SCHAUM'S OUTLINE OF

THEORY AND PROBLEMS

of

ELECTROMAGNETICS

•

by

JOSEPH A. EDMINISTER, M.S.E.
Associate Professor of Electrical Engineering
The University of Akron

SCHAUM'S OUTLINE SERIES
McGRAW-HILL BOOK COMPANY

New York St. Louis San Francisco Auckland Bogotá Düsseldorf Johannesburg
London Madrid Mexico Montreal New Delhi Panama Paris
São Paulo Singapore Sydney Tokyo Toronto

0-07-018990-0

5 6 7 8 9 10 11 12 13 14 15 16 17 18 19 20 SH SH 8 4 3 2

Library of Congress Cataloging in Publication Data

Edminister, Joseph.
Schaum's outline of electromagnetics.

(Schaum's outline series)
Includes index.
1. Electromagnetism. I. Title.
QC760.E35 537'.02'02 78-9482
ISBN 0-07-018990-0

Preface

This book is intended to serve as a supplement to any of the introductory textbooks in electromagnetic field theory for engineers; it may also be used by itself as the text for a brief first course. As in other Schaum's Outlines the emphasis is on how to solve problems. Each chapter consists of an ample set of problems with detailed solutions, and a further set of problems with answers, preceded by a simplified outline of the principles and facts needed to understand the problems and their solutions. Although electromagnetic problems of the physical world tend to be quite elaborate, it was decided in this book to present mostly short, single-concept problems. It is felt that this will prove advantageous to the student who seeks help on a particular point, as well as to those who may use the book for review purposes.

Throughout the book the mathematics has been kept as simple as possible, and an abstract approach has been avoided. Concrete examples are liberally used and numerous graphs and sketches are given. I have found in many years of teaching that the solution of most problems begins with a carefully drawn sketch.

This book is dedicated to my students, who have shown me where the difficulties in the subject lie. For editorial assistance I want to express my gratitude to the staff of McGraw-Hill. Sincere thanks to Thomas R. Connell for his great care in checking all the problems and offering suggestions. Eileen Kerns deserves thanks for her capable typing of the manuscript. And finally, thanks are due to my family, in particular my wife Nina, for constant support and encouragement, without which the book could not have been written.

JOSEPH A. EDMINISTER

Contents

Chapter 1

Vector Analysis

1.1 VECTOR NOTATION

In order to distinguish *vectors* (quantities having magnitude and direction) from *scalars* (quantities having magnitude only) the vectors are denoted by boldface symbols. A *unit vector*, one of absolute value (or magnitude or length) 1, will in this book always be indicated by a boldface, lowercase **a**. The unit vector in the direction of a vector **A** is determined by dividing **A** by its absolute value:

$$\mathbf{a}_A = \frac{\mathbf{A}}{|\mathbf{A}|} \quad \text{or} \quad \frac{\mathbf{A}}{A}$$

where $|\mathbf{A}| = A = \sqrt{\mathbf{A} \cdot \mathbf{A}}$ (see Section 1.2).

By use of the unit vectors $\mathbf{a}_x$, $\mathbf{a}_y$, $\mathbf{a}_z$ along the x, y, and z axes of a cartesian coordinate system, an arbitrary vector can be written in *component form*:

$$\mathbf{A} = A_x \mathbf{a}_x + A_y \mathbf{a}_y + A_z \mathbf{a}_z$$

1.2 VECTOR ALGEBRA

1. Vectors may be added and subtracted.

$$\begin{aligned} \mathbf{A} \pm \mathbf{B} &= (A_x \mathbf{a}_x + A_y \mathbf{a}_y + A_z \mathbf{a}_z) \pm (B_x \mathbf{a}_x + B_y \mathbf{a}_y + B_z \mathbf{a}_z) \\ &= (A_x \pm B_x)\mathbf{a}_x + (A_y \pm B_y)\mathbf{a}_y + (A_z \pm B_z)\mathbf{a}_z \end{aligned}$$

2. The associative, distributive, and commutative laws apply.

$$\mathbf{A} + (\mathbf{B} + \mathbf{C}) = (\mathbf{A} + \mathbf{B}) + \mathbf{C}$$

$$k(\mathbf{A} + \mathbf{B}) = k\mathbf{A} + k\mathbf{B} \qquad (k_1 + k_2)\mathbf{A} = k_1\mathbf{A} + k_2\mathbf{A}$$

$$\mathbf{A} + \mathbf{B} = \mathbf{B} + \mathbf{A}$$

3. The *dot product* of two vectors is, by definition,

$$\mathbf{A} \cdot \mathbf{B} = AB\cos\theta \qquad \text{(read “A dot B”)}$$

where θ is the smaller angle between **A** and **B**. From the component form it can be shown that

$$\mathbf{A} \cdot \mathbf{B} = A_x B_x + A_y B_y + A_z B_z$$

In particular,

$$\mathbf{A} \cdot \mathbf{A} = |\mathbf{A}|^2 = A_x^2 + A_y^2 + A_z^2$$

4. The *cross product* of two vectors is, by definition,

$$\mathbf{A} \times \mathbf{B} = (AB\sin\theta)\mathbf{a}_n \qquad \text{(read “A cross B”)}$$

where θ is the smaller angle between **A** and **B**, and $\mathbf{a}_n$ is a unit vector normal to the plane determined by **A** and **B** when they are drawn from a common point. There are two normals to the plane, so further specification is needed. The normal selected is the one in the direction of advance of a right-hand screw when **A** is turned toward **B** (Fig. 1-1). Because of this direction requirement, the commutative law does not apply to the cross product; instead,

$$\mathbf{A} \times \mathbf{B} = -\mathbf{B} \times \mathbf{A}$$

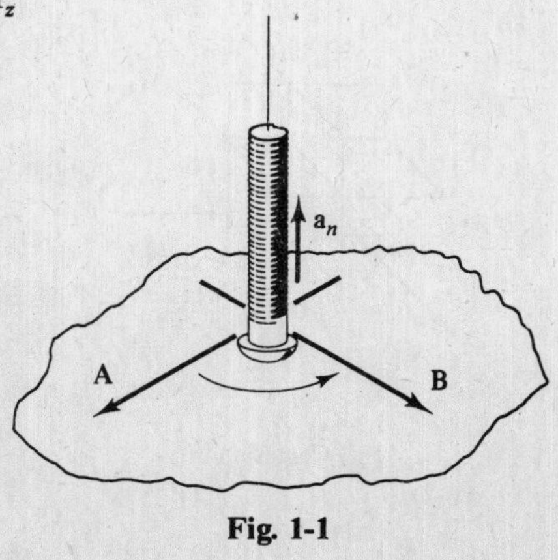

Fig. 1-1

Expanding the cross product in component form,

$$\mathbf{A} \times \mathbf{B} = (A_x \mathbf{a}_x + A_y \mathbf{a}_y + A_z \mathbf{a}_z) \times (B_x \mathbf{a}_x + B_y \mathbf{a}_y + B_z \mathbf{a}_z)$$
$$= (A_y B_z - A_z B_y)\mathbf{a}_x + (A_z B_x - A_x B_z)\mathbf{a}_y + (A_x B_y - A_y B_x)\mathbf{a}_z$$

which is conveniently expressed as a determinant:

$$\mathbf{A} \times \mathbf{B} = \begin{vmatrix} \mathbf{a}_x & \mathbf{a}_y & \mathbf{a}_z \\ A_x & A_y & A_z \\ B_x & B_y & B_z \end{vmatrix}$$

1.3 COORDINATE SYSTEMS

A problem which has cylindrical or spherical symmetry could be expressed and solved in the familiar cartesian coordinate system. However, the solution would fail to show the symmetry and in most cases would be needlessly complex. Therefore, throughout this book, in addition to the cartesian coordinate system, the circular cylindrical and the spherical coordinate systems will be used. All three will be examined together in order to illustrate the similarities and the differences.

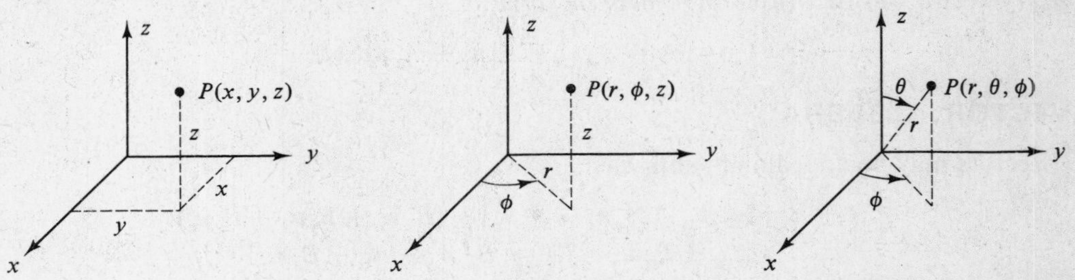

(a) Cartesian (b) Cylindrical (c) Spherical

Fig. 1-2

A point P is described by three coordinates, in cartesian (x, y, z), in circular cylindrical (r, ϕ, z) and in spherical (r, θ, ϕ), as shown in Fig. 1-2. The order of specifying the coordinates is important and should be carefully followed. The angle ϕ is the same angle in both the cylindrical and spherical systems. But, in the order of the coordinates, ϕ appears in the second position in cylindrical, (r, ϕ, z), and the third position in spherical, (r, θ, ϕ). The same symbol, r, is used in both cylindrical and

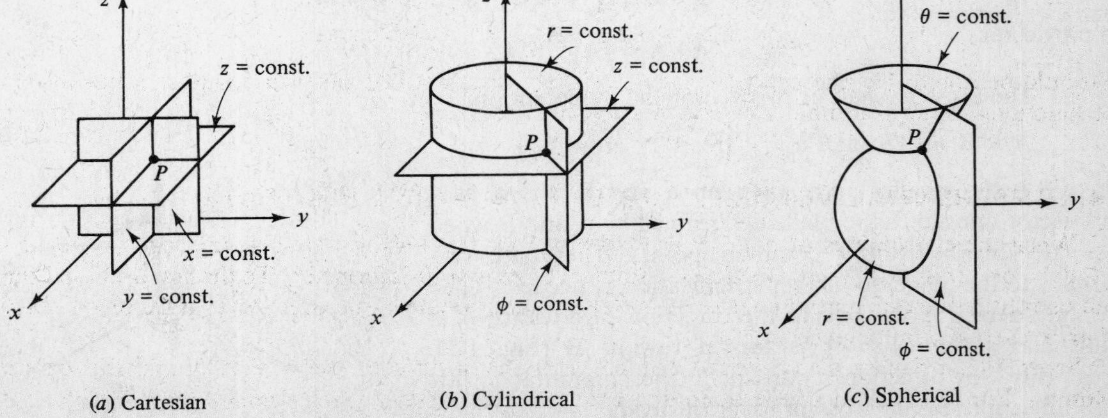

(a) Cartesian (b) Cylindrical (c) Spherical

Fig. 1-3

spherical for two quite different things. In cylindrical coordinates r measures the distance from the z axis in a plane normal to the z axis, while in the spherical system r measures the distance from the origin to the point. It should be clear from the context of the problem which r is intended.

A point is also defined by the intersection of three orthogonal surfaces, as shown in Fig. 1-3. In cartesian coordinates the surfaces are the infinite planes $x = \text{const.}$, $y = \text{const.}$, and $z = \text{const.}$ In cylindrical coordinates, $z = \text{const.}$ is the same infinite plane as in cartesian; $\phi = \text{const.}$ is a half plane with its edge along the z axis; $r = \text{const.}$ is a right circular cylinder. These three surfaces are orthogonal and their intersection locates point P. In spherical coordinates, $\phi = \text{const.}$ is the same half plane as in cylindrical; $r = \text{const.}$ is a sphere with its center at the origin; $\theta = \text{const.}$ is a right circular cone whose axis is the z axis and whose vertex is at the origin. Note that θ is limited to the range $0 \le \theta \le \pi$.

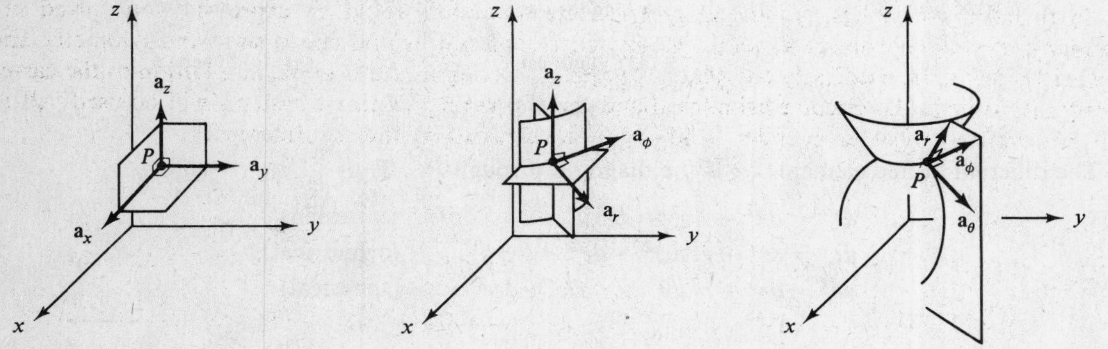

(a) Cartesian (b) Cylindrical (c) Spherical

Fig. 1-4

Figure 1-4 shows the three unit vectors at point P. In the cartesian system the unit vectors have fixed directions, independent of the location of P. This is not true for the other two systems (except in the case of $\mathbf{a}_z$). Each unit vector is normal to its coordinate surface and is in the direction in which the coordinate increases. Notice that all these systems are right-handed:

$$\mathbf{a}_x \times \mathbf{a}_y = \mathbf{a}_z \qquad \mathbf{a}_r \times \mathbf{a}_\phi = \mathbf{a}_z \qquad \mathbf{a}_r \times \mathbf{a}_\theta = \mathbf{a}_\phi$$

The component forms of a vector in the three systems are

$$\mathbf{A} = A_x \mathbf{a}_x + A_y \mathbf{a}_y + A_z \mathbf{a}_z \qquad \text{(cartesian)}$$
$$\mathbf{A} = A_r \mathbf{a}_r + A_\phi \mathbf{a}_\phi + A_z \mathbf{a}_z \qquad \text{(cylindrical)}$$
$$\mathbf{A} = A_r \mathbf{a}_r + A_\theta \mathbf{a}_\theta + A_\phi \mathbf{a}_\phi \qquad \text{(spherical)}$$

It should be noted that the components A_x, A_r, A_ϕ, etc., are not generally constants but more often are functions of the coordinates in that particular system.

1.4 DIFFERENTIAL VOLUME, SURFACE, AND LINE ELEMENTS

When the coordinates of point P are expanded to $(x + dx, y + dy, z + dz)$ or $(r + dr, \phi + d\phi, z + dz)$ or $(r + dr, \theta + d\theta, \phi + d\phi)$ a differential volume dv is formed. To the first order in infinitesimal quantities the differential volume is, in all three coordinate systems, a rectangular box. The value of dv in each system is given in Fig. 1-5.

From Fig. 1-5 may also be read the areas of the surface elements that bound the differential volume. For instance, in spherical coordinates, the differential surface element perpendicular to $\mathbf{a}_r$ is

$$dS = (r\,d\theta)(r \sin\theta\,d\phi) = r^2 \sin\theta\,d\theta\,d\phi$$

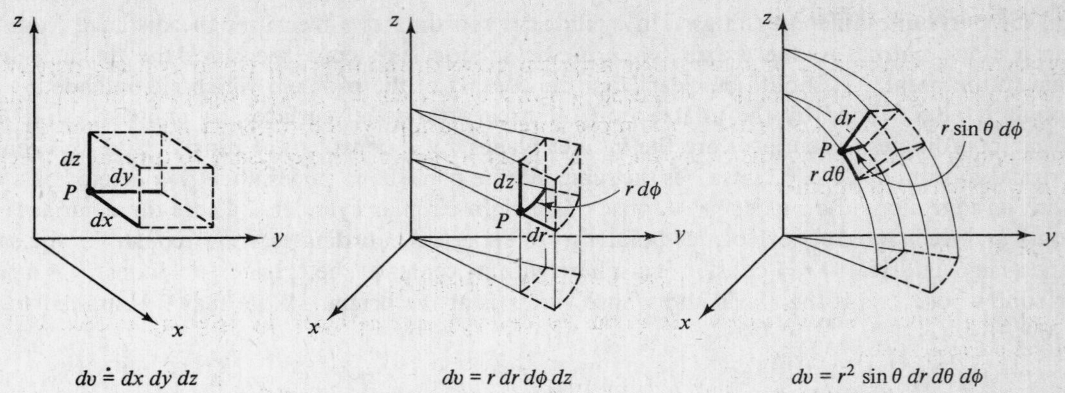

(*a*) Cartesian (*b*) Cylindrical (*c*) Spherical

Fig. 1-5

The differential line element, $d\ell$, is the diagonal through P. Thus

$$d\ell^2 = dx^2 + dy^2 + dz^2 \qquad \text{(cartesian)}$$
$$d\ell^2 = dr^2 + r^2\,d\phi^2 + dz^2 \qquad \text{(cylindrical)}$$
$$d\ell^2 = dr^2 + r^2\,d\theta^2 + r^2 \sin^2\theta\,d\phi^2 \qquad \text{(spherical)}$$

1.5 VECTOR FIELDS

The vector expressions in electromagnetics are generally such that the coefficients of the unit vectors contain the variables. Therefore, the expression changes its magnitude and direction from point to point throughout the region of interest.

Consider, for example, the vector

$$\mathbf{E} = -x\mathbf{a}_x + y\mathbf{a}_y$$

Values of x and y may be substituted into the expression to give $\mathbf{E}$ at the various locations. After a number of points are examined, the pattern becomes evident. Figure 1-6 shows this field.

In addition, a vector field may vary with time. Thus, the two-dimensional field examined above could be given a time variation such as

$$\mathbf{E} = (-x\mathbf{a}_x + y\mathbf{a}_y)\sin\omega t$$

or

$$\mathbf{E} = (-x\mathbf{a}_x + y\mathbf{a}_y)e^{j\omega t}$$

The electric and magnetic fields of the later chapters are all time-variable. And, as might be expected, they will be differentiated with respect to time and also integrated with respect to time. However, both operations will follow naturally and seldom cause any great difficulty.

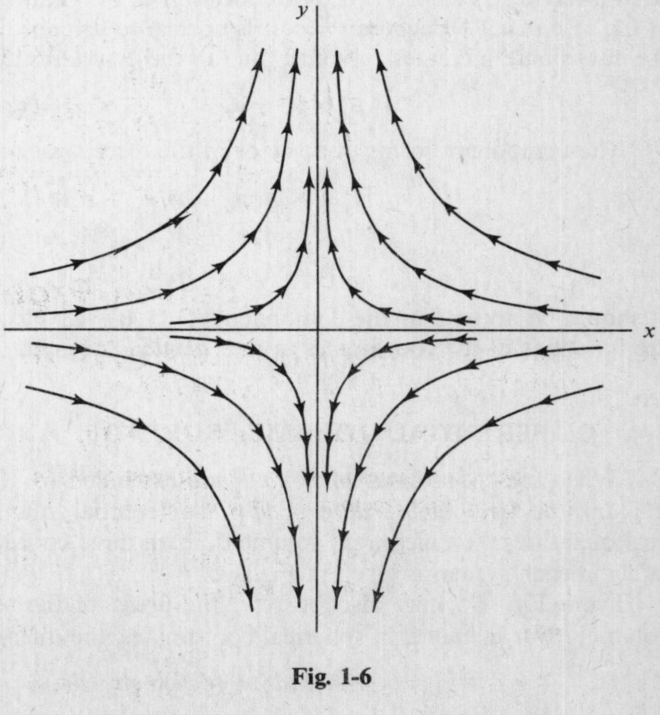

Fig. 1-6

1.6 TRANSFORMATIONS

The vector or vector field in a particular problem exists in the physical world and the coordinate system which is employed to express it is merely a frame of reference. A wise choice of the coordinate system at the outset will often result in a more direct solution to the problem and a concise final expression which shows the symmetry present. At times, however, it is necessary to transform a vector field in one system into another.

EXAMPLE 1 Consider

$$\mathbf{A} = 5r\mathbf{a}_r + 2\sin\phi\,\mathbf{a}_\theta + 2\cos\theta\,\mathbf{a}_\phi$$

in spherical coordinates. The variables r, θ, ϕ can be changed into cartesian by referring to Fig. 1-2 and applying basic trigonometry. Thus

$$r = \sqrt{x^2+y^2+z^2} \qquad \cos\theta = \frac{z}{\sqrt{x^2+y^2+z^2}} \qquad \tan\phi = \frac{y}{x}$$

Now the spherical components of the vector field $\mathbf{A}$ can be written in terms of x, y, and z:

$$\mathbf{A} = 5\sqrt{x^2+y^2+z^2}\,\mathbf{a}_r + \frac{2y}{\sqrt{x^2+y^2}}\mathbf{a}_\theta + \frac{2z}{\sqrt{x^2+y^2+z^2}}\mathbf{a}_\phi$$

The unit vectors $\mathbf{a}_r$, $\mathbf{a}_\theta$, and $\mathbf{a}_\phi$ can also be transformed into their cartesian equivalents by referring to Fig. 1-4 and applying basic trigonometry. Thus

$$\mathbf{a}_r = \frac{x}{\sqrt{x^2+y^2+z^2}}\mathbf{a}_x + \frac{y}{\sqrt{x^2+y^2+z^2}}\mathbf{a}_y + \frac{z}{\sqrt{x^2+y^2+z^2}}\mathbf{a}_z$$

$$\mathbf{a}_\theta = \frac{xz}{\sqrt{x^2+y^2+z^2}\sqrt{x^2+y^2}}\mathbf{a}_x + \frac{yz}{\sqrt{x^2+y^2+z^2}\sqrt{x^2+y^2}}\mathbf{a}_y - \frac{\sqrt{x^2+y^2}}{\sqrt{x^2+y^2+z^2}}\mathbf{a}_z$$

$$\mathbf{a}_\phi = \frac{-y}{\sqrt{x^2+y^2}}\mathbf{a}_x + \frac{x}{\sqrt{x^2+y^2}}\mathbf{a}_y$$

Combining these with the transformed components results in

$$\mathbf{A} = \left(5x + \frac{2xyz}{\sqrt{x^2+y^2+z^2}\,(x^2+y^2)} - \frac{2yz}{\sqrt{x^2+y^2+z^2}\sqrt{x^2+y^2}}\right)\mathbf{a}_x$$
$$+ \left(5y + \frac{2y^2z}{(x^2+y^2)\sqrt{x^2+y^2+z^2}} + \frac{2xy}{\sqrt{x^2+y^2+z^2}\sqrt{x^2+y^2}}\right)\mathbf{a}_y$$
$$+ \left(5z - \frac{2y}{\sqrt{x^2+y^2+z^2}}\right)\mathbf{a}_z$$

Solved Problems

1.1 Show that the vector directed from $M(x_1, y_1, z_1)$ to $N(x_2, y_2, z_2)$ in Fig. 1-7 is given by

$$(x_2 - x_1)\mathbf{a}_x + (y_2 - y_1)\mathbf{a}_y + (z_2 - z_1)\mathbf{a}_z$$

The coordinates of M and N are used to write the two position vectors $\mathbf{A}$ and $\mathbf{B}$ in Fig. 1-7.

$$\mathbf{A} = x_1\mathbf{a}_x + y_1\mathbf{a}_y + z_1\mathbf{a}_z$$
$$\mathbf{B} = x_2\mathbf{a}_x + y_2\mathbf{a}_y + z_2\mathbf{a}_z$$

Then

$$\mathbf{B} - \mathbf{A} = (x_2 - x_1)\mathbf{a}_x + (y_2 - y_1)\mathbf{a}_y + (z_2 - z_1)\mathbf{a}_z$$

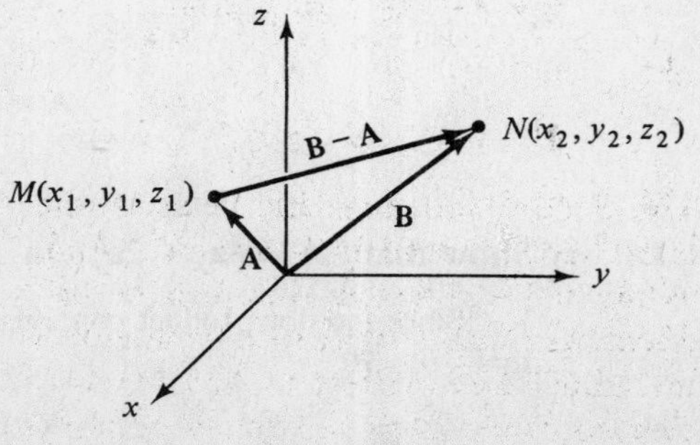

Fig. 1-7

1.2. Find the vector **A** directed from $(2, -4, 1)$ to $(0, -2, 0)$ in cartesian coordinates and find the unit vector along **A**.

$$\mathbf{A} = (0-2)\mathbf{a}_x + (-2-(-4))\mathbf{a}_y + (0-1)\mathbf{a}_z = -2\mathbf{a}_x + 2\mathbf{a}_y - \mathbf{a}_z$$

$$|\mathbf{A}|^2 = (-2)^2 + (2)^2 + (-1)^2 = 9$$

$$\mathbf{a}_A = \frac{\mathbf{A}}{|\mathbf{A}|} = -\frac{2}{3}\mathbf{a}_x + \frac{2}{3}\mathbf{a}_y - \frac{1}{3}\mathbf{a}_z$$

1.3. Find the distance between $(5, 3\pi/2, 0)$ and $(5, \pi/2, 10)$ in cylindrical coordinates.

First, obtain the *cartesian* position vectors **A** and **B** (see Fig. 1-8).

$$\mathbf{A} = -5\mathbf{a}_y \qquad \mathbf{B} = 5\mathbf{a}_y + 10\mathbf{a}_z$$

Then $\mathbf{B} - \mathbf{A} = 10\mathbf{a}_y + 10\mathbf{a}_z$ and the required distance between the points is

$$|\mathbf{B} - \mathbf{A}| = 10\sqrt{2}$$

The cylindrical coordinates of the points cannot be used to obtain a vector between the points in the same manner as was employed in Problem 1.1 in cartesian coordinates.

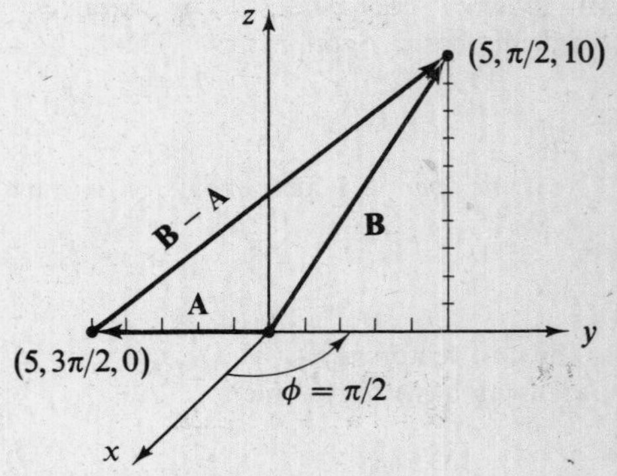

Fig. 1-8

1.4. Show that $\mathbf{A}\cdot\mathbf{B} = A_x B_x + A_y B_y + A_z B_z$.

Express the dot product in component form.

$$\begin{aligned}\mathbf{A}\cdot\mathbf{B} &= (A_x\mathbf{a}_x + A_y\mathbf{a}_y + A_z\mathbf{a}_z)\cdot(B_x\mathbf{a}_x + B_y\mathbf{a}_y + B_z\mathbf{a}_z)\\ &= (A_x\mathbf{a}_x)\cdot(B_x\mathbf{a}_x) + (A_x\mathbf{a}_x)\cdot(B_y\mathbf{a}_y) + (A_x\mathbf{a}_x)\cdot(B_z\mathbf{a}_z)\\ &\quad + (A_y\mathbf{a}_y)\cdot(B_x\mathbf{a}_x) + (A_y\mathbf{a}_y)\cdot(B_y\mathbf{a}_y) + (A_y\mathbf{a}_y)\cdot(B_z\mathbf{a}_z)\\ &\quad + (A_z\mathbf{a}_z)\cdot(B_x\mathbf{a}_x) + (A_z\mathbf{a}_z)\cdot(B_y\mathbf{a}_y) + (A_z\mathbf{a}_z)\cdot(B_z\mathbf{a}_z)\end{aligned}$$

However, $\mathbf{a}_x\cdot\mathbf{a}_x = \mathbf{a}_y\cdot\mathbf{a}_y = \mathbf{a}_z\cdot\mathbf{a}_z = 1$ because the $\cos\theta$ in the dot product is unity when the angle is zero. And when $\theta = 90°$, $\cos\theta$ is zero. Hence all other dot products of the unit vectors are zero. Thus

$$\mathbf{A}\cdot\mathbf{B} = A_x B_x + A_y B_y + A_z B_z$$

1.5. Given $\mathbf{A} = 2\mathbf{a}_x + 4\mathbf{a}_y - 3\mathbf{a}_z$ and $\mathbf{B} = \mathbf{a}_x - \mathbf{a}_y$, find $\mathbf{A}\cdot\mathbf{B}$ and $\mathbf{A}\times\mathbf{B}$.

$$\mathbf{A}\cdot\mathbf{B} = (2)(1) + (4)(-1) + (-3)(0) = -2$$

$$\mathbf{A}\times\mathbf{B} = \begin{vmatrix}\mathbf{a}_x & \mathbf{a}_y & \mathbf{a}_z\\ 2 & 4 & -3\\ 1 & -1 & 0\end{vmatrix} = -3\mathbf{a}_x - 3\mathbf{a}_y - 6\mathbf{a}_z$$

1.6. Show that $\mathbf{A} = 4\mathbf{a}_x - 2\mathbf{a}_y - \mathbf{a}_z$ and $\mathbf{B} = \mathbf{a}_x + 4\mathbf{a}_y - 4\mathbf{a}_z$ are perpendicular.

Since the dot product contains $\cos\theta$, a dot product of zero from any two nonzero vectors implies that $\theta = 90°$.

$$\mathbf{A}\cdot\mathbf{B} = (4)(1) + (-2)(4) + (-1)(-4) = 0$$

1.7. Given $\mathbf{A} = 2\mathbf{a}_x + 4\mathbf{a}_y$ and $\mathbf{B} = 6\mathbf{a}_y - 4\mathbf{a}_z$, find the smaller angle between them using (*a*) the cross product, (*b*) the dot product.

(*a*)
$$\mathbf{A} \times \mathbf{B} = \begin{vmatrix} \mathbf{a}_x & \mathbf{a}_y & \mathbf{a}_z \\ 2 & 4 & 0 \\ 0 & 6 & -4 \end{vmatrix} = -16\mathbf{a}_x + 8\mathbf{a}_y + 12\mathbf{a}_z$$

$$|\mathbf{A}| = \sqrt{(2)^2 + (4)^2 + (0)^2} = 4.47$$
$$|\mathbf{B}| = \sqrt{(0)^2 + (6)^2 + (-4)^2} = 7.21$$
$$|\mathbf{A} \times \mathbf{B}| = \sqrt{(-16)^2 + (8)^2 + (12)^2} = 21.54$$

Then, since $|\mathbf{A} \times \mathbf{B}| = |\mathbf{A}|\,|\mathbf{B}| \sin\theta$,

$$\sin\theta = \frac{21.54}{(4.47)(7.21)} = 0.668 \qquad \text{or} \qquad \theta = 41.9°$$

(*b*)
$$\mathbf{A} \cdot \mathbf{B} = (2)(0) + (4)(6) + (0)(-4) = 24$$

$$\cos\theta = \frac{\mathbf{A} \cdot \mathbf{B}}{|\mathbf{A}||\mathbf{B}|} = \frac{24}{(4.47)(7.21)} = 0.745 \qquad \text{or} \qquad \theta = 41.9°$$

1.8. Given $\mathbf{F} = (y - 1)\mathbf{a}_x + 2x\mathbf{a}_y$, find the vector at $(2, 2, 1)$ and its projection on **B**, where $\mathbf{B} = 5\mathbf{a}_x - \mathbf{a}_y + 2\mathbf{a}_z$.

$$\mathbf{F}(2, 2, 1) = (2 - 1)\mathbf{a}_x + (2)(2)\mathbf{a}_y = \mathbf{a}_x + 4\mathbf{a}_y$$

As indicated in Fig. 1-9, the projection of one vector on a second vector is obtained by expressing the unit vector in the direction of the second vector and taking the dot product.

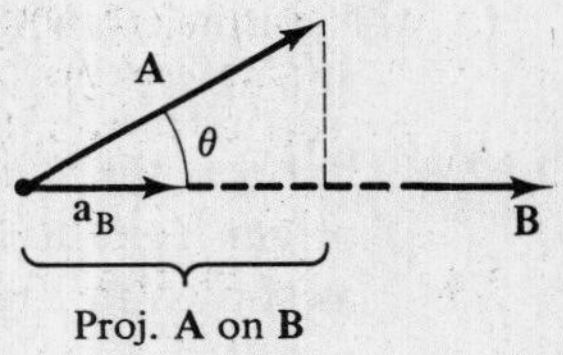

Fig. 1-9

$$\text{Proj. } \mathbf{A} \text{ on } \mathbf{B} = \mathbf{A} \cdot \mathbf{a}_B = \frac{\mathbf{A} \cdot \mathbf{B}}{|\mathbf{B}|}$$

Thus, at $(2, 2, 1)$,

$$\text{Proj. } \mathbf{F} \text{ on } \mathbf{B} = \frac{\mathbf{F} \cdot \mathbf{B}}{|\mathbf{B}|} = \frac{(1)(5) + (4)(-1) + (0)(2)}{\sqrt{30}} = \frac{1}{\sqrt{30}}$$

1.9. Given $\mathbf{A} = \mathbf{a}_x + \mathbf{a}_y$, $\mathbf{B} = \mathbf{a}_x + 2\mathbf{a}_z$, and $\mathbf{C} = 2\mathbf{a}_y + \mathbf{a}_z$, find $(\mathbf{A} \times \mathbf{B}) \times \mathbf{C}$ and compare it with $\mathbf{A} \times (\mathbf{B} \times \mathbf{C})$.

$$\mathbf{A} \times \mathbf{B} = \begin{vmatrix} \mathbf{a}_x & \mathbf{a}_y & \mathbf{a}_z \\ 1 & 1 & 0 \\ 1 & 0 & 2 \end{vmatrix} = 2\mathbf{a}_x - 2\mathbf{a}_y - \mathbf{a}_z$$

Then
$$(\mathbf{A} \times \mathbf{B}) \times \mathbf{C} = \begin{vmatrix} \mathbf{a}_x & \mathbf{a}_y & \mathbf{a}_z \\ 2 & -2 & -1 \\ 0 & 2 & 1 \end{vmatrix} = -2\mathbf{a}_y + 4\mathbf{a}_z$$

A similar calculation gives $\mathbf{A} \times (\mathbf{B} \times \mathbf{C}) = 2\mathbf{a}_x - 2\mathbf{a}_y + 3\mathbf{a}_z$. Thus the parentheses that indicate which cross product is to be taken first are essential in the vector triple product.

1.10. Using the vectors **A**, **B**, and **C** of Problem 1.9, find $\mathbf{A} \cdot \mathbf{B} \times \mathbf{C}$ and compare it with $\mathbf{A} \times \mathbf{B} \cdot \mathbf{C}$.

From Problem 1.9, $\mathbf{B} \times \mathbf{C} = -4\mathbf{a}_x - \mathbf{a}_y + 2\mathbf{a}_z$. Then

$$\mathbf{A} \cdot \mathbf{B} \times \mathbf{C} = (1)(-4) + (1)(-1) + (0)(2) = -5$$

Also from Problem 1.9, $\mathbf{A} \times \mathbf{B} = 2\mathbf{a}_x - 2\mathbf{a}_y - \mathbf{a}_z$. Then

$$\mathbf{A} \times \mathbf{B} \cdot \mathbf{C} = (2)(0) + (-2)(2) + (-1)(1) = -5$$

Parentheses are not needed in the scalar triple product since it has meaning only when the cross product is taken first. In general, it can be shown that

$$\mathbf{A} \cdot \mathbf{B} \times \mathbf{C} = \begin{vmatrix} A_x & A_y & A_z \\ B_x & B_y & B_z \\ C_x & C_y & C_z \end{vmatrix}$$

As long as the vectors appear in the same cyclic order the result is the same. The scalar triple products out of this cyclic order have a change in sign.

1.11. Express the unit vector which points from $z = h$ on the z axis toward $(r, \phi, 0)$ in cylindrical coordinates. See Fig. 1-10.

The vector $\mathbf{R}$ is the difference of two vectors:

$$\mathbf{R} = r\mathbf{a}_r - h\mathbf{a}_z$$

$$\mathbf{a}_R = \frac{\mathbf{R}}{|\mathbf{R}|} = \frac{r\mathbf{a}_r - h\mathbf{a}_z}{\sqrt{r^2 + h^2}}$$

The angle ϕ does not appear explicitly in these expressions. Nevertheless, both $\mathbf{R}$ and $\mathbf{a}_R$ vary with ϕ through $\mathbf{a}_r$.

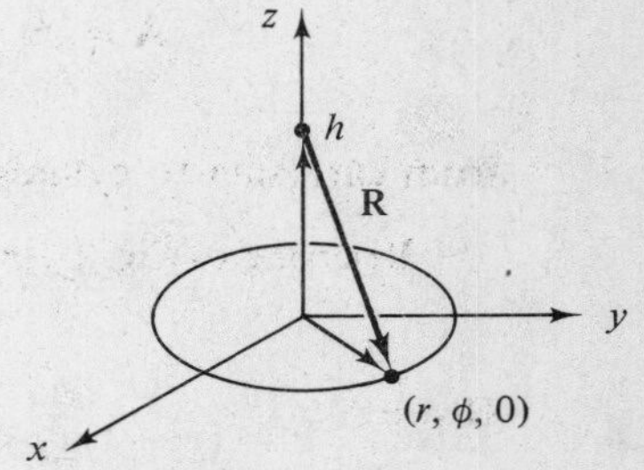

Fig. 1-10

1.12. Express the unit vector which is directed toward the origin from an arbitrary point on the plane $z = -5$, as shown in Fig. 1-11.

Since the problem is in cartesian coordinates, the two-point formula of Problem 1.1 applies.

$$\mathbf{R} = -x\mathbf{a}_x - y\mathbf{a}_y + 5\mathbf{a}_z$$

$$\mathbf{a}_R = \frac{-x\mathbf{a}_x - y\mathbf{a}_y + 5\mathbf{a}_z}{\sqrt{x^2 + y^2 + 25}}$$

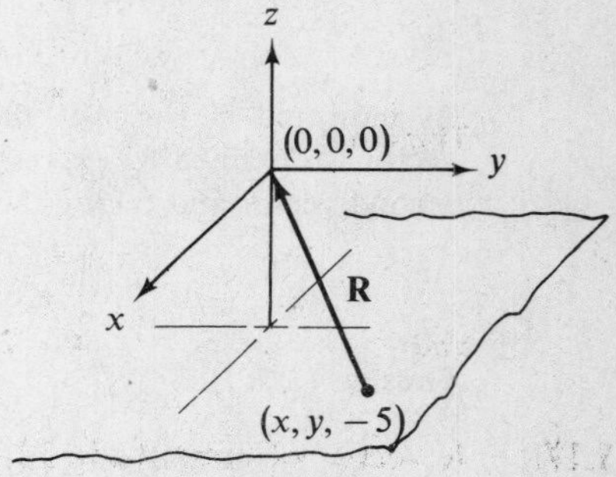

Fig. 1-11

1.13. Use the spherical coordinate system to find the area of the strip $\alpha \le \theta \le \beta$ on the spherical shell of radius a (Fig. 1-12). What results when $\alpha = 0$ and $\beta = \pi$?

The differential surface element is [see Fig. 1-5(*c*)]

$$dS = r^2 \sin\theta \, d\theta \, d\phi$$

Then

$$A = \int_0^{2\pi} \int_\alpha^\beta a^2 \sin\theta \, d\theta \, d\phi$$
$$= 2\pi a^2(\cos\alpha - \cos\beta)$$

When $\alpha = 0$ and $\beta = \pi$, $A = 4\pi a^2$, the surface area of the entire sphere.

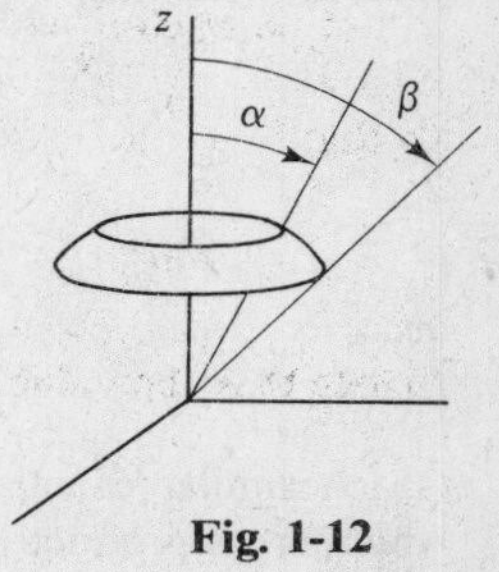

Fig. 1-12

1.14. Develop the equation for the volume of a sphere of radius a from the differential volume.

From Fig. 1-5(*c*), $dv = r^2 \sin\theta \, dr \, d\theta \, d\phi$. Then

$$v = \int_0^{2\pi} \int_0^{\pi} \int_0^{a} r^2 \sin\theta \, dr \, d\theta \, d\phi = \frac{4}{3}\pi a^3$$

1.15. Use the cylindrical coordinate system to find the area of the curved surface of a right circular cylinder where $r = 2$ m, $h = 5$ m, and $30° \le \phi \le 120°$ (see Fig. 1-13).

The differential surface element is $dS = r\,d\phi\,dz$. Then

$$A = \int_0^5 \int_{\pi/6}^{2\pi/3} 2\,d\phi\,dz$$
$$= 5\pi \text{ m}^2$$

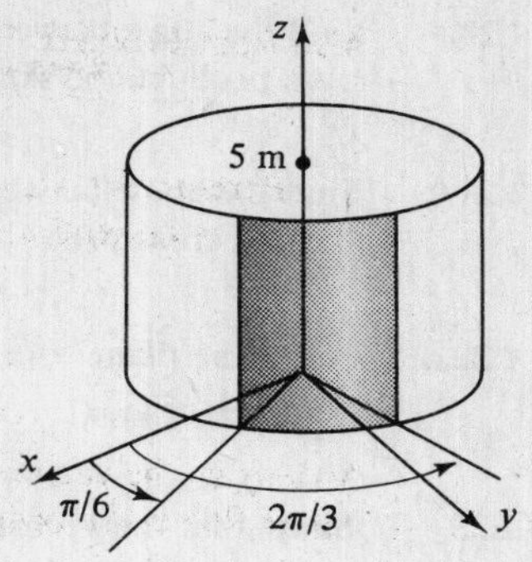

Fig. 1-13

1.16. Transform

$$\mathbf{A} = y\mathbf{a}_x + x\mathbf{a}_y + \frac{x^2}{\sqrt{x^2 + y^2}}\mathbf{a}_z$$

from cartesian to cylindrical coordinates.

Referring to Fig. 1-2(*b*),

$$x = r\cos\phi \qquad y = r\sin\phi \qquad r = \sqrt{x^2 + y^2}$$

Hence $$\mathbf{A} = r\sin\phi\,\mathbf{a}_x + r\cos\phi\,\mathbf{a}_y + r\cos^2\phi\,\mathbf{a}_z$$

Now the projections of the cartesian unit vectors on $\mathbf{a}_r$, $\mathbf{a}_\phi$, and $\mathbf{a}_z$ are obtained:

$$\mathbf{a}_x \cdot \mathbf{a}_r = \cos\phi \qquad \mathbf{a}_x \cdot \mathbf{a}_\phi = -\sin\phi \qquad \mathbf{a}_x \cdot \mathbf{a}_z = 0$$
$$\mathbf{a}_y \cdot \mathbf{a}_r = \sin\phi \qquad \mathbf{a}_y \cdot \mathbf{a}_\phi = \cos\phi \qquad \mathbf{a}_y \cdot \mathbf{a}_z = 0$$
$$\mathbf{a}_z \cdot \mathbf{a}_r = 0 \qquad \mathbf{a}_z \cdot \mathbf{a}_\phi = 0 \qquad \mathbf{a}_z \cdot \mathbf{a}_z = 1$$

Therefore

$$\mathbf{a}_x = \cos\phi\,\mathbf{a}_r - \sin\phi\,\mathbf{a}_\phi$$
$$\mathbf{a}_y = \sin\phi\,\mathbf{a}_r + \cos\phi\,\mathbf{a}_\phi$$
$$\mathbf{a}_z = \mathbf{a}_z$$

and $$\mathbf{A} = 2r\sin\phi\cos\phi\,\mathbf{a}_r + (r\cos^2\phi - r\sin^2\phi)\mathbf{a}_\phi + r\cos^2\phi\,\mathbf{a}_z$$

1.17. A vector of magnitude 10 points from $(5, 5\pi/4, 0)$ in cylindrical coordinates toward the origin (Fig. 1-14). Express the vector in cartesian coordinates.

In cylindrical coordinates, the vector may be expressed as $10\mathbf{a}_r$, where $\phi = \pi/4$. Hence

$$A_x = 10\cos\frac{\pi}{4} = \frac{10}{\sqrt{2}} \qquad A_y = 10\sin\frac{\pi}{4} = \frac{10}{\sqrt{2}} \qquad A_z = 0$$

so that

$$\mathbf{A} = \frac{10}{\sqrt{2}}\mathbf{a}_x + \frac{10}{\sqrt{2}}\mathbf{a}_y$$

Notice that the value of the radial coordinate, 5, is immaterial.

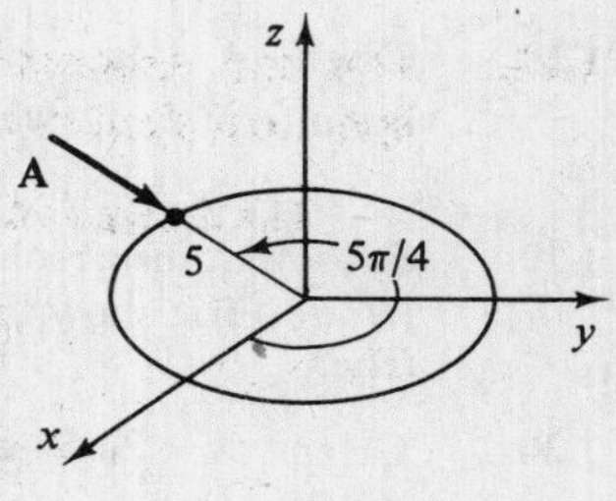

Fig. 1-14

Supplementary Problems

1.18. Given $\mathbf{A} = 4\mathbf{a}_y + 10\mathbf{a}_z$ and $\mathbf{B} = 2\mathbf{a}_x + 3\mathbf{a}_y$, find the projection of **A** on **B**. *Ans.* $12/\sqrt{13}$

1.19. Given $\mathbf{A} = (10/\sqrt{2})(\mathbf{a}_x + \mathbf{a}_z)$ and $\mathbf{B} = 3(\mathbf{a}_y + \mathbf{a}_z)$, express the projection of **B** on **A** as a vector in the direction of **A**. *Ans.* $1.50\,(\mathbf{a}_x + \mathbf{a}_z)$

1.20. Find the angle between $\mathbf{A} = 10\mathbf{a}_y + 2\mathbf{a}_z$ and $\mathbf{B} = -4\mathbf{a}_y + 0.5\,\mathbf{a}_z$ using both the dot product and the cross product. *Ans.* 161.5°

1.21. Find the angle between $\mathbf{A} = 5.8\,\mathbf{a}_y + 1.55\,\mathbf{a}_z$ and $\mathbf{B} = -6.93\,\mathbf{a}_y + 4.0\,\mathbf{a}_z$ using both the dot product and the cross product. *Ans.* 135°

1.22. Given the plane $4x + 3y + 2z = 12$, find the unit vector normal to the surface in the direction away from the origin. *Ans.* $(4\mathbf{a}_x + 3\mathbf{a}_y + 2\mathbf{a}_z)/\sqrt{29}$

1.23. Show that the vector fields **A** and **B** are everywhere perpendicular if $A_x B_x + A_y B_y + A_z B_z = 0$.

1.24. Find the relationship which the cartesian components of **A** and **B** must satisfy if the vector fields are everywhere parallel.

Ans. $\dfrac{A_x}{B_x} = \dfrac{A_y}{B_y} = \dfrac{A_z}{B_z}$

1.25. Express the unit vector directed toward the origin from an arbitrary point on the line described by $x = 0$, $y = 3$.

Ans. $\mathbf{a} = \dfrac{-3\mathbf{a}_y - z\mathbf{a}_z}{\sqrt{9 + z^2}}$

1.26. Express the unit vector directed toward the point (x_1, y_1, z_1) from an arbitrary point in the plane $y = -5$.

Ans. $\mathbf{a} = \dfrac{(x_1 - x)\mathbf{a}_x + (y_1 + 5)\mathbf{a}_y + (z_1 - z)\mathbf{a}_z}{\sqrt{(x_1 - x)^2 + (y_1 + 5)^2 + (z_1 - z)^2}}$

1.27. Express the unit vector directed toward the point $(0, 0, h)$ from an arbitrary point in the plane $z = -2$. Explain the result as h approaches -2.

Ans. $\mathbf{a} = \dfrac{-x\mathbf{a}_x - y\mathbf{a}_y + (h + 2)\mathbf{a}_z}{\sqrt{x^2 + y^2 + (h + 2)^2}}$

1.28. Given $\mathbf{A} = 5\mathbf{a}_x$ and $\mathbf{B} = 4\mathbf{a}_x + B_y\mathbf{a}_y$, find B_y such that the angle between **A** and **B** is 45°. If **B** also has a term $B_z\mathbf{a}_z$, what relationship must exist between B_y and B_z? *Ans.* $B_y = \pm 4$, $\sqrt{B_y^2 + B_z^2} = 4$

1.29. Show that the absolute value of $\mathbf{A} \cdot \mathbf{B} \times \mathbf{C}$ is the volume of the parallelepiped with edges **A**, **B**, and **C**. (*Hint*: First show that the base has area $|\mathbf{B} \times \mathbf{C}|$.)

1.30. Given $\mathbf{A} = 2\mathbf{a}_x - \mathbf{a}_z$, $\mathbf{B} = 3\mathbf{a}_x + \mathbf{a}_y$, and $\mathbf{C} = -2\mathbf{a}_x + 6\mathbf{a}_y - 4\mathbf{a}_z$, show that **C** is ⊥ to both **A** and **B**.

1.31. Given $\mathbf{A} = \mathbf{a}_x - \mathbf{a}_y$, $\mathbf{B} = 2\mathbf{a}_z$, and $\mathbf{C} = -\mathbf{a}_x + 3\mathbf{a}_y$, find $\mathbf{A} \cdot \mathbf{B} \times \mathbf{C}$. Examine other variations of the scalar triple product. *Ans.* -4

1.32. Using the vectors of Problem 1.31 find $(\mathbf{A} \times \mathbf{B}) \times \mathbf{C}$. *Ans.* $-8\mathbf{a}_z$

1.33. Find the unit vector directed from $(2, -5, -2)$ toward $(14, -5, 3)$.

Ans. $\mathbf{a} = \dfrac{12}{13}\mathbf{a}_x + \dfrac{5}{13}\mathbf{a}_z$

1.34. Show why the method of Problem 1.1 cannot be used in cylindrical coordinates for the points (r_1, ϕ_1, z_1) and (r_2, ϕ_2, z_2). Examine the same question for spherical coordinates.

1.35. Verify that the distance d between the two points of Problem 1.34 is given by

$$d^2 = r_1^2 + r_2^2 - 2r_1 r_2 \cos(\phi_2 - \phi_1) + (z_2 - z_1)^2$$

1.36. Find the vector directed from $(10, 3\pi/4, \pi/6)$ to $(5, \pi/4, \pi)$, where the points are given in spherical coordinates.
Ans. $-9.66\,\mathbf{a}_x - 3.54\,\mathbf{a}_y + 10.61\,\mathbf{a}_z$

1.37. Find the distance between $(2, \pi/6, 0)$ and $(1, \pi, 2)$, where the points are given in cylindrical coordinates.
Ans. 3.53

1.38. Find the distance between $(1, \pi/4, 0)$ and $(1, 3\pi/4, \pi)$, where the points are given in spherical coordinates.
Ans. 2.0

1.39. Use spherical coordinates and integrate to find the area of the region $0 \le \phi \le \alpha$ on the spherical shell of radius a. What is the result when $\alpha = 2\pi$? *Ans.* $2\alpha a^2$, $A = 4\pi a^2$

1.40. Use cylindrical coordinates to find the area of the curved surface of a right circular cylinder of radius a and height h. *Ans.* $2\pi ah$

1.41. Use cylindrical coordinates and integrate to obtain the volume of the right circular cylinder of Problem 1.40.
Ans. $\pi a^2 h$

1.42. Use spherical coordinates to write the differential surface areas dS_1 and dS_2 and then integrate to obtain the areas of the surfaces marked *1* and *2* in Fig. 1-15.
Ans. $\pi/4$, $\pi/6$

1.43. Use spherical coordinates to find the volume of a hemispherical shell of inner radius 2.00 m and outer radius 2.02 m. *Ans.* 0.162π m^3

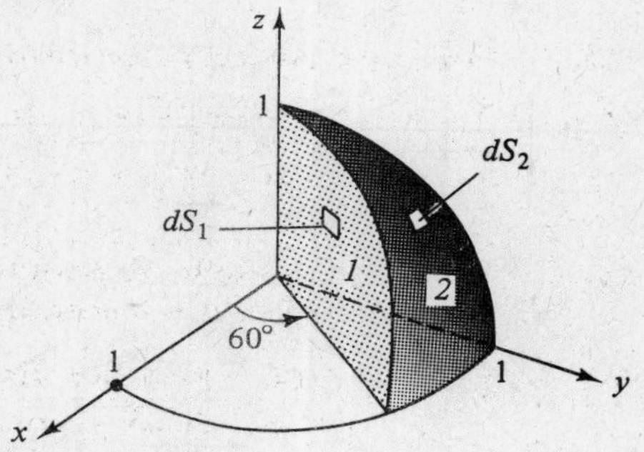

Fig. 1-15

1.44. Using spherical coordinates to express the differential volume, integrate to obtain the volume defined by $1 \le r \le 2$ m, $0 \le \theta \le \pi/2$, and $0 \le \phi \le \pi/2$. *Ans.* $\dfrac{7\pi}{6}$ m^3

1.45. Transform the vector $\mathbf{A} = A_x\mathbf{a}_x + A_y\mathbf{a}_y + A_z\mathbf{a}_z$ into cylindrical coordinates.
Ans. $\mathbf{A} = (A_x \cos\phi + A_y \sin\phi)\mathbf{a}_r + (-A_x \sin\phi + A_y \cos\phi)\mathbf{a}_\phi + A_z\mathbf{a}_z$

1.46. Transform the vector $\mathbf{A} = A_r\mathbf{a}_r + A_\theta\mathbf{a}_\theta + A_\phi\mathbf{a}_\phi$ into cartesian coordinates.

Ans.
$$\mathbf{A} = \left(\frac{A_r x}{\sqrt{x^2+y^2+z^2}} + \frac{A_\theta xz}{\sqrt{x^2+y^2+z^2}\sqrt{x^2+y^2}} - \frac{A_\phi y}{\sqrt{x^2+y^2}}\right)\mathbf{a}_x$$
$$+ \left(\frac{A_r y}{\sqrt{x^2+y^2+z^2}} + \frac{A_\theta yz}{\sqrt{x^2+y^2+z^2}\sqrt{x^2+y^2}} + \frac{A_\phi x}{\sqrt{x^2+y^2}}\right)\mathbf{a}_y$$
$$+ \left(\frac{A_r z}{\sqrt{x^2+y^2+z^2}} - \frac{A_\theta \sqrt{x^2+y^2}}{\sqrt{x^2+y^2+z^2}}\right)\mathbf{a}_z$$

1.47. Transform the vector $\mathbf{F} = r^{-1}\mathbf{a}_r$ in spherical coordinates into cartesian coordinates.

Ans. $\mathbf{F} = \dfrac{x\mathbf{a}_x + y\mathbf{a}_y + z\mathbf{a}_z}{x^2 + y^2 + z^2}$

1.48. In cylindrical coordinates $r = \text{const.}$ defines a right circular cylinder and $\mathbf{F} = F\mathbf{a}_r$ describes a force everywhere normal to the surface. Express the surface and the force in cartesian coordinates.

Ans. $x^2 + y^2 = \text{const.}$, $\mathbf{F} = \dfrac{x\mathbf{a}_x + y\mathbf{a}_y}{\sqrt{x^2 + y^2}}$

1.49. Transform the vector field $\mathbf{F} = 2\cos\theta\,\mathbf{a}_r + \sin\theta\,\mathbf{a}_\theta$ into cartesian coordinates.

Ans. $\mathbf{F} = \dfrac{3xz\mathbf{a}_x + 3yz\mathbf{a}_y + (2z^2 - x^2 - y^2)\mathbf{a}_z}{x^2 + y^2 + z^2}$

1.50. Sketch the vector field $\mathbf{F} = y\mathbf{a}_x + x\mathbf{a}_y$. *Ans.* See Fig. 1-16.

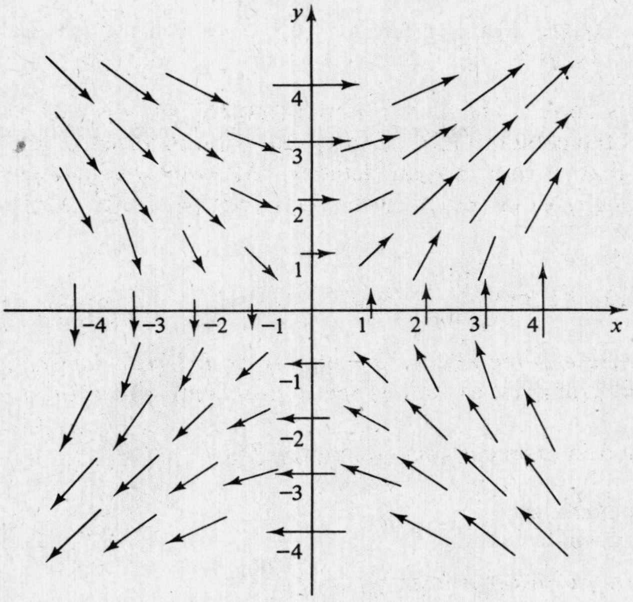

Fig. 1-16

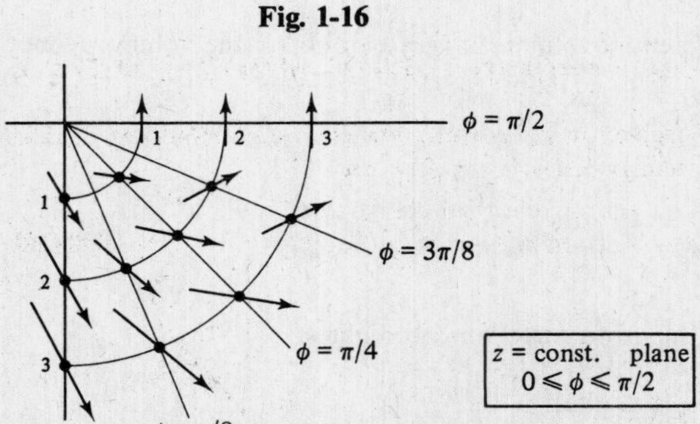

Fig. 1-17

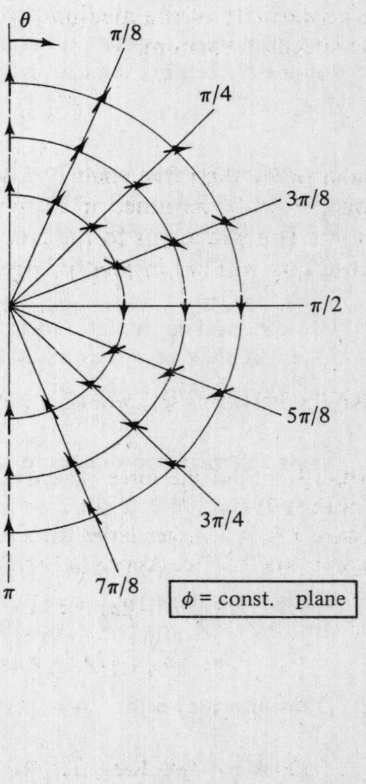

Fig. 1-18

1.51. Sketch the cylindrical coordinate field $\mathbf{F} = 2r\cos\phi\,\mathbf{a}_r + r\mathbf{a}_\phi$. *Ans.* See Fig. 1-17.

1.52. Sketch the vector field of Problem 1.49, using spherical coordinates. *Ans.* See Fig. 1-18.

Chapter 2

Coulomb Forces and Electric Field Intensity

2.1 COULOMB'S LAW

There is a force between two charges which is directly proportional to the charge magnitudes and inversely proportional to the square of the separation distance. This is *Coulomb's law*, which was developed from work with small charged bodies and a delicate torsion balance. In vector form, it is stated thus,

$$\mathbf{F} = \frac{Q_1 Q_2}{4\pi\epsilon d^2}\mathbf{a}$$

Rationalized SI units will be used throughout this book. The force is in newtons (N), the distance is in meters (m), and the (derived) unit of charge is the coulomb (C). The system is rationalized by the factor 4π, introduced in this law in order that it not appear later in Maxwell's equations. ϵ is the *permittivity* of the medium, with the units $C^2/N \cdot m^2$ or, equivalently, farads per meter (F/m). For free space or vacuum,

$$\epsilon = \epsilon_0 = 8.854 \times 10^{-12}\,\text{F/m} \approx \frac{10^{-9}}{36\pi}\ \text{F/m}$$

For media other than free space, $\epsilon = \epsilon_0 \epsilon_r$, where ϵ_r is the *relative permittivity* or *dielectric constant*. Free space is to be assumed in all problems and examples, as well as the approximate value for ϵ_0, unless there is a statement to the contrary.

Subscripts will aid in identifying the force and in expressing its direction. Thus,

$$\mathbf{F}_1 = \frac{Q_1 Q_2}{4\pi\epsilon_0 d^2}\mathbf{a}_{21}$$

describes the force on Q_1, where the unit vector $\mathbf{a}_{21}$ is directed from Q_2 to Q_1.

EXAMPLE 1 Find the force on charge Q_1, 20 μC, due to charge Q_2, -300 μC, where Q_1 is at (0, 1, 2) m and Q_2 at (2, 0, 0) m.

Because 1 C is a rather large unit, charges are often given in microcoulombs (μC), nanocoulombs (nC), or picocoulombs (pC). (See Appendix for the SI prefix system.) Referring to Fig. 2-1,

$$\mathbf{R}_{21} = -2\mathbf{a}_x + \mathbf{a}_y + 2\mathbf{a}_z$$

$$\mathbf{a}_{21} = \frac{1}{3}(-2\mathbf{a}_x + \mathbf{a}_y + 2\mathbf{a}_z)$$

Then

$$\mathbf{F}_1 = \frac{(20 \times 10^{-6})(-300 \times 10^{-6})}{4\pi(10^{-9}/36\pi)(3)^2}\left(\frac{-2\mathbf{a}_x + \mathbf{a}_y + 2\mathbf{a}_z}{3}\right)$$

$$= 6\left(\frac{2\mathbf{a}_x - \mathbf{a}_y - 2\mathbf{a}_z}{3}\right)\ \text{N}$$

The force magnitude is 6 N and the direction is such that Q_1 is attracted to Q_2.

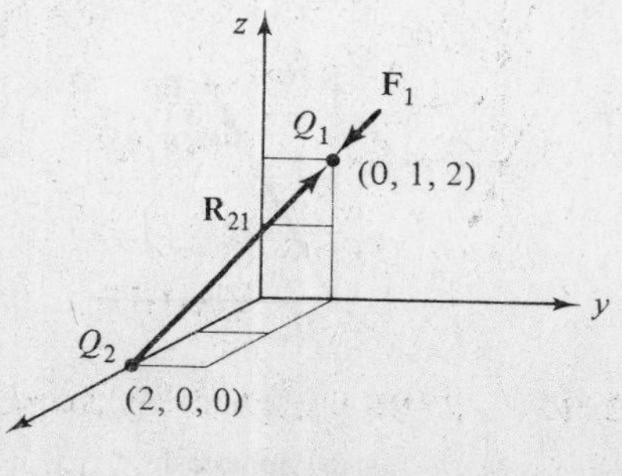

Fig. 2-1

In the region around an isolated point charge there is a spherically symmetrical *force field*. This is made evident when charge Q is fixed at the origin, as in Fig. 2-2, and a second charge, Q_T, is moved about in the region. At each location a force acts along the line joining the two charges, directed away from the origin if the charges are of like sign. This can be expressed in spherical coordinates by

$$\mathbf{F}_T = \frac{QQ_T}{4\pi\epsilon_0 r^2}\mathbf{a}_r$$

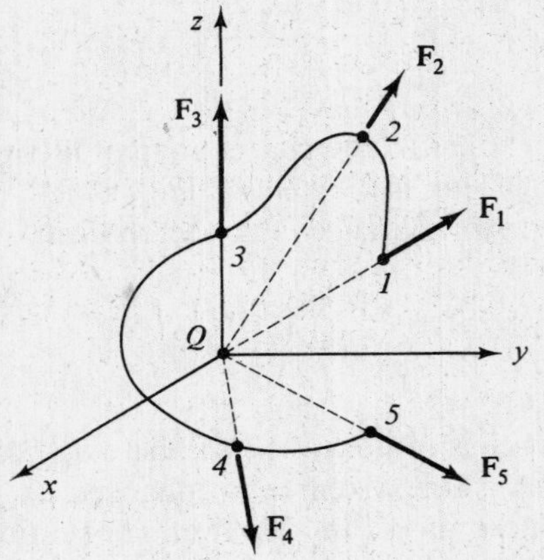

Fig. 2-2

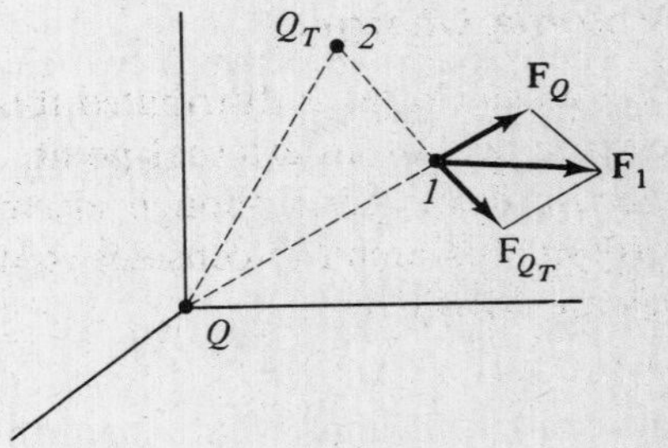

Fig. 2-3

It should be noted that unless $Q_T \ll Q$ the symmetrical field about Q is disturbed by Q_T. At location *1* in Fig. 2-3 the force $\mathbf{F}_1$ is seen to be the vector sum

$$\mathbf{F}_1 = \mathbf{F}_{Q_T} + \mathbf{F}_Q$$

This should come as no surprise, since if Q has a force field so also must Q_T. When the two charges are in the same region, the resulting field will of necessity be the point-by-point vector sum of the two fields. This is the *superposition principle* for coulomb forces; it extends to any number of charges.

2.2 ELECTRIC FIELD INTENSITY

Suppose that, in the above situation, the test charge Q_T is sufficiently small so as not to disturb significantly the field of the fixed point charge Q. Then the *electric field intensity*, $\mathbf{E}$, due to Q is defined to be the force per unit charge on Q_T:

$$\mathbf{E} = \frac{1}{Q_T}\mathbf{F}_T = \frac{Q}{4\pi\epsilon_0 r^2}\mathbf{a}_r$$

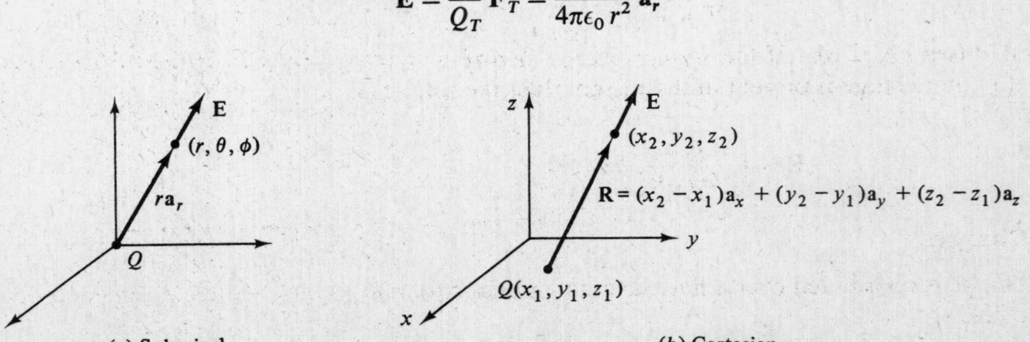

(*a*) Spherical (*b*) Cartesian

Fig. 2-4

This expression for **E** is in spherical coordinates with origin at the location of Q [Fig. 2-4(a)]; it may be transformed to other coordinate systems by the method of Section 1.6. In an arbitrary cartesian coordinate system,

$$\mathbf{E} = \frac{Q}{4\pi\epsilon_0 R^2}\mathbf{a}_R$$

where the separation vector **R** is as given in Fig. 2-4(b).

The units of **E** are newtons per coulomb (N/C) or the equivalent, volts per meter (V/m).

2.3 CHARGE DISTRIBUTIONS

Volume Charge.

When charge is distributed throughout a specified volume, each charge element contributes to the electric field at an external point. A summation or integration is then required to obtain the total electric field. Even though electric charge in its smallest division is found to be an electron or proton, it is useful to consider continuous (in fact, differentiable) charge distributions and to define a *charge density* by

$$\rho = \frac{dQ}{dv} \quad (\text{C/m}^3)$$

Note the units in parentheses, which is meant to signify that ρ will be in C/m^3 provided that the variables are expressed in proper SI units (C for Q and m^3 for v). This convention will be used throughout this book.

With reference to volume v in Fig. 2-5, each differential charge dQ produces a differential electric field

$$d\mathbf{E} = \frac{dQ}{4\pi\epsilon_0 R^2}\mathbf{a}_R$$

at the observation point P. Assuming that the only charge in the region is contained within the volume, the total electric field at P is obtained by integration over the volume:

$$\mathbf{E} = \int_v \frac{\rho\mathbf{a}_R}{4\pi\epsilon_0 R^2}\,dv$$

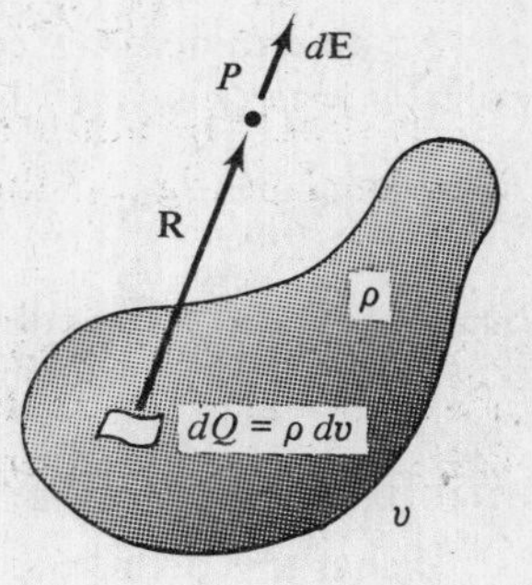

Fig. 2-5

Sheet Charge.

Charge may also be distributed over a surface or a sheet. Then each differential charge dQ on the sheet results in a differential electric field

$$d\mathbf{E} = \frac{dQ}{4\pi\epsilon_0 R^2}\mathbf{a}_R$$

at point P (see Fig. 2-6). If the *surface charge density* is ρ_s (C/m^2) and if no other charge is present in the region, then the total electric field at P is

$$\mathbf{E} = \int_S \frac{\rho_s\mathbf{a}_R}{4\pi\epsilon_0 R^2}\,dS$$

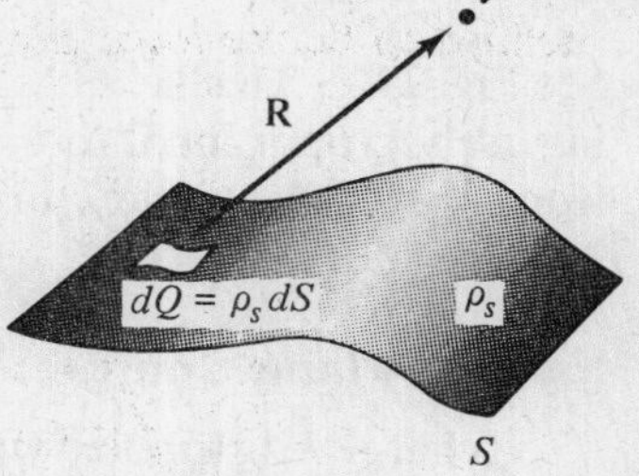

Fig. 2-6

Line Charge.

If charge is distributed over a line, each differential charge dQ along the line produces a differential electric field

$$d\mathbf{E} = \frac{dQ}{4\pi\epsilon_0 R^2}\mathbf{a}_R$$

at P (see Fig. 2-7). And if the *line charge density* is ρ_ℓ (C/m), and no other charge is in the region, then the total electric field at P is

$$\mathbf{E} = \int_L \frac{\rho_\ell \, \mathbf{a}_R}{4\pi\epsilon_0 R^2} \, d\ell$$

It should be emphasized that in all three of the above charge distributions and corresponding integrals for **E**, the unit vector $\mathbf{a}_R$ is variable, depending on the coordinates of the charge element dQ. Thus $\mathbf{a}_R$ cannot be removed from the integrand.

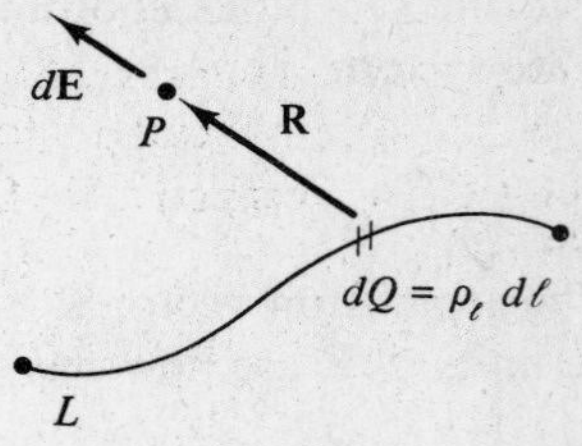

Fig. 2-7

2.4 STANDARD CHARGE CONFIGURATIONS

In three special cases the integration discussed in Section 2.3 is either unnecessary or easily carried out. In regard to these standard configurations (and to others which will be covered in this chapter) it should be noted that the charge is not "on a conductor." When a problem states that charge is distributed in the form of a disk, for example, it does not mean a disk-shaped conductor with charge on the surface. (In Chapter 6, conductors with surface charge will be examined.) Although it may now require a stretch of the imagination, these charges should be thought of as somehow suspended in space, fixed in the specified configuration.

Point Charge.

As determined in Section 2.3, the field of a single point charge Q is given by

$$\mathbf{E} = \frac{Q}{4\pi\epsilon_0 r^2} \mathbf{a}_r \qquad \text{(spherical coordinates)}$$

See Fig. 2-4(a). This is a spherically symmetric field that follows an *inverse-square law* (like gravitation).

Infinite Line Charge.

If charge is distributed with *uniform* density ρ_ℓ (C/m) along an *infinite* straight line—which will be chosen as the z axis—then the field is given by

$$\mathbf{E} = \frac{\rho_\ell}{2\pi\epsilon_0 r} \mathbf{a}_r \qquad \text{(cylindrical coordinates)}$$

See Fig. 2-8. This field has cylindrical symmetry and is inversely proportional to the *first power* of the distance from the line charge. For a derivation of **E**, see Problem 2.9.

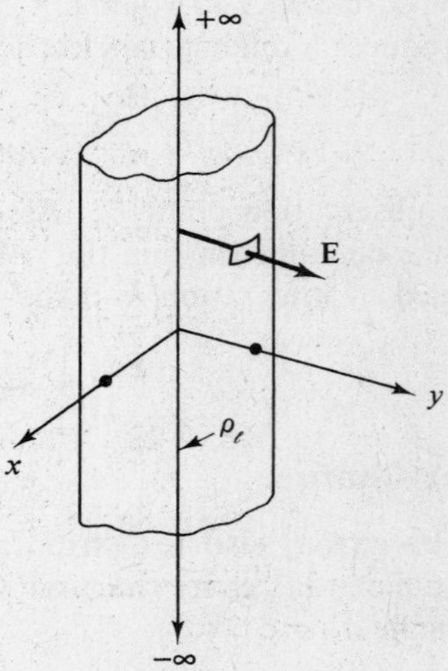

Fig. 2-8

Infinite Plane Charge.

If charge is distributed with *uniform* density ρ_s (C/m²) over an *infinite* plane, then the field is given by

$$\mathbf{E} = \frac{\rho_s}{2\epsilon_0} \mathbf{a}_n$$

See Fig. 2-9. This field is of constant magnitude and has mirror symmetry about the plane charge. For a derivation of **E**, see Problem 2.12.

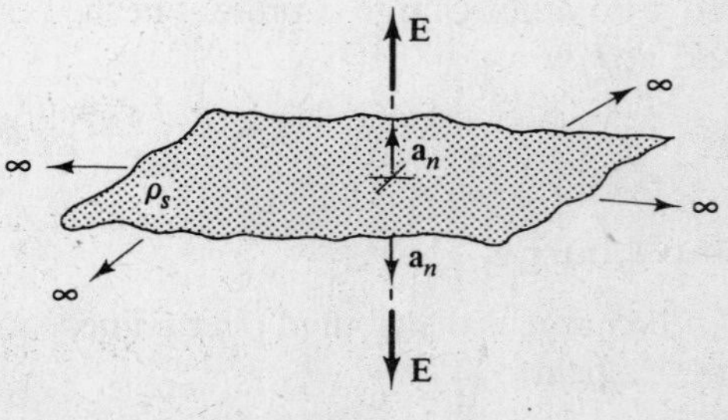

Fig. 2-9

Solved Problems

2.1. Two point charges, $Q_1 = 50\ \mu C$ and $Q_2 = 10\ \mu C$, are located at $(-1, 1, -3)$ m and $(3, 1, 0)$ m respectively (Fig. 2-10). Find the force on Q_1.

$$\mathbf{R}_{21} = -4\mathbf{a}_x - 3\mathbf{a}_z$$

$$\mathbf{a}_{21} = \frac{-4\mathbf{a}_x - 3\mathbf{a}_z}{5}$$

$$\mathbf{F}_1 = \frac{Q_1 Q_2}{4\pi\epsilon_0 R_{21}^2}\mathbf{a}_{21}$$

$$= \frac{(50 \times 10^{-6})(10^{-5})}{4\pi(10^{-9}/36\pi)(5)^2}\left(\frac{-4\mathbf{a}_x - 3\mathbf{a}_z}{5}\right)$$

$$= (0.18)(-0.8\mathbf{a}_x - 0.6\mathbf{a}_z)\ \text{N}$$

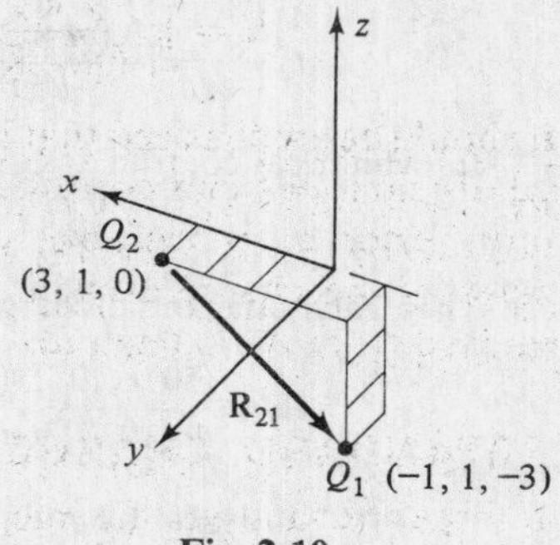

Fig. 2-10

The force has a magnitude of 0.18 N and a direction given by the unit vector $-0.8\mathbf{a}_x - 0.6\mathbf{a}_z$. In component form,

$$\mathbf{F}_1 = -0.144\mathbf{a}_x - 0.108\mathbf{a}_z\ \text{N}$$

2.2. Refer to Fig. 2-11. Find the force on a $100\ \mu C$ charge at $(0, 0, 3)$ m if four like charges of $20\ \mu C$ are located on the x and y axes at ± 4 m.

Consider the force due to the charge at $y = 4$,

$$\frac{(10^{-4})(20 \times 10^{-6})}{4\pi(10^{-9}/36\pi)(5)^2}\left(\frac{-4\mathbf{a}_y + 3\mathbf{a}_z}{5}\right)$$

The y component will be canceled by the charge at $y = -4$. Similarly, the x components due to the other two charges will cancel. Hence

$$\mathbf{F} = 4\left(\frac{18}{25}\right)\left(\frac{3}{5}\mathbf{a}_z\right) = 1.73\mathbf{a}_z\ \text{N}$$

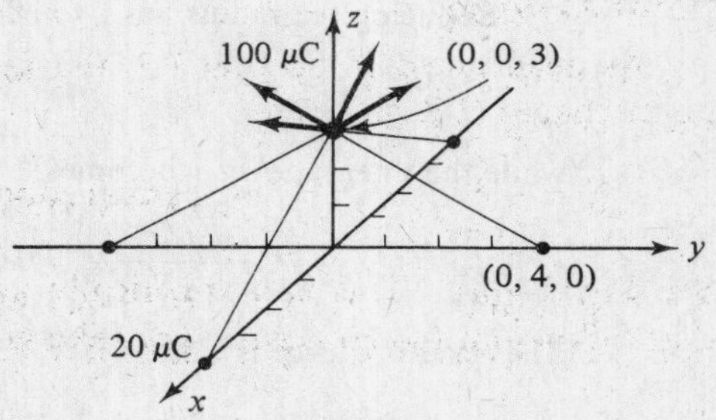

Fig. 2-11

2.3. Refer to Fig. 2-12. Point charge $Q_1 = 300\ \mu C$, located at $(1, -1, -3)$ m, experiences a force

$$\mathbf{F}_1 = 8\mathbf{a}_x - 8\mathbf{a}_y + 4\mathbf{a}_z\ \ \text{N}$$

due to point charge Q_2 at $(3, -3, -2)$ m. Determine Q_2.

$$\mathbf{R}_{21} = -2\mathbf{a}_x + 2\mathbf{a}_y - \mathbf{a}_z$$

Note that, because

$$\frac{8}{-2} = \frac{-8}{2} = \frac{4}{-1}$$

the given force is along $\mathbf{R}_{21}$ (see Problem 1.24), as it must be.

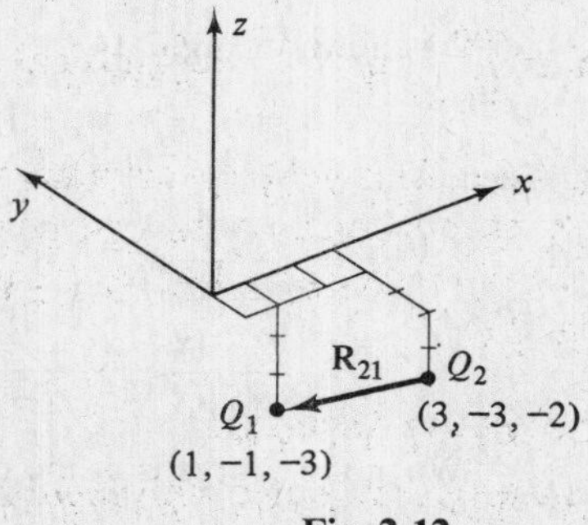

Fig. 2-12

$$\mathbf{F}_1 = \frac{Q_1 Q_2}{4\pi\epsilon_0 R^2}\mathbf{a}_R$$

$$8\mathbf{a}_x - 8\mathbf{a}_y + 4\mathbf{a}_z = \frac{(300 \times 10^{-6})Q_2}{4\pi(10^{-9}/36\pi)(3)^2}\left(\frac{-2\mathbf{a}_x + 2\mathbf{a}_y - \mathbf{a}_z}{3}\right)$$

Solving, $Q_2 = -40\ \mu C$.

2.4. Find the force on a point charge of 50 μC at (0, 0, 5) m due to a charge of 500π μC that is uniformly distributed over the circular disk $r \leq 5$ m, $z = 0$ m (see Fig. 2-13).

The charge density is

$$\rho_s = \frac{Q}{A} = \frac{500\pi \times 10^{-6}}{\pi(5)^2} = 0.2 \times 10^{-4}\ \text{C/m}^2$$

In cylindrical coordinates,

$$\mathbf{R} = -r\mathbf{a}_r + 5\mathbf{a}_z$$

Then each differential charge results in a differential force

$$d\mathbf{F} = \frac{(50 \times 10^{-6})(\rho_s\, r\, dr\, d\phi)}{4\pi(10^{-9}/36\pi)(r^2 + 25)}\left(\frac{-r\mathbf{a}_r + 5\mathbf{a}_z}{\sqrt{r^2 + 25}}\right)$$

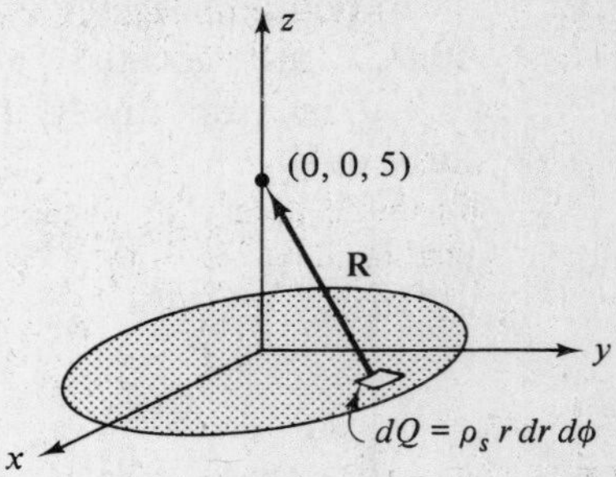

Fig. 2-13

Before integrating, note that the radial components will cancel and that $\mathbf{a}_z$ is constant. Hence,

$$\mathbf{F} = \int_0^{2\pi}\int_0^5 \frac{(50 \times 10^{-6})(0.2 \times 10^{-4})5r\,dr\,d\phi}{4\pi(10^{-9}/36\pi)(r^2 + 25)^{3/2}}\mathbf{a}_z$$

$$= 90\pi \int_0^5 \frac{r\,dr}{(r^2 + 25)^{3/2}}\mathbf{a}_z = 90\pi\left[\frac{-1}{\sqrt{r^2 + 25}}\right]_0^5 \mathbf{a}_z = 16.56\mathbf{a}_z\ \text{N}$$

2.5. Repeat Problem 2.4 for a disk of radius 2 m.

Reducing the radius has two effects: the charge density is increased by a factor

$$\frac{\rho_2}{\rho_1} = \frac{(5)^2}{(2)^2} = 6.25$$

while the integral over r becomes

$$\int_0^2 \frac{r\,dr}{(r^2 + 25)^{3/2}} = 0.0143 \qquad \text{instead of} \qquad \int_0^5 \frac{r\,dr}{(r^2 + 25)^{3/2}} = 0.0586$$

The resulting force is

$$\mathbf{F} = (6.25)\left(\frac{0.0143}{0.0586}\right)(16.56\mathbf{a}_z\ \text{N}) = 25.27\mathbf{a}_z\ \text{N}$$

2.6. Find the expression for the electric field at P due to a point charge Q at (x_1, y_1, z_1). Repeat with the charge placed at the origin.

As shown in Fig. 2-14,

$$\mathbf{R} = (x - x_1)\mathbf{a}_x + (y - y_1)\mathbf{a}_y + (z - z_1)\mathbf{a}_z$$

Then

$$\mathbf{E} = \frac{Q}{4\pi\epsilon_0 R^2}\mathbf{a}_R$$

$$= \frac{Q}{4\pi\epsilon_0}\frac{(x - x_1)\mathbf{a}_x + (y - y_1)\mathbf{a}_y + (z - z_1)\mathbf{a}_z}{[(x - x_1)^2 + (y - y_1)^2 + (z - z_1)^2]^{3/2}}$$

When the charge is at the origin,

$$\mathbf{E} = \frac{Q}{4\pi\epsilon_0}\frac{x\mathbf{a}_x + y\mathbf{a}_y + z\mathbf{a}_z}{(x^2 + y^2 + z^2)^{3/2}}$$

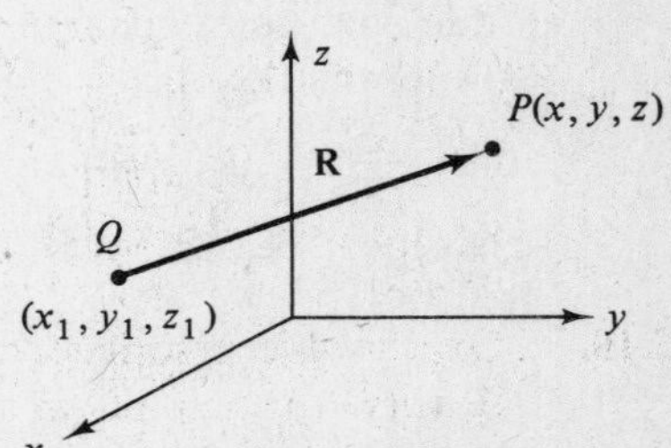

Fig. 2-14

but this expression fails to show the symmetry of the field. In spherical coordinates with Q at the origin,

$$\mathbf{E} = \frac{Q}{4\pi\epsilon_0 r^2}\mathbf{a}_r$$

and now the symmetry is apparent.

2.7. Find **E** at the origin due to a point charge of 64.4 nC located at $(-4, 3, 2)$ m in cartesian coordinates.

The electric field intensity due to a point charge Q at the origin in spherical coordinates is

$$\mathbf{E} = \frac{Q}{4\pi\epsilon_0 r^2}\mathbf{a}_r$$

In this problem the distance is $\sqrt{29}$ m and the vector from the charge to the origin, where **E** is to be evaluated, is $\mathbf{R} = 4\mathbf{a}_x - 3\mathbf{a}_y - 2\mathbf{a}_z$.

$$\mathbf{E} = \frac{64.4 \times 10^{-9}}{4\pi(10^{-9}/36\pi)(29)}\left(\frac{4\mathbf{a}_x - 3\mathbf{a}_y - 2\mathbf{a}_z}{\sqrt{29}}\right) = (20.0)\left(\frac{4\mathbf{a}_x - 3\mathbf{a}_y - 2\mathbf{a}_z}{\sqrt{29}}\right) \quad \text{V/m}$$

2.8. Find **E** at $(0, 0, 5)$ m due to $Q_1 = 0.35\ \mu\text{C}$ at $(0, 4, 0)$ m and $Q_2 = -0.55\ \mu\text{C}$ at $(3, 0, 0)$ m (see Fig. 2-15).

$$\mathbf{R}_1 = -4\mathbf{a}_y + 5\mathbf{a}_z$$

$$\mathbf{R}_2 = -3\mathbf{a}_x + 5\mathbf{a}_z$$

$$\mathbf{E}_1 = \frac{0.35 \times 10^{-6}}{4\pi(10^{-9}/36\pi)(41)}\left(\frac{-4\mathbf{a}_y + 5\mathbf{a}_z}{\sqrt{41}}\right)$$

$$= -48.0\mathbf{a}_y + 60.0\mathbf{a}_z \quad \text{V/m}$$

$$\mathbf{E}_2 = \frac{-0.55 \times 10^{-6}}{4\pi(10^{-9}/36\pi)(34)}\left(\frac{-3\mathbf{a}_x + 5\mathbf{a}_z}{\sqrt{34}}\right)$$

$$= 74.9\mathbf{a}_x - 124.9\mathbf{a}_z \quad \text{V/m}$$

and $\mathbf{E} = \mathbf{E}_1 + \mathbf{E}_2 = 74.9\mathbf{a}_x - 48.0\mathbf{a}_y - 64.9\mathbf{a}_z$ V/m

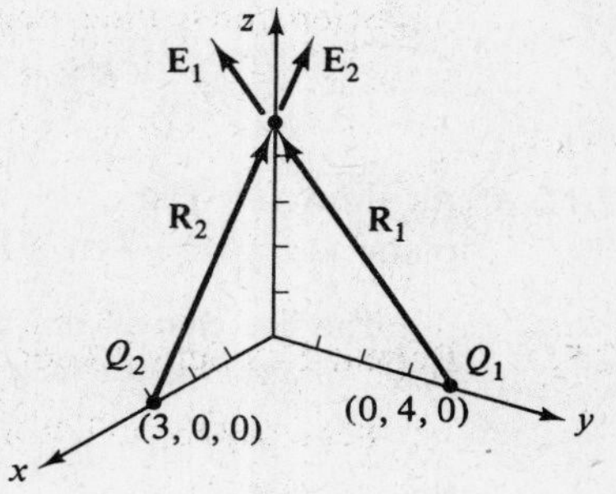

Fig. 2-15

2.9. Charge is distributed uniformly along an infinite straight line with density ρ_ℓ. Develop the expression for **E** at the general point P.

Cylindrical coordinates will be used, with the line charge as the z axis (see Fig. 2-16). At P,

$$d\mathbf{E} = \frac{dQ}{4\pi\epsilon_0 R^2}\left(\frac{r\mathbf{a}_r - z\mathbf{a}_z}{\sqrt{r^2 + z^2}}\right)$$

Since for every dQ at z there is another charge dQ at $-z$, the z components cancel. Then

$$\mathbf{E} = \int_{-\infty}^{\infty} \frac{\rho_\ell r\,dz}{4\pi\epsilon_0(r^2 + z^2)^{3/2}}\mathbf{a}_r$$

$$= \frac{\rho_\ell r}{4\pi\epsilon_0}\left[\frac{z}{r^2\sqrt{r^2 + z^2}}\right]_{-\infty}^{\infty}\mathbf{a}_r = \frac{\rho_\ell}{2\pi\epsilon_0 r}\mathbf{a}_r$$

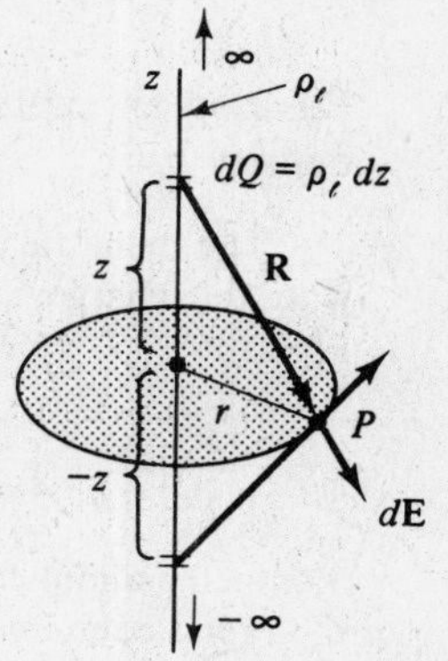

Fig. 2-16

2.10. On the line described by $x = 2$ m, $y = -4$ m there is a uniform charge distribution of density $\rho_\ell = 20$ nC/m. Determine the electric field **E** at $(-2, -1, 4)$ m.

With some modification for cartesian coordinates the expression obtained in Problem 2.9 can be used with this uniform line charge. Since the line is parallel to $\mathbf{a}_z$, the field has no z component. Referring to Fig. 2-17,

$$\mathbf{R} = -4\mathbf{a}_x + 3\mathbf{a}_y$$

and

$$\mathbf{E} = \frac{20 \times 10^{-9}}{2\pi\epsilon_0(5)}\left(\frac{-4\mathbf{a}_x + 3\mathbf{a}_y}{5}\right) = -57.6\mathbf{a}_x + 43.2\mathbf{a}_y \quad \text{V/m}$$

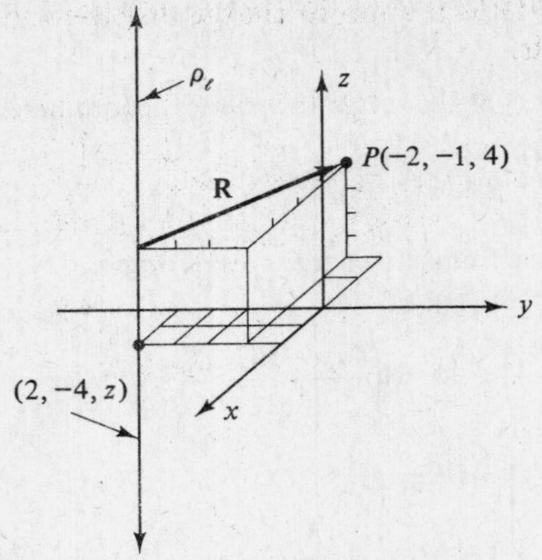

Fig. 2-17

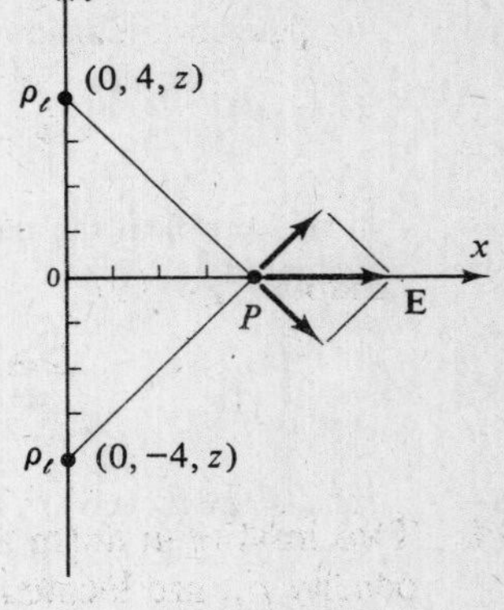

Fig. 2-18

2.11. As shown in Fig. 2-18, two uniform line charges of density $\rho_\ell = 4$ nC/m lie in the $x = 0$ plane at $y = \pm 4$ m. Find **E** at (4, 0, 10) m.

The line charges are both parallel to $\mathbf{a}_z$; their fields are radial and parallel to the xy plane. For either line charge the magnitude of the field at P would be

$$E = \frac{\rho_\ell}{2\pi\epsilon_0 r} = \frac{18}{\sqrt{2}} \text{ V/m}$$

The field due to both line charges is, by superposition,

$$\mathbf{E} = 2\left(\frac{18}{\sqrt{2}} \cos 45°\right)\mathbf{a}_x = 18\mathbf{a}_x \text{ V/m}$$

2.12. Develop an expression for **E** due to charge uniformly distributed over an infinite plane with density ρ_s.

The cylindrical coordinate system will be used, with the charge in the $z = 0$ plane as shown in Fig. 2-19.

$$d\mathbf{E} = \frac{\rho_s\, r\, dr\, d\phi}{4\pi\epsilon_0(r^2 + z^2)}\left(\frac{-r\mathbf{a}_r + z\mathbf{a}_z}{\sqrt{r^2 + z^2}}\right)$$

Symmetry about the z axis results in cancellation of the radial components.

$$\mathbf{E} = \int_0^{2\pi}\int_0^{\infty} \frac{\rho_s\, rz\, dr\, d\phi}{4\pi\epsilon_0(r^2 + z^2)^{3/2}}\,\mathbf{a}_z = \frac{\rho_s z}{2\epsilon_0}\left[\frac{-1}{\sqrt{r^2 + z^2}}\right]_0^{\infty}\mathbf{a}_z = \frac{\rho_s}{2\epsilon_0}\,\mathbf{a}_z$$

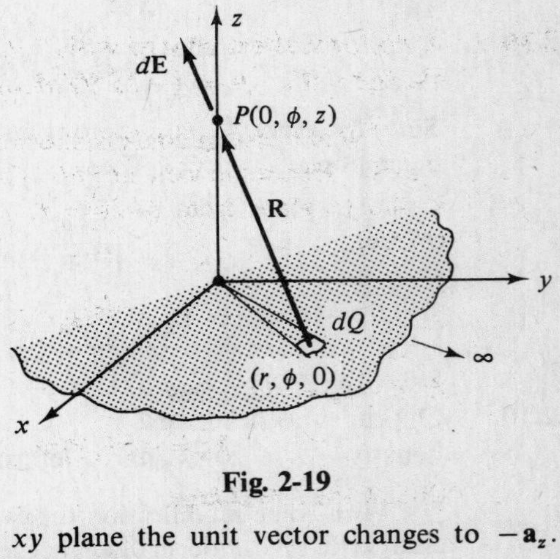

Fig. 2-19

This result is for points above the xy plane. Below the xy plane the unit vector changes to $-\mathbf{a}_z$. The generalized form may be written using $\mathbf{a}_n$, the unit normal vector:

$$E = \frac{\rho_s}{2\epsilon_0}\,\mathbf{a}_n$$

The electric field is everywhere normal to the plane of the charge and its magnitude is independent of the distance from the plane.

2.13. As shown in Fig. 2-20, the plane $y = 3$ m contains a uniform charge distribution of density $\rho_s = (10^{-8}/6\pi)$ C/m². Determine **E** at all points.

For $y > 3$ m,

$$\mathbf{E} = \frac{\rho_s}{2\epsilon_0}\mathbf{a}_n = 30\mathbf{a}_y \text{ V/m}$$

and for $y < 3$ m,

$$\mathbf{E} = -30\mathbf{a}_y \text{ V/m}$$

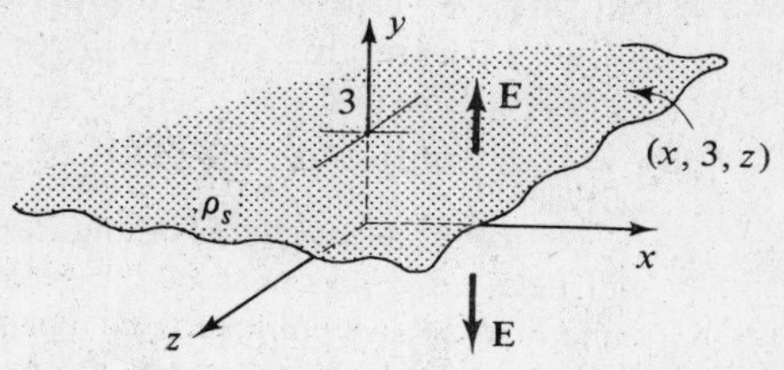

Fig. 2-20

2.14. Two infinite uniform sheets of charge, each with density ρ_s, are located at $x = \pm 1$ (Fig. 2-21). Determine **E** in all regions.

Only parts of the two sheets of charge are shown in Fig. 2-21. Both sheets result in **E** fields that are directed along x, independent of the distance. Then

$$\mathbf{E}_1 + \mathbf{E}_2 = \begin{cases} -(\rho_s/\epsilon_0)\mathbf{a}_x & x < -1 \\ 0 & -1 < x < 1 \\ (\rho_s/\epsilon_0)\mathbf{a}_x & x > 1 \end{cases}$$

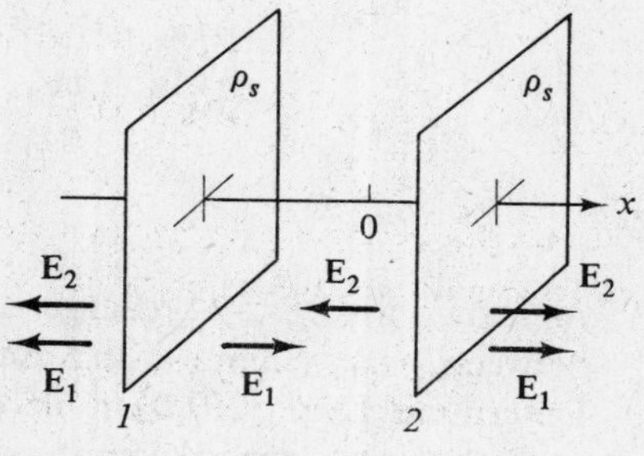

Fig. 2-21

2.15. Repeat Problem 2.14 with ρ_s on $x = -1$ and $-\rho_s$ on $x = 1$.

$$\mathbf{E}_1 + \mathbf{E}_2 = \begin{cases} 0 & x < -1 \\ (\rho_s/\epsilon_0)\mathbf{a}_x & -1 < x < 1 \\ 0 & x > 1 \end{cases}$$

2.16. A uniform sheet charge with $\rho_s = (1/3\pi)$ nC/m² is located at $z = 5$ m and a uniform line charge with $\rho_\ell = (-25/9)$ nC/m at $z = -3$ m, $y = 3$ m. Find **E** at $(x, -1, 0)$ m.

The two charge configurations are parallel to the x axis. Hence the view in Fig. 2-22 is taken looking at the yz plane from positive x. Due to the sheet charge,

$$\mathbf{E}_s = \frac{\rho_s}{2\epsilon_0}\mathbf{a}_n$$

At P, $\mathbf{a}_n = -\mathbf{a}_z$ and

$$\mathbf{E}_s = -6\mathbf{a}_z \text{ V/m}$$

Due to the line charge,

$$\mathbf{E}_\ell = \frac{\rho_\ell}{2\pi\epsilon_0 r}\mathbf{a}_r$$

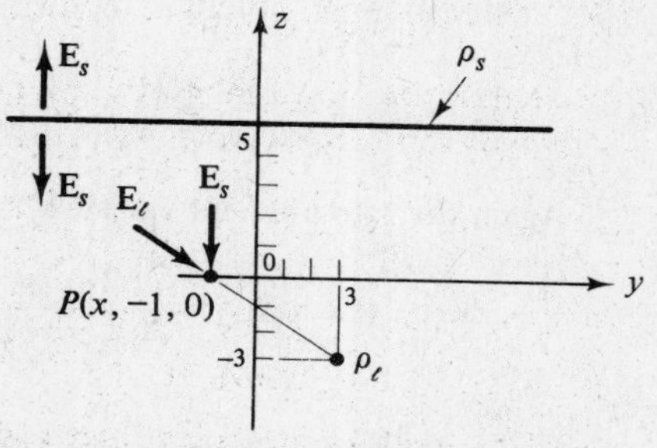

Fig. 2-22

and at P

$$\mathbf{E}_\ell = 8\mathbf{a}_y - 6\mathbf{a}_z \quad \text{V/m}$$

The total electric field is the sum, $\mathbf{E} = \mathbf{E}_\ell + \mathbf{E}_s = 8\mathbf{a}_y - 12\mathbf{a}_z$ V/m.

2.17. Determine **E** at (2, 0, 2) m due to three standard charge distributions as follows: a uniform sheet at $x = 0$ m with $\rho_{s1} = (1/3\pi)\ \text{nC/m}^2$, a uniform sheet at $x = 4$ m with $\rho_{s2} = (-1/3\pi)\ \text{nC/m}^2$, and a uniform line at $x = 6$ m, $y = 0$ m with $\rho_\ell = -2$ nC/m.

Since the three charge configurations are parallel with $\mathbf{a}_z$, there will be no z component of the field. Point (2, 0, 2) will have the same field as any point $(2, 0, z)$. In Fig. 2-23, P is located between the two sheet charges, where the fields add due to the difference in sign.

$$\mathbf{E} = \frac{\rho_{s1}}{2\epsilon_0}\mathbf{a}_n + \frac{\rho_{s2}}{2\epsilon_0}\mathbf{a}_n + \frac{\rho_\ell}{2\pi\epsilon_0 r}\mathbf{a}_r$$
$$= 6\mathbf{a}_x + 6\mathbf{a}_x + 9\mathbf{a}_x$$
$$= 21\mathbf{a}_x \text{ V/m}$$

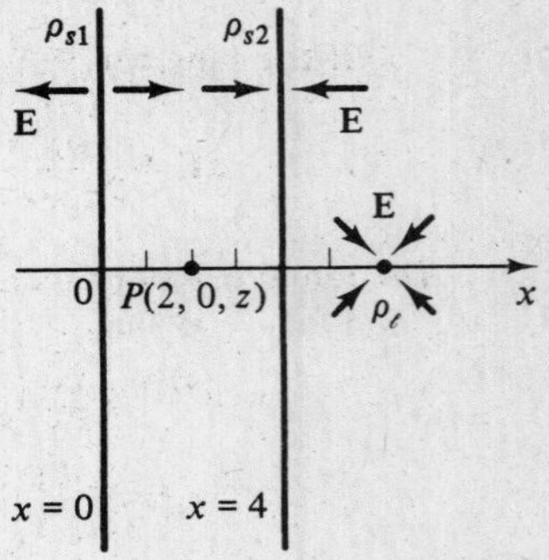

Fig. 2-23

2.18. As shown in Fig. 2-24, charge is distributed along the z axis between $z = \pm 5$ m with a uniform density $\rho_\ell = 20$ nC/m. Determine **E** at (2, 0, 0) m in cartesian coordinates. Also express the answer in cylindrical coordinates.

$$d\mathbf{E} = \frac{20 \times 10^{-9}\, dz}{4\pi(10^{-9}/36\pi)(4 + z^2)}\left(\frac{2\mathbf{a}_x - z\mathbf{a}_z}{\sqrt{4 + z^2}}\right) \quad \text{(V/m)}$$

Symmetry with respect to the $z = 0$ plane eliminates any z component in the result.

$$\mathbf{E} = 180 \int_{-5}^{5} \frac{2\, dz}{(4 + z^2)^{3/2}}\mathbf{a}_x = 167\mathbf{a}_x \text{ V/m}$$

In cylindrical coordinates, $\mathbf{E} = 167\mathbf{a}_r$ V/m.

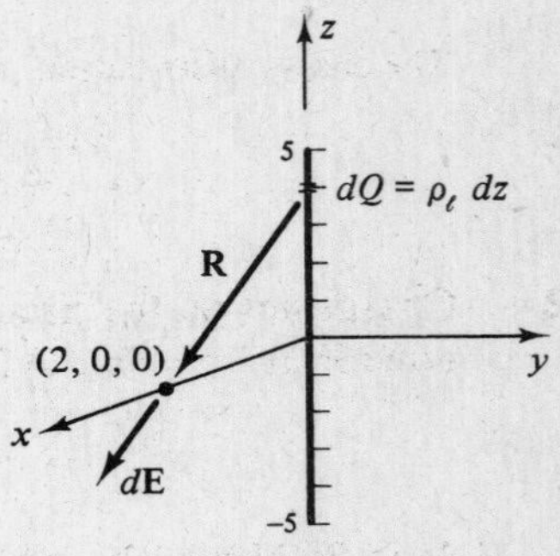

Fig. 2-24

2.19. Charge is distributed along the z axis from $z = 5$ m to ∞ and from $z = -5$ m to $-\infty$ (see Fig. 2-25) with the same density as in Problem 2.18, 20 nC/m. Find **E** at (2, 0, 0) m.

$$d\mathbf{E} = \frac{20 \times 10^{-9}\, dz}{4\pi(10^{-9}/36\pi)(4 + z^2)}\left(\frac{2\mathbf{a}_x - z\mathbf{a}_z}{\sqrt{4 + z^2}}\right) \quad \text{(V/m)}$$

Again the z component vanishes.

$$\mathbf{E} = 180\left[\int_5^\infty \frac{2\, dz}{(4 + z^2)^{3/2}} + \int_{-\infty}^{-5} \frac{2\, dz}{(4 + z^2)^{3/2}}\right]\mathbf{a}_x$$
$$= 13\mathbf{a}_x \text{ V/m}$$

In cylindrical coordinates, $\mathbf{E} = 13\mathbf{a}_r$ V/m.

When the charge configurations of Problems 2.18 and 2.19 are superimposed, the result is a uniform line charge.

$$\mathbf{E} = \frac{\rho_\ell}{2\pi\epsilon_0 r}\mathbf{a}_r = 180\mathbf{a}_r \text{ V/m}$$

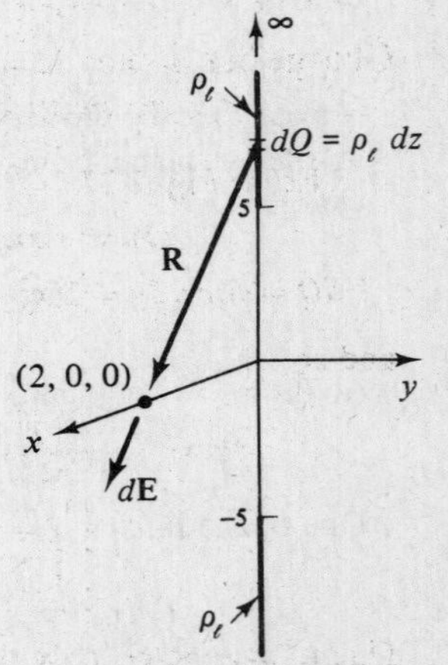

Fig. 2-25

2.20. Find the electric field intensity $\mathbf{E}$ at $(0, \phi, h)$ in cylindrical coordinates due to the uniformly charged disk $r \le a, z = 0$ (see Fig. 2-26).

If the constant charge density is ρ_s,

$$d\mathbf{E} = \frac{\rho_s\, r\, dr\, d\phi}{4\pi\epsilon_0(r^2 + h^2)}\left(\frac{-r\mathbf{a}_r + h\mathbf{a}_z}{\sqrt{r^2 + h^2}}\right)$$

The radial components cancel. Therefore

$$\mathbf{E} = \frac{\rho_s h}{4\pi\epsilon_0}\int_0^{2\pi}\int_0^a \frac{r\, dr\, d\phi}{(r^2 + h^2)^{3/2}}\,\mathbf{a}_z = \frac{\rho_s h}{2\epsilon_0}\left(\frac{-1}{\sqrt{a^2 + h^2}} + \frac{1}{h}\right)\mathbf{a}_z$$

Note that as $a \to \infty$, $\mathbf{E} \to (\rho_s/2\epsilon_0)\mathbf{a}_z$, the field due to a uniform plane sheet.

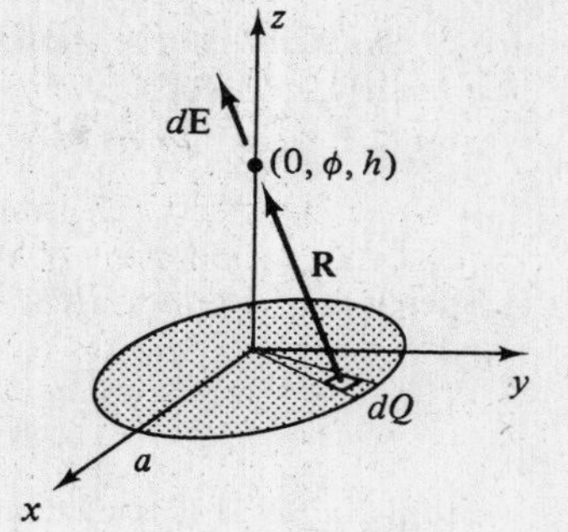

Fig. 2-26

2.21. Charge lies on the circular disk $r \le a, z = 0$ with density $\rho_s = \rho_0 \sin^2 \phi$. Determine $\mathbf{E}$ at $(0, \phi, h)$.

$$d\mathbf{E} = \frac{\rho_0(\sin^2 \phi) r\, dr\, d\phi}{4\pi\epsilon_0(r^2 + h^2)}\left(\frac{-r\mathbf{a}_r + h\mathbf{a}_z}{\sqrt{r^2 + h^2}}\right)$$

The charge distribution, though not uniform, still is symmetrical such that all radial components cancel.

$$\mathbf{E} = \frac{\rho_0 h}{4\pi\epsilon_0}\int_0^{2\pi}\int_0^a \frac{(\sin^2 \phi) r\, dr\, d\phi}{(r^2 + h^2)^{3/2}}\,\mathbf{a}_z = \frac{\rho_0 h}{4\epsilon_0}\left(\frac{-1}{\sqrt{a^2 + h^2}} + \frac{1}{h}\right)\mathbf{a}_z$$

2.22. Charge lies on the circular disk $r \le 4$ m, $z = 0$ with density $\rho_s = (10^{-4}/r)$ (C/m^2). Determine $\mathbf{E}$ at $r = 0$, $z = 3$ m.

$$d\mathbf{E} = \frac{(10^{-4}/r) r\, dr\, d\phi}{4\pi\epsilon_0(r^2 + 9)}\left(\frac{-r\mathbf{a}_r + 3\mathbf{a}_z}{\sqrt{r^2 + 9}}\right) \quad \text{(V/m)}$$

As in Problems 2.20 and 2.21 the radial component vanishes by symmetry.

$$\mathbf{E} = (2.7 \times 10^6)\int_0^{2\pi}\int_0^4 \frac{dr\, d\phi}{(r^2 + 9)^{3/2}}\,\mathbf{a}_z = 1.51 \times 10^6\mathbf{a}_z \text{ V/m} \quad \text{or} \quad 1.51\mathbf{a}_z \text{ MV/m}$$

2.23. Charge lies in the $z = -3$ m plane in the form of a square sheet defined by $-2 \le x \le 2$ m, $-2 \le y \le 2$ m with charge density $\rho_s = 2(x^2 + y^2 + 9)^{3/2}$ nC/m^2. Find $\mathbf{E}$ at the origin.

From Fig. 2-27,

$$\mathbf{R} = -x\mathbf{a}_x - y\mathbf{a}_y + 3\mathbf{a}_z \quad \text{(m)}$$

$$dQ = \rho_s\, dx\, dy = 2(x^2 + y^2 + 9)^{3/2} \times 10^{-9}\, dx\, dy \quad \text{(C)}$$

and so

$$d\mathbf{E} = \frac{2(x^2 + y^2 + 9)^{3/2} \times 10^{-9}\, dx\, dy}{4\pi\epsilon_0(x^2 + y^2 + 9)} \times \left(\frac{-x\mathbf{a}_x - y\mathbf{a}_y + 3\mathbf{a}_z}{\sqrt{x^2 + y^2 + 9}}\right) \quad \text{(V/m)}$$

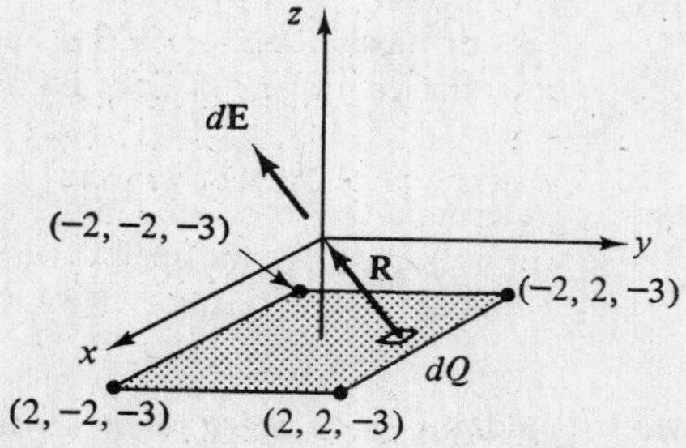

Fig. 2-27

Due to symmetry, only the z component of $\mathbf{E}$ exists.

$$\mathbf{E} = \int_{-2}^{2}\int_{-2}^{2} \frac{6 \times 10^{-9}\, dx\, dy}{4\pi\epsilon_0}\,\mathbf{a}_z = 864\mathbf{a}_z \text{ V/m}$$

2.24. A charge of uniform density $\rho_s = 0.3\ \text{nC/m}^2$ covers the plane $2x - 3y + z = 6$ m. Find **E** on the side of the plane containing the origin.

Since this charge configuration is a uniform sheet, $E = \rho_s/2\epsilon_0$ and $\mathbf{E} = (17.0)\mathbf{a}_n$ V/m. The unit normal vectors for a plane $Ax + By + Cz = D$ are

$$\mathbf{a}_n = \pm\frac{A\mathbf{a}_x + B\mathbf{a}_y + C\mathbf{a}_z}{\sqrt{A^2 + B^2 + C^2}}$$

Therefore, the unit normal vectors for this plane are

$$\mathbf{a}_n = \pm\frac{2\mathbf{a}_x - 3\mathbf{a}_y + \mathbf{a}_z}{\sqrt{14}}$$

From Fig. 2-28 it is evident that the unit vector on the side of the plane containing the origin is produced by the negative sign. The electric field at the origin is

$$\mathbf{E} = (17.0)\left(\frac{-2\mathbf{a}_x + 3\mathbf{a}_y - \mathbf{a}_z}{\sqrt{14}}\right)\ \text{V/m}$$

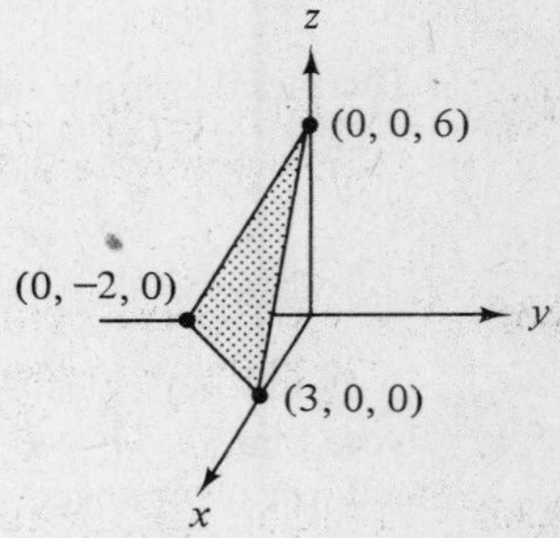

Fig. 2-28

Supplementary Problems

2.25. Two point charges, $Q_1 = 250\ \mu\text{C}$ and $Q_2 = -300\ \mu\text{C}$, are located at $(5, 0, 0)$ m and $(0, 0, -5)$ m, respectively. Find the force on Q_2. *Ans.* $\mathbf{F}_2 = (13.5)\left(\frac{\mathbf{a}_x + \mathbf{a}_z}{\sqrt{2}}\right)$ N

2.26. Two point charges, $Q_1 = 30\ \mu\text{C}$ and $Q_2 = -100\ \mu\text{C}$, are located at $(2, 0, 5)$ m and $(-1, 0, -2)$ m, respectively. Find the force on Q_1. *Ans.* $\mathbf{F}_1 = (0.465)\left(\frac{-3\mathbf{a}_x - 7\mathbf{a}_z}{\sqrt{58}}\right)$ N

2.27. In Problem 2.26 find the force on Q_2. *Ans* $-\mathbf{F}_1$

2.28. Four point charges, each 20 μC, are on the x and y axes at ± 4 m. Find the force on a 100 μC point charge at $(0, 0, 3)$ m. *Ans.* $1.73\,\mathbf{a}_z$ N

2.29. Ten identical charges of 500 μC each are spaced equally around a circle of radius 2 m. Find the force on a charge of $-20\ \mu$C located on the axis, 2 m from the plane of the circle. *Ans.* $(79.5)(-\mathbf{a}_n)$ N

2.30. Determine the force on a point charge of 50 μC at $(0, 0, 5)$ m due to a point charge of $500\pi\ \mu$C at the origin. Compare the answer with Problems 2.4 and 2.5, where this same total charge is distributed over a circular disk. *Ans.* $28.3\,\mathbf{a}_z$ N

2.31. Find the force on a point charge of 30 μC at $(0, 0, 5)$ m due to a 4 m square in the $z = 0$ plane between $x = \pm 2$ m and $y = \pm 2$ m with a total charge of 500 μC, distributed uniformly. *Ans.* $4.66\,\mathbf{a}_z$ N

2.32. Show that the force on a point charge anywhere within a circular ring of uniform charge density is zero, provided the point charge remains in the plane of the ring.

2.33. Two identical point charges of Q (C) each are separated by a distance d (m). Express the electric field $\mathbf{E}$ for points along the line joining the two charges.
Ans. If the charges are at $x = 0$ and $x = d$, then, for $0 < x < d$,

$$\mathbf{E} = \frac{Q}{4\pi\epsilon_0}\left[\frac{1}{x^2} - \frac{1}{(d-x)^2}\right]\mathbf{a}_x \quad \text{(V/m)}$$

2.34. Identical charges of Q (C) are located at the eight corners of a cube with a side ℓ (m). Show that the coulomb force on each charge has magnitude $(3.29\,Q^2/4\pi\epsilon_0\ell^2)$ N.

2.35. Show that the electric field $\mathbf{E}$ outside a spherical shell of uniform charge density ρ_s is the same as $\mathbf{E}$ due to the total charge on the shell located at the center.

2.36. Develop the expression in cartesian coordinates for $\mathbf{E}$ due to an infinitely long, straight charge configuration of uniform density ρ_ℓ. *Ans.* $\mathbf{E} = \dfrac{\rho_\ell}{2\pi\epsilon_0}\dfrac{x\mathbf{a}_x + y\mathbf{a}_y}{x^2 + y^2}$

2.37. A uniform charge distribution, infinite in extent, lies along the z axis with $\rho_\ell = 20$ nC/m. Find the electric field $\mathbf{E}$ at $(6, 8, 3)$ m, expressing it in both cartesian and cylindrical coordinates.
Ans. $21.6\mathbf{a}_x + 28.8\mathbf{a}_y$ V/m, $36\mathbf{a}_r$ V/m

2.38. Two identical uniform line charges of $\rho_\ell = 4$ nC/m are parallel to the z axis at $x = 0$, $y = \pm 4$ m. Determine the electric field $\mathbf{E}$ at $(\pm 4, 0, z)$ m. *Ans.* $\pm 18\mathbf{a}_x$ V/m

2.39. Two identical uniform line charges of $\rho_\ell = 5$ nC/m are parallel to the x axis, one at $z = 0$, $y = -2$ m and the other at $z = 0$, $y = 4$ m. Find $\mathbf{E}$ at $(4, 1, 3)$ m. *Ans.* $30\mathbf{a}_z$ V/m

2.40. Determine $\mathbf{E}$ at the origin due to a uniform line charge distribution with $\rho_\ell = 3.30$ nC/m located at $x = 3$ m, $y = 4$ m. *Ans.* $-7.13\mathbf{a}_x - 9.50\mathbf{a}_y$ V/m

2.41. Referring to Problem 2.40, at what other points will the value of $\mathbf{E}$ be the same? *Ans.* $(0, 0, z)$

2.42. Two meters from the z axis, E due to a uniform line charge along the z axis is known to be 1.80×10^4 V/m. Find the uniform charge density ρ_ℓ. *Ans.* 2.0 μC/m

2.43. The plane $-x + 3y - 6z = 6$ m contains a uniform charge distribution $\rho_s = 0.53$ nC/m². Find $\mathbf{E}$ on the side containing the origin. *Ans.* $30\left(\dfrac{\mathbf{a}_x - 3\mathbf{a}_y + 6\mathbf{a}_z}{\sqrt{46}}\right)$ V/m

2.44. Two infinite sheets of uniform charge density $\rho_s = (10^{-9}/6\pi)$ C/m² are located at $z = -5$ m and $y = -5$ m. Determine the uniform line charge density ρ_ℓ necessary to produce the same value of $\mathbf{E}$ at $(4, 2, 2)$ m, if the line charge is located at $z = 0$, $y = 0$. *Ans.* 0.667 nC/m

2.45. Two uniform charge distributions are as follows: a sheet of uniform charge density $\rho_s = -50$ nC/m² at $y = 2$ m and a uniform line of $\rho_\ell = 0.2$ μC/m at $z = 2$ m, $y = -1$ m. At what points in the region will $\mathbf{E}$ be zero? *Ans.* $(x, -2.273, 2.0)$ m

2.46. A uniform sheet of charge with $\rho_s = (-1/3\pi)$ nC/m^2 is located at $z = 5$ m and a uniform line of charge with $\rho_\ell = (-25/9)$ nC/m is located at $z = -3$ m, $y = 3$ m. Find the electric field **E** at $(0, -1, 0)$ m. *Ans.* $8\mathbf{a}_y$ V/m

2.47. A uniform line charge of $\rho_\ell = (\sqrt{2} \times 10^{-8}/6)$ C/m lies along the x axis and a uniform sheet of charge is located at $y = 5$ m. Along the line $y = 3$ m, $z = 3$ m the electric field **E** has only a z component. What is ρ_s for the sheet? *Ans.* 125 pC/m^2

2.48. A uniform line charge of $\rho_\ell = 3.30$ nC/m is located at $x = 3$ m, $y = 4$ m. A point charge Q is 2 m from the origin. Find the charge Q and its location such that the electric field is zero at the origin. *Ans.* 5.28 nC at $(-1.2, -1.6, 0)$ m

2.49. A circular ring of charge with radius 2 m lies in the $z = 0$ plane, with center at the origin. If the uniform charge density is $\rho_\ell = 10$ nC/m, find the point charge Q at the origin which would produce the same electric field **E** at $(0, 0, 5)$ m. *Ans.* 100.5 nC

2.50. The circular disk $r \le 2$ m in the $z = 0$ plane has a charge density $\rho_s = 10^{-8}/r$ (C/m^2). Determine the electric field **E** for the point $(0, \phi, h)$. *Ans.* $\dfrac{1.13 \times 10^3}{h\sqrt{4 + h^2}}\,\mathbf{a}_z$ (V/m)

2.51. Examine the result in Problem 2.50 as h becomes much greater than 2 m and compare it to the field at h which results when the total charge on the disk is concentrated at the origin.

2.52. A finite sheet of charge, of density $\rho_s = 2x(x^2 + y^2 + 4)^{3/2}$ (C/m^2), lies in the $z = 0$ plane for $0 \le x \le 2$ m and $0 \le y \le 2$ m. Determine **E** at $(0, 0, 2)$ m.

Ans. $(18 \times 10^9)\left(-\dfrac{16}{3}\mathbf{a}_x - 4\mathbf{a}_y + 8\mathbf{a}_z\right)$ V/m $= 18\left(-\dfrac{16}{3}\mathbf{a}_x - 4\mathbf{a}_y + 8\mathbf{a}_z\right)$ GV/m

2.53. Determine the electric field **E** at $(8, 0, 0)$ m due to a charge of 10 nC distributed uniformly along the x axis between $x = -5$ m and $x = 5$ m. Repeat for the same total charge distributed between $x = -1$ m and $x = 1$ m. *Ans.* $2.31\mathbf{a}_x$ V/m, $1.43\mathbf{a}_x$ V/m

2.54. The circular disk $r \le 1$ m, $z = 0$ has a charge density $\rho_s = 2(r^2 + 25)^{3/2}e^{-10r}$ (C/m^2). Find **E** at $(0, 0, 5)$ m. *Ans.* $5.66\mathbf{a}_z$ GV/m

2.55. Show that the electric field is zero everywhere inside a uniformly charged spherical shell.

2.56. Charge is distributed with constant density ρ throughout a spherical volume of radius a. By using the results of Problems 2.35 and 2.55, show that

$$\mathbf{E} = \begin{cases} \dfrac{r\rho}{3\epsilon_0}\,\mathbf{a}_r & r \le a \\[2ex] \dfrac{a^3\rho}{3\epsilon_0 r^2}\,\mathbf{a}_r & r \ge a \end{cases}$$

where r is the distance from the center of the sphere.

Chapter 3

Electric Flux and Gauss's Law

3.1 NET CHARGE IN A REGION

With charge density defined as in Section 2.3, it is possible to obtain the net charge contained in a specified volume by integration. From

$$dQ = \rho\, dv \quad \text{(C)}$$

it follows that

$$Q = \int_v \rho\, dv \quad \text{(C)}$$

Of course, ρ need not be constant throughout the volume v.

3.2 ELECTRIC FLUX AND FLUX DENSITY

By definition, *electric flux*, Ψ, originates on positive charge and terminates on negative charge. In the absence of negative charge, the flux Ψ terminates at infinity. Also by definition, one coulomb of electric charge gives rise to one coulomb of electric flux. Hence

$$\Psi = Q \quad \text{(C)}$$

In Fig. 3-1(a) the flux lines leave $+Q$ and terminate on $-Q$. This assumes that the two charges are of equal magnitude. The case of positive charge with no negative charge in the region is illustrated in Fig. 3-1(b). Here the flux lines are equally spaced throughout the solid angle, and reach out toward infinity.

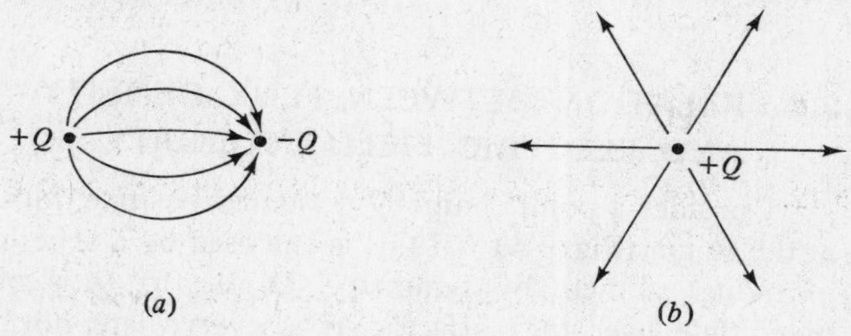

Fig. 3-1

While the electric flux Ψ is a scalar quantity, the *density of electric flux*, **D**, is a vector field which takes its direction from the lines of flux. If in the neighborhood of point P the lines of flux have the direction of the unit vector **a** (see Fig. 3-2) and if an amount of flux $d\Psi$ crosses the differential area dS, which is normal to **a**, then the electric flux density at P is

$$\mathbf{D} = \frac{d\Psi}{dS}\mathbf{a} \quad (\text{C/m}^2)$$

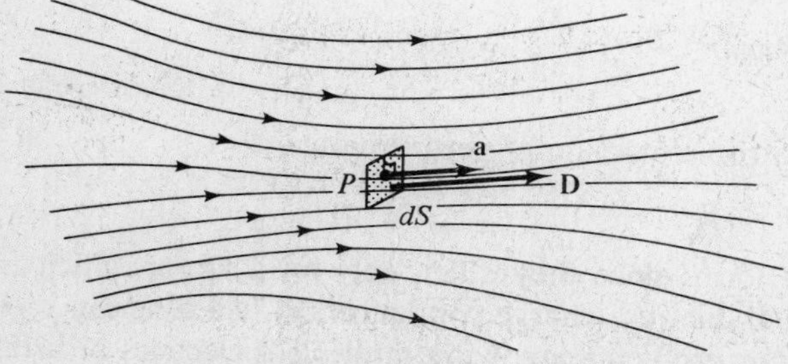

Fig. 3-2

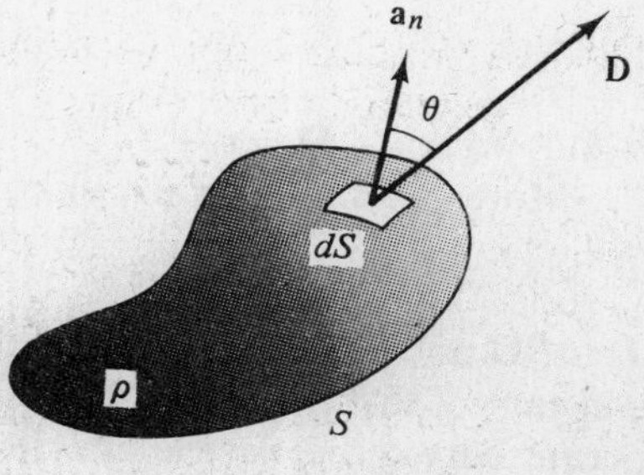

Fig. 3-3

A volume charge distribution of density ρ (C/m^3) is shown enclosed by surface S in Fig. 3-3. Since each coulomb of charge Q has, by definition, one coulomb of flux Ψ, it follows that the net flux crossing the closed surface S is an exact measure of the net charge enclosed. However, the density **D** may vary in magnitude and direction from point to point of S; in general, **D** will not be along the normal to S. If, at the surface element dS, **D** makes an angle θ with the normal, then the differential flux crossing dS is given by

$$\begin{aligned} d\Psi &= D\,dS\cos\theta \\ &= \mathbf{D}\cdot dS\,\mathbf{a}_n \\ &= \mathbf{D}\cdot d\mathbf{S} \end{aligned}$$

where $d\mathbf{S}$ is the vector surface element, of magnitude dS and direction $\mathbf{a}_n$. The unit vector $\mathbf{a}_n$ is always taken to point out of S, so that $d\Psi$ is the amount of flux passing from the interior of S to the exterior of S through dS.

3.3 GAUSS'S LAW

Integration of the above expression for $d\Psi$ over the closed surface S gives, since $\Psi = Q$,

$$\oint \mathbf{D}\cdot d\mathbf{S} = Q_{\text{enc}}$$

This is Gauss's law, which states that *the total flux out of a closed surface is equal to the net charge within that surface.* It will be seen that a great deal of valuable information can be obtained from the application of Gauss's law without actually carrying out the integration.

3.4 RELATION BETWEEN FLUX DENSITY AND ELECTRIC FIELD INTENSITY

Consider a point charge Q (assumed positive, for simplicity) at the origin (Fig. 3-4). If this is enclosed by a spherical surface of radius r, then, by symmetry, **D** due to Q is of constant magnitude over the surface and is everywhere normal to the surface. Gauss's law then gives

$$Q = \oint \mathbf{D}\cdot d\mathbf{S} = D\oint dS = D(4\pi r^2)$$

from which $D = Q/4\pi r^2$. Therefore

$$\mathbf{D} = \frac{Q}{4\pi r^2}\,\mathbf{a}_n = \frac{Q}{4\pi r^2}\,\mathbf{a}_r$$

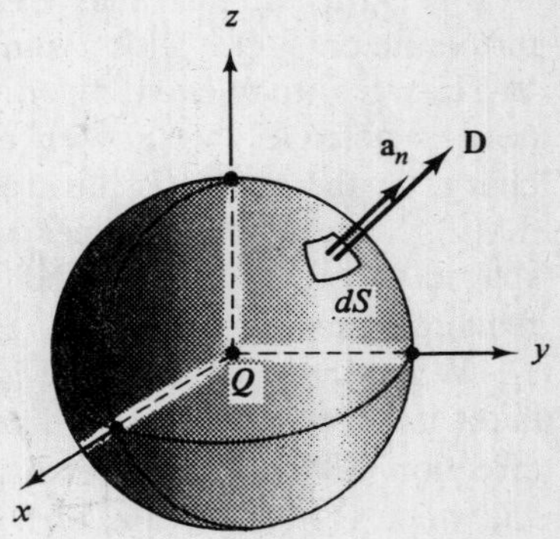

Fig. 3-4

But, from Section 2.2, the electric field intensity due to Q is

$$\mathbf{E} = \frac{Q}{4\pi\epsilon_0 r^2}\,\mathbf{a}_r$$

It follows that $\mathbf{D} = \epsilon_0\mathbf{E}$.

More generally, for any electric field in an isotropic medium of permittivity ϵ,

$$\mathbf{D} = \epsilon\mathbf{E}$$

Thus, **D** and **E** fields will have exactly the same form, since they differ only by a factor which is a constant of the medium. While the electric field **E** due to a charge configuration is a function of the permittivity ϵ, the electric flux density **D** is not. In problems involving multiple dielectrics a distinct advantage will be found in first obtaining **D**, then converting to **E** within each dielectric.

3.5 SPECIAL GAUSSIAN SURFACES

The spherical surface used in the derivation of Section 3.4 was a *special gaussian surface* in that it satisfied the following defining conditions:

1. the surface is closed
2. at each point of the surface **D** is either normal or tangential to the surface
3. D has the same value at all points of the surface where **D** is normal

EXAMPLE 1 Use a special gaussian surface to find **D** due to a uniform line change ρ_ℓ (C/m).

Take the line charge as the z axis of cylindrical coordinates (Fig. 3-5). By cylindrical symmetry, **D** can only have an r component, and this component can only depend on r. Thus, the special gaussian surface for this problem is a closed right circular cylinder whose axis is the z axis (Fig. 3-6). Applying Gauss's law,

$$Q = \int_1 \mathbf{D} \cdot d\mathbf{S} + \int_2 \mathbf{D} \cdot d\mathbf{S} + \int_3 \mathbf{D} \cdot d\mathbf{S}$$

Over surfaces 1 and 3, **D** and $d\mathbf{S}$ are orthogonal, and so the integrals vanish. Over 2, **D** and $d\mathbf{S}$ are parallel (or antiparallel, if ρ_ℓ is negative), and D is constant because r is constant. Thus,

$$Q = D \int_2 dS = D(2\pi r L)$$

where L is the length of the cylinder. But the enclosed charge is $Q = \rho_\ell L$. Hence,

$$D = \frac{\rho_\ell}{2\pi r} \quad \text{and} \quad \mathbf{D} = \frac{\rho_\ell}{2\pi r}\mathbf{a}_r$$

Observe the simplicity of the above derivation as compared to Problem 2.9.

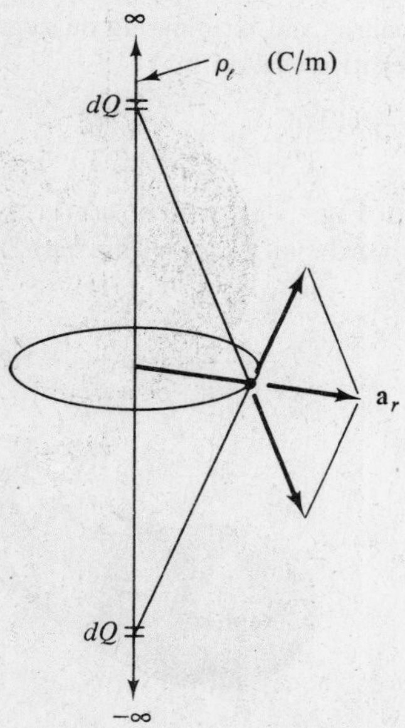

Fig. 3-5

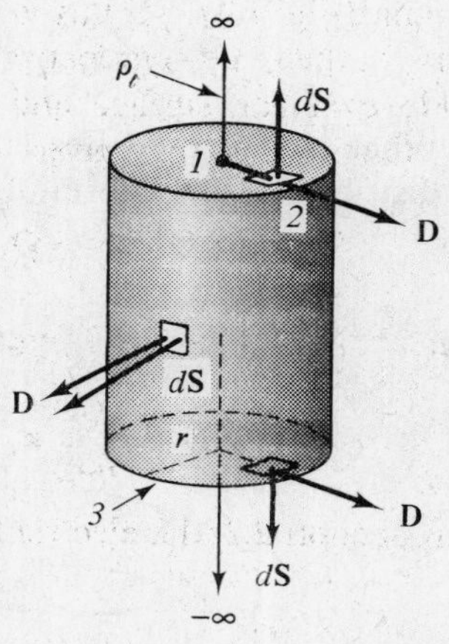

Fig. 3-6

The one serious limitation to the method of special gaussian surfaces is that it can be utilized only for highly symmetrical charge configurations. However, for other configurations, the method can still provide quick approximations to the field at locations very close to or very far from the charges. See Problem 3.40.

Solved Problems

3.1. Find the charge in the volume defined by $0 \le x \le 1$ m, $0 \le y \le 1$ m and $0 \le z \le 1$ m if $\rho = 30x^2y$ (μC/m^3). What change occurs for the limits $-1 \le y \le 0$ m?

Since $dQ = \rho\, dv$,

$$Q = \int_0^1 \int_0^1 \int_0^1 30x^2y\,dx\,dy\,dz = 5\ \mu\text{C}$$

For the change in limits on y,

$$Q = \int_0^1 \int_{-1}^0 \int_0^1 30x^2y\,dx\,dy\,dz = -5\ \mu\text{C}$$

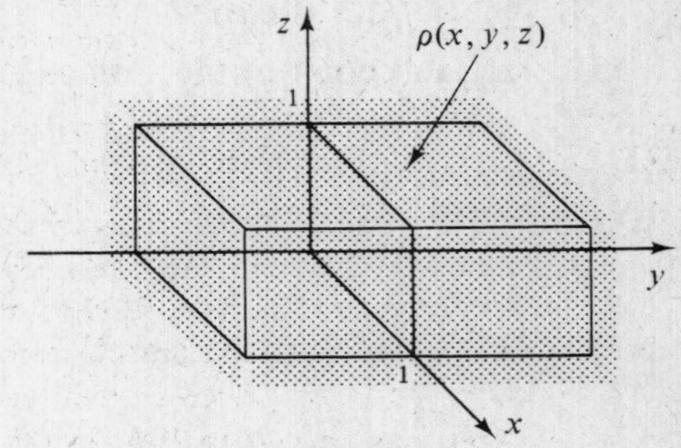

Fig. 3-7

3.2. Find the charge in the volume defined by $1 \le r \le 2$ m in spherical coordinates if

$$\rho = \frac{5\cos^2\phi}{r^4} \quad (\text{C/m}^3)$$

By integration,

$$Q = \int_0^{2\pi} \int_0^{\pi} \int_1^2 \left(\frac{5\cos^2\phi}{r^4}\right) r^2 \sin\theta\,dr\,d\theta\,d\phi = 5\pi\ \text{C}$$

3.3. Three point charges, $Q_1 = 30$ nC, $Q_2 = 150$ nC and $Q_3 = -70$ nC, are enclosed by surface S. What net flux crosses S?

Since electric flux was defined as originating on positive charge and terminating on negative charge, part of the flux from the positive charges terminates on the negative charge.

$$\Psi_{\text{net}} = Q_{\text{net}} = 30 + 150 - 70 = 110\ \text{nC}$$

3.4. What net flux crosses the closed surface S shown in Fig. 3-8, which contains a charge distribution in the form of a plane disk of radius 4 m with a density $\rho_s = (\sin^2\phi)/2r$ (C/m^2)?

$$\Psi = Q = \int_0^{2\pi} \int_0^4 \left(\frac{\sin^2\phi}{2r}\right) r\,dr\,d\phi = 2\pi\ \text{C}$$

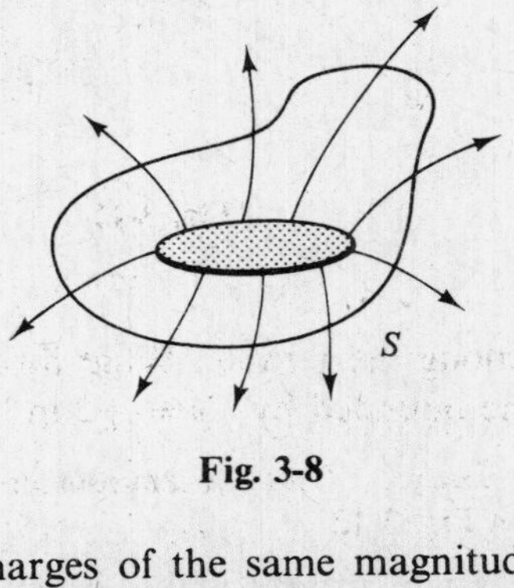

Fig. 3-8

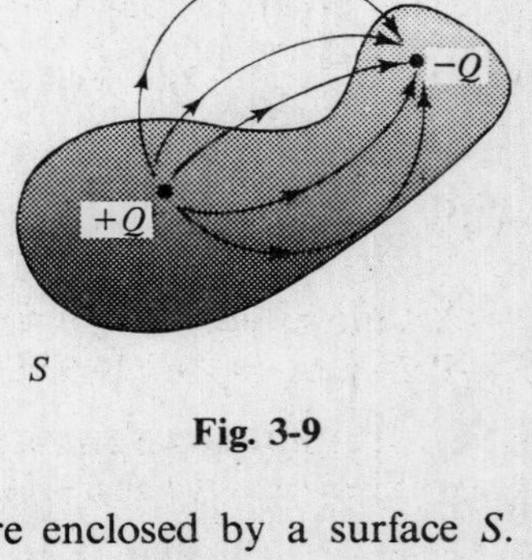

Fig. 3-9

3.5. Two charges of the same magnitude but opposite signs are enclosed by a surface S. Can flux Ψ cross the surface?

While flux can cross the surface, as shown in Fig. 3-9, the *net* flux *out of* S will be zero so long as the charges are of the same magnitude.

3.6. A circular disk of radius 4 m with a charge density $\rho_s = 12 \sin \phi \quad \mu\text{C/m}^2$ is enclosed by surface S. What net flux crosses S?

$$\Psi = Q = \int_0^{2\pi} \int_0^4 (12 \sin \phi) r \, dr \, d\phi = 0 \ \mu\text{C}$$

Since the disk contains equal amounts of positive and negative charge $[\sin(\phi + \pi) = -\sin \phi]$, no net flux crosses S.

3.7. Charge in the form of a plane sheet with density $\rho_s = 40 \ \mu\text{C/m}^2$ is located at $z = -0.5$ m. A uniform line charge of $\rho_\ell = -6 \ \mu\text{C/m}$ lies along the y axis. What net flux crosses the surface of a cube 2 m on an edge, centered at the origin, as shown in Fig. 3-10?

$$\Psi = Q_{\text{enc}}$$

The charge enclosed from the plane is

$$Q = (4 \text{ m}^2)(40 \ \mu\text{C/m}^2) = 160 \ \mu\text{C}$$

and from the line

$$Q = (2 \text{ m})(-6 \ \mu\text{C/m}) = -12 \ \mu\text{C}$$

Thus, $Q_{\text{enc}} = \Psi = 160 - 12 = 148 \ \mu\text{C}$.

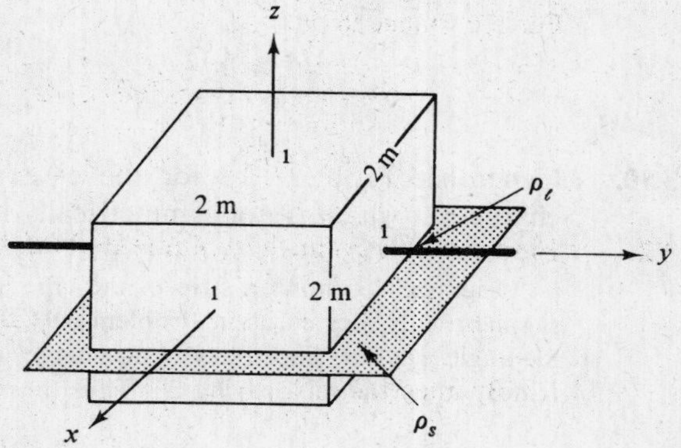

Fig. 3-10

3.8. A point charge Q is at the origin of a spherical coordinate system. Find the flux which crosses the portion of a spherical shell described by $\alpha \leq \theta \leq \beta$ (Fig. 3-11). What is the result if $\alpha = 0$ and $\beta = \pi/2$?

The total flux $\Psi = Q$ crosses a complete spherical shell of area $4\pi r^2$. The area of the strip is given by

$$A = \int_0^{2\pi} \int_\alpha^\beta r^2 \sin \theta \, d\theta \, d\phi$$
$$= 2\pi r^2 (-\cos \beta + \cos \alpha)$$

Then the flux through the strip is

$$\Psi_{\text{net}} = \frac{A}{4\pi r^2} Q = \frac{Q}{2} (-\cos \beta + \cos \alpha)$$

For $\alpha = 0$, $\beta = \pi/2$ (a hemisphere), this becomes $\Psi_{\text{net}} = Q/2$.

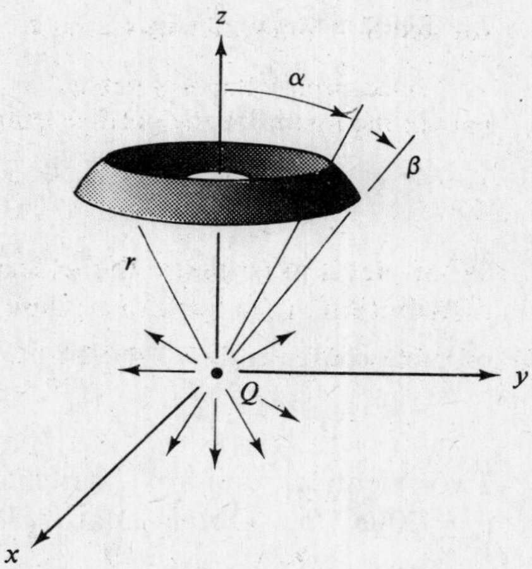

Fig. 3-11

3.9. A uniform line charge with $\rho_\ell = 50 \ \mu\text{C/m}$ lies along the x axis. What flux per unit length, Ψ/L, crosses the portion of the $z = -3$ m plane bounded by $y = \pm 2$ m?

The flux is uniformly distributed around the line charge. Thus the amount crossing the strip is obtained from the angle subtended compared to 2π. In Fig. 3-12,

$$\alpha = 2 \arctan\left(\frac{2}{3}\right) = 1.176 \text{ rad}$$

Then

$$\frac{\Psi}{L} = 50\left(\frac{1.176}{2\pi}\right) = 9.36 \ \mu\text{C/m}$$

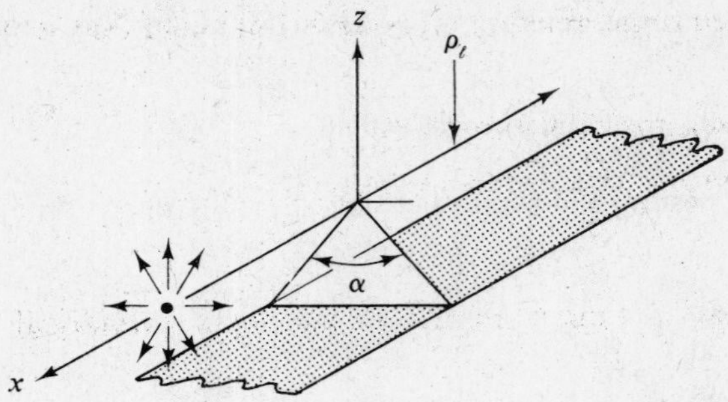

Fig. 3-12

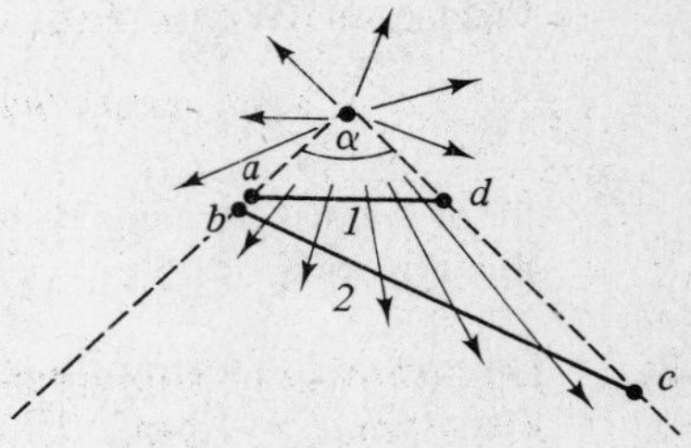

Fig. 3-13

3.10. Generalize Problem 3.9 for the case of a plane strip whose edges are parallel to a line charge but which is not symmetrically located with respect to the line charge.

Figure 3-13 shows a strip of the type in question, labeled *2*, and another strip, labeled *1*, which is symmetrically located as in Problem 3.9. From Problem 3.9 the flux across strip *1* is determined by the angle α. But, since there is no charge in the region *abcd*, Gauss's law shows that the flux entering *1* must equal the flux leaving *2*. Thus the flux across *2* is also determined by the subtended angle α.

3.11. A point charge, $Q = 30$ nC, is located at the origin in cartesian coordinates. Find the electric flux density **D** at $(1, 3, -4)$ m.

Referring to Fig. 3-14,

$$\mathbf{D} = \frac{Q}{4\pi R^2}\mathbf{a}_R$$

$$= \frac{30 \times 10^{-9}}{4\pi(26)}\left(\frac{\mathbf{a}_x + 3\mathbf{a}_y - 4\mathbf{a}_z}{\sqrt{26}}\right)$$

$$= (9.18 \times 10^{-11})\left(\frac{\mathbf{a}_x + 3\mathbf{a}_y - 4\mathbf{a}_z}{\sqrt{26}}\right) \quad \text{C/m}^2$$

or, more conveniently, $D = 91.8\ \text{pC/m}^2$.

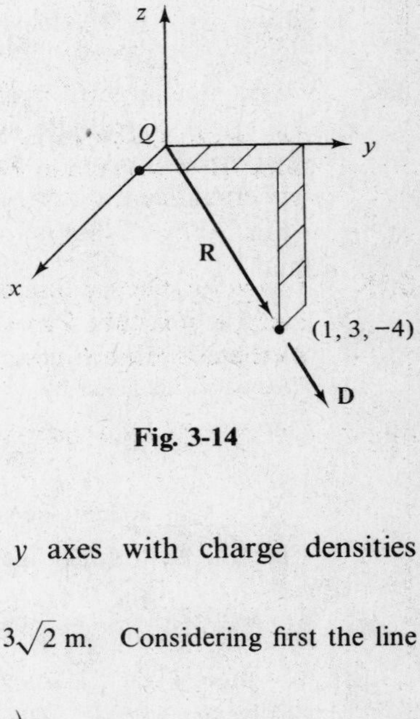

Fig. 3-14

3.12. Two identical uniform line charges lie along the x and y axes with charge densities $\rho_\ell = 20\ \mu$C/m. Obtain **D** at $(3, 3, 3)$ m.

The distance from the observation point to either line charge is $3\sqrt{2}$ m. Considering first the line charge on the x axis,

$$\mathbf{D}_1 = \frac{\rho_\ell}{2\pi r_1}\mathbf{a}_{r1} = \frac{20\ \mu\text{C/m}}{2\pi(3\sqrt{2}\ \text{m})}\left(\frac{\mathbf{a}_y + \mathbf{a}_z}{\sqrt{2}}\right)$$

and now the y axis line charge,

$$\mathbf{D}_2 = \frac{\rho_\ell}{2\pi r_2}\mathbf{a}_{r2} = \frac{20\ \mu\text{C/m}}{2\pi(3\sqrt{2}\ \text{m})}\left(\frac{\mathbf{a}_x + \mathbf{a}_z}{\sqrt{2}}\right)$$

The total flux density is the vector sum,

$$\mathbf{D} = \frac{20}{2\pi(3\sqrt{2})}\left(\frac{\mathbf{a}_x + \mathbf{a}_y + 2\mathbf{a}_z}{\sqrt{2}}\right) = (1.30)\left(\frac{\mathbf{a}_x + \mathbf{a}_y + 2\mathbf{a}_z}{\sqrt{2}}\right) \quad \mu\text{C/m}^2$$

3.13. Given that $\mathbf{D} = 10x\mathbf{a}_x$ (C/m²), determine the flux crossing a 1 m² area that is normal to the x axis at $x = 3$ m.

Since $\mathbf{D}$ is constant over the area and perpendicular to it,

$$\Psi = DA = (30\ \text{C/m}^2)(1\ \text{m}^2) = 30\ \text{C}$$

3.14. Determine the flux crossing an area of 1 mm² on the surface of a cylindrical shell at $r = 10$ m, $z = 2$ m, $\phi = 53.2°$ if

$$\mathbf{D} = 2x\mathbf{a}_x + 2(1 - y)\mathbf{a}_y + 4z\mathbf{a}_z \quad (\text{C/m}^2)$$

At point P (see Fig. 3-15),

$$x = 10\cos 53.2° = 6$$
$$y = 10\sin 53.2° = 8$$

Then, at P,

$$\mathbf{D} = 12\mathbf{a}_x - 14\mathbf{a}_y + 8\mathbf{a}_z \quad \text{C/m}^2$$

The $1\ \text{mm}^2 = 10^{-6}\ \text{m}^2$ area, which is very small compared to the units on $\mathbf{D}$, can be approximated as

$$d\mathbf{S} = 10^{-6}(0.6\mathbf{a}_x + 0.8\mathbf{a}_y) \quad \text{m}^2$$

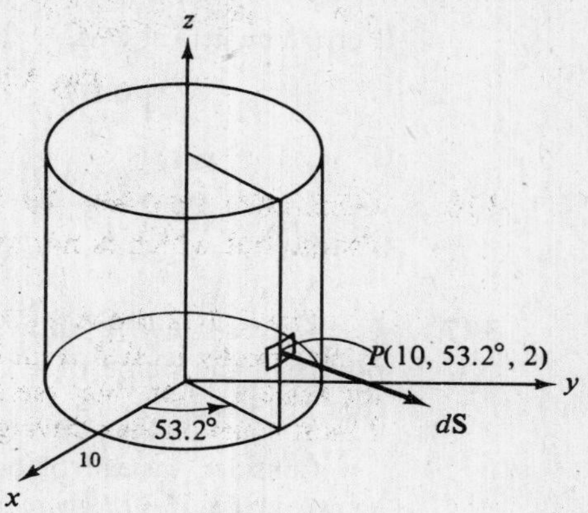

Fig. 3-15

Then

$$d\Psi = \mathbf{D}\cdot d\mathbf{S} = (12\mathbf{a}_x - 14\mathbf{a}_y + 8\mathbf{a}_z)\cdot 10^{-6}(0.6\mathbf{a}_x + 0.8\mathbf{a}_y) = -4.0\ \mu\text{C}$$

The negative sign indicates that flux crosses this differential surface in a direction toward the z axis rather than outward in the direction of $d\mathbf{S}$.

3.15. Given an electric flux density $\mathbf{D} = 2x\mathbf{a}_x + 3\mathbf{a}_y$ (C/m²), determine the net flux crossing the surface of a cube 2 m on an edge centered at the origin. (The edges of the cube are parallel to the coordinate axes.)

$$\begin{aligned}
\Psi = \oint \mathbf{D}\cdot d\mathbf{S} &= \int_{x=1} (2\mathbf{a}_x + 3\mathbf{a}_y)\cdot(dS\,\mathbf{a}_x) + \int_{x=-1} (-2\mathbf{a}_x + 3\mathbf{a}_y)\cdot(-dS\,\mathbf{a}_x) \\
&+ \int_{y=1} (2x\mathbf{a}_x + 3\mathbf{a}_y)\cdot(dS\,\mathbf{a}_y) + \int_{y=-1} (2x\mathbf{a}_x + 3\mathbf{a}_y)\cdot(-dS\,\mathbf{a}_y) \\
&+ \int_{z=1} (2x\mathbf{a}_x + 3\mathbf{a}_y)\cdot(dS\,\mathbf{a}_z) + \int_{z=-1} (2x\mathbf{a}_x + 3\mathbf{a}_y)\cdot(-dS\,\mathbf{a}_z) \\
&= 2\int_{x=1} dS + 2\int_{x=-1} dS + 3\int_{y=1} dS - 3\int_{y=-1} dS + 0 + 0 \\
&= (2 + 2 + 3 - 3)(2^2) = 16\ \text{C}
\end{aligned}$$

3.16. A uniform line charge of $\rho_\ell = 3\ \mu\text{C/m}$ lies along the z axis, and a concentric circular cylinder of radius 2 m has $\rho_s = (-1.5/4\pi)\ \mu\text{C/m}^2$. Both distributions are infinite in extent with z. Use Gauss's law to find $\mathbf{D}$ in all regions.

Using the special gaussian surface A in Fig. 3-16 and proceeding as in Example 1, Section 3.5,

$$\mathbf{D} = \frac{\rho_\ell}{2\pi r}\mathbf{a}_r \qquad 0 < r < 2$$

Using the special gaussian surface B,

$$Q_{\text{enc}} = \oint \mathbf{D} \cdot d\mathbf{S}$$

$$(\rho_\ell + 4\pi\rho_s)L = D(2\pi r L)$$

from which

$$\mathbf{D} = \frac{\rho_\ell + 4\pi\rho_s}{2\pi r}\mathbf{a}_r \qquad r > 2$$

For the numerical data,

$$\mathbf{D} = \begin{cases} \dfrac{0.477}{r}\mathbf{a}_r \quad (\mu\text{C/m}^2) & 0 < r < 2\text{ m} \\ \dfrac{0.239}{r}\mathbf{a}_r \quad (\mu\text{C/m}^2) & r > 2\text{ m} \end{cases}$$

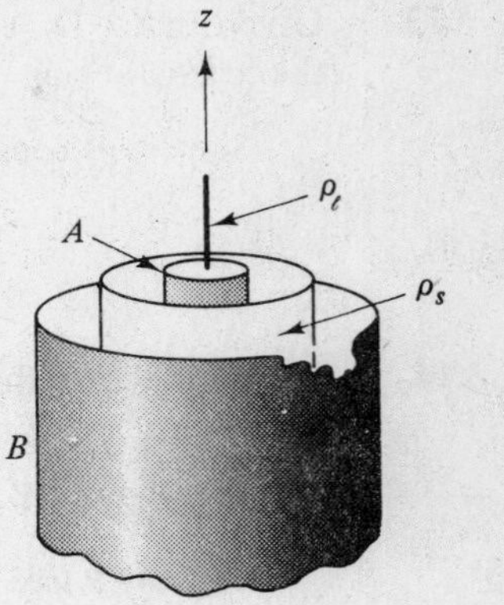

Fig. 3-16

3.17. Use Gauss's law to show that **D** and **E** are zero at all points in the plane of a uniformly charged circular ring that are inside the ring.

Consider, instead of one ring, the charge configuration shown in Fig. 3-17, where the uniformly charged cylinder is infinite in extent, made up of many rings. For gaussian surface *1*,

$$Q_{\text{enc}} = 0 = D \oint dS$$

Hence $\mathbf{D} = 0$ for $r < R$. Since Ψ is completely in the radial direction, a slice dz can be taken from the cylinder of charge and the result found above will still apply to this ring. For all points within the ring, in the plane of the ring, **D** and **E** are zero.

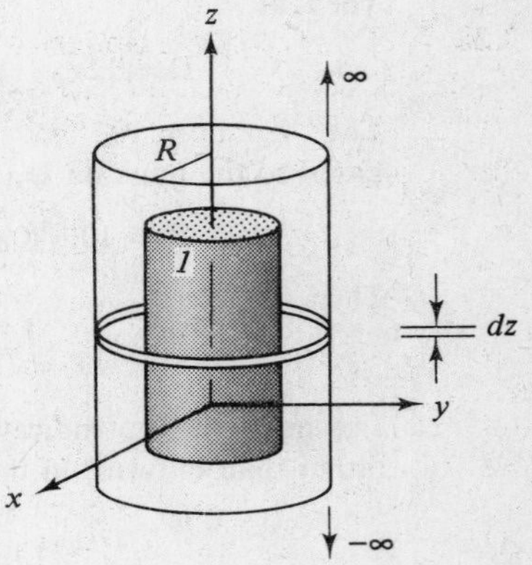

Fig. 3-17

3.18. A charge configuration in cylindrical coordinates is given by $\rho = 5re^{-2r}$ (C/m^3). Use Gauss's law to find **D**.

Since ρ is not a function of ϕ or z, the flux Ψ is completely radial. It is also true that, for r constant, the flux density **D** must be of constant magnitude. Then a proper special gaussian surface is a closed right circular cylinder. The integrals over the plane ends vanish, so that Gauss's law becomes

$$Q_{\text{enc}} = \int_{\substack{\text{lateral}\\\text{surface}}} \mathbf{D} \cdot d\mathbf{S}$$

$$\int_0^L \int_0^{2\pi} \int_0^r 5re^{-2r}\, r\, dr\, d\phi\, dz = D(2\pi r L)$$

$$5\pi L[e^{-2r}(-r^2 - r - \tfrac{1}{2}) + \tfrac{1}{2}] = D(2\pi r L)$$

Hence
$$\mathbf{D} = \frac{2.5}{r}[\tfrac{1}{2} - e^{-2r}(r^2 + r + \tfrac{1}{2})]\mathbf{a}_r \quad (\text{C/m}^2)$$

3.19. The volume in cylindrical coordinates between $r = 2$ m and $r = 4$ m contains a uniform charge density ρ (C/m^3). Use Gauss's law to find **D** in all regions.

From Fig. 3-18, for $0 < r < 2$ m,

$$Q_{enc} = D(2\pi r L)$$
$$\mathbf{D} = 0$$

For $2 \le r \le 4$ m,

$$\pi\rho L(r^2 - 4) = D(2\pi r L)$$
$$\mathbf{D} = \frac{\rho}{2r}(r^2 - 4)\mathbf{a}_r \quad (\text{C/m}^2)$$

For $r > 4$ m,

$$12\pi\rho L = D(2\pi r L)$$
$$\mathbf{D} = \frac{6\rho}{r}\mathbf{a}_r \quad (\text{C/m}^2)$$

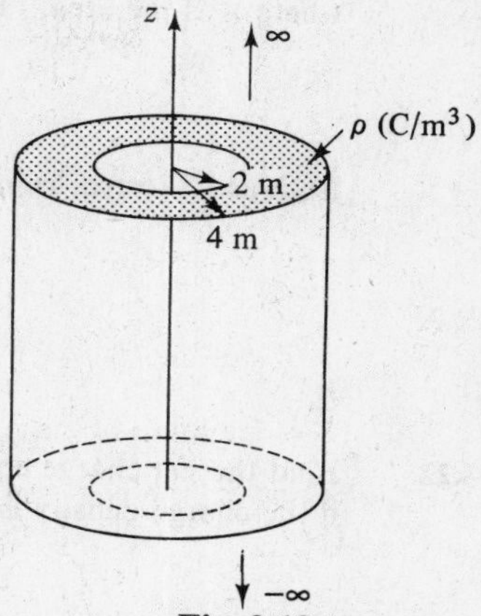

Fig. 3-18

3.20. The volume in spherical coordinates described by $r \le a$ contains a uniform charge density ρ. Use Gauss's law to determine **D** and compare your results with those for the corresponding **E** field, found in Problem 2.56. What point charge at the origin will result in the same **D** field for $r > a$?

For a gaussian surface such as Σ in Fig. 3-19,

$$Q_{enc} = \oint \mathbf{D} \cdot d\mathbf{S}$$
$$\frac{4}{3}\pi r^3 \rho = D(4\pi r^2)$$

and

$$\mathbf{D} = \frac{\rho r}{3}\mathbf{a}_r \qquad r \le a$$

For points outside the charge distribution,

$$\frac{4}{3}\pi a^3 \rho = D(4\pi r^2) \qquad \text{whence} \qquad \mathbf{D} = \frac{\rho a^3}{3r^2}\mathbf{a}_r \qquad r > a$$

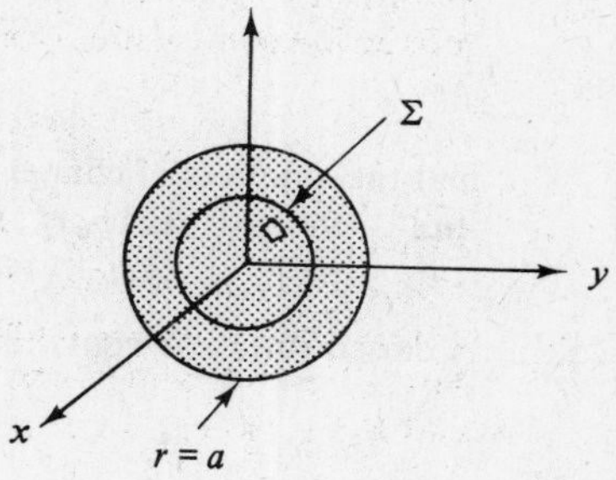

Fig. 3-19

If a point charge $Q = (4/3)\pi a^3 \rho$ is placed at the origin, the **D** field for $r > a$ will be the same. This point charge is the same as the total charge contained in the volume.

3.21. A parallel-plate capacitor has a surface charge on the lower side of the upper plate of $+\rho_s$ (C/m²). The upper surface of the lower plate contains $-\rho_s$ (C/m²). Neglect fringing and use Gauss's law to find **D** and **E** in the region between the plates.

All flux leaving the positive charge on the upper plate terminates on the equal negative charge on the lower plate. The statement *neglect fringing* insures that all flux is normal to the plates. For the special gaussian surface shown in Fig. 3-20,

$$Q_{enc} = \int_{top} \mathbf{D} \cdot d\mathbf{S} + \int_{bottom} \mathbf{D} \cdot d\mathbf{S} + \int_{side} \mathbf{D} \cdot d\mathbf{S}$$
$$= 0 + \int_{bottom} \mathbf{D} \cdot d\mathbf{S} + 0$$

or

$$\rho_s A = D \int dS = DA$$

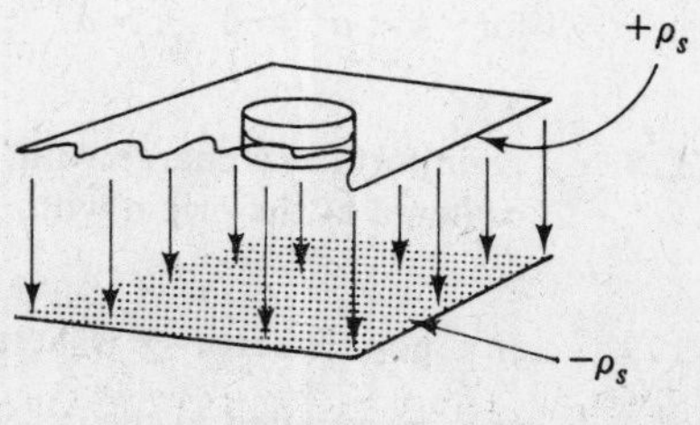

Fig. 3-20

where A is the area. Consequently,

$$\mathbf{D} = \rho_s \mathbf{a}_n \ (\text{C/m}^2) \qquad \text{and} \qquad \mathbf{E} = \frac{\rho_s}{\epsilon_0} \mathbf{a}_n \ (\text{V/m})$$

Both are directed from the positive to the negative plate.

Supplementary Problems

3.22. Find the net charge enclosed in a cube 2 m on an edge, parallel to the axes and centered at the origin, if the charge density is

$$\rho = 50x^2 \cos\left(\frac{\pi}{2} y\right) \quad (\mu\text{C/m}^3)$$

Ans. 84.9 μC

3.23. Find the charge enclosed in the volume $1 \leq r \leq 3$ m, $0 \leq \phi \leq \pi/3$, $0 \leq z \leq 2$ m given the charge density $\rho = 2z \sin^2 \phi$ (C/m³). *Ans.* 4.91 C

3.24. Given a charge density in spherical coordinates

$$\rho = \frac{\rho_0}{(r/r_0)^2} e^{-r/r_0} \cos^2 \phi$$

find the amounts of charge in the spherical volumes enclosed by $r = r_0$, $r = 5r_0$, and $r = \infty$.
Ans. $3.97 \rho_0 r_0^3$, $6.24 \rho_0 r_0^3$, $6.28 \rho_0 r_0^3$

3.25. A closed surface S contains a finite line charge distribution, $0 \leq \ell \leq \pi$ m, with charge density

$$\rho_\ell = -\rho_0 \sin \frac{\ell}{2} \quad (\text{C/m})$$

What net flux crosses the surface S? *Ans.* $-2\rho_0$ (C)

3.26. Charge is distributed in the spherical region $r \leq 2$ m with density

$$\rho = \frac{-200}{r^2} \quad (\mu\text{C/m}^3)$$

What net flux crosses the surfaces $r = 1$ m, $r = 4$ m, and $r = 500$ m?
Ans. -800π μC, -1600π μC, -1600π μC

3.27. A point charge Q is at the origin of spherical coordinates and a spherical shell charge distribution at $r = a$ has a total charge of $Q' - Q$, uniformly distributed. What flux crosses the surfaces $r = k$ for $k < a$ and $k > a$? *Ans.* Q, Q'

3.28. A uniform line charge with $\rho_\ell = 3$ μC/m lies along the x axis. What flux crosses a spherical surface centered at the origin with $r = 3$ m? *Ans.* 18 μC

3.29. If a point charge Q is at the origin, find an expression for the flux which crosses the portion of a sphere, centered at the origin, described by $\alpha \leq \phi \leq \beta$. *Ans.* $\dfrac{\beta - \alpha}{2\pi} Q$

3.30. A point charge of Q (C) is at the center of a spherical coordinate system. Find the flux Ψ which crosses an area of 4π m^2 on a concentric spherical shell of radius 3 m. *Ans.* $Q/9$ (C)

3.31. An area of 40.2 m^2 on the surface of a spherical shell of radius 4 m is crossed by 10 μC of flux in an inward direction. What point charge at the origin is indicated? *Ans.* $-50\ \mu$C

3.32. A uniform line charge ρ_ℓ lies along the x axis. What percent of the flux from the line crosses the strip of the $y = 6$ plane having $-1 \le z \le 1$? *Ans.* 5.26%

3.33. A point charge, $Q = 3$ nC, is located at the origin of a cartesian coordinate system. What flux Ψ crosses the portion of the $z = 2$ m plane for which $-4 \le x \le 4$ m and $-4 \le y \le 4$ m?
Ans. 0.5 nC

3.34. A uniform line charge with $\rho_\ell = 5\ \mu$C/m lies along the x axis. Find **D** at (3, 2, 1) m.
Ans. $(0.356)\left(\dfrac{2\mathbf{a}_y + \mathbf{a}_z}{\sqrt{5}}\right)$ μC/m^2

3.35. A point charge of $+Q$ is at the origin of a spherical coordinate system, surrounded by a concentric uniform distribution of charge on a spherical shell at $r = a$ for which the total charge is $-Q$. Find the flux Ψ crossing spherical surfaces at $r < a$ and $r > a$. Obtain D in all regions.

Ans. $\Psi = 4\pi r^2 D = \begin{cases} +Q & r < a \\ 0 & r > a \end{cases}$

3.36. Given that $\mathbf{D} = 500e^{-0.1x}\mathbf{a}_x$ (μC/m^2), find the flux Ψ crossing surfaces of area 1 m^2 normal to the x axis and located at $x = 1$ m, $x = 5$ m, and $x = 10$ m. *Ans.* 452 μC, 303 μC, 184 μC

3.37. Given that $\mathbf{D} = 5x^2\mathbf{a}_x + 10z\mathbf{a}_z$ (C/m^2), find the net outward flux crossing the surface of a cube 2 m on an edge centered at the origin. The edges of the cube are parallel to the axes.
Ans. 80 C

3.38. Given that

$$\mathbf{D} = 30e^{-r/b}\mathbf{a}_r - 2\frac{z}{b}\mathbf{a}_z \quad (\text{C/m}^2)$$

in cylindrical coordinates, find the outward flux crossing the right circular cylinder described by $r = 2b$, $z = 0$, and $z = 5b$ (m). *Ans.* $129b^2$ (C)

3.39. Given that

$$\mathbf{D} = 2r\cos\phi\,\mathbf{a}_\phi - \frac{\sin\phi}{3r}\mathbf{a}_z$$

in cylindrical coordinates, find the flux crossing the portion of the $z = 0$ plane defined by $r \le a$, $0 \le \phi \le \pi/2$. Repeat for $3\pi/2 \le \phi \le 2\pi$. Assume positive flux in the $\mathbf{a}_z$ direction.

Ans. $-\dfrac{a}{3}, \dfrac{a}{3}$

3.40. In cylindrical coordinates, the disk $r \le a$, $z = 0$ carries charge with nonuniform density $\rho_s(r, \phi)$. Use appropriate special gaussian surfaces to find approximate values of D on the z axis (a) very close to the disk ($0 < z \ll a$), (b) very far from the disk ($z \gg a$).

Ans. (a) $\dfrac{\rho_s(0, \phi)}{2}$; ($b$) $\dfrac{Q}{4\pi z^2}$ where $Q = \displaystyle\int_0^{2\pi}\int_0^a \rho_s(r, \phi)\, r\, dr\, d\phi$

3.41. A point charge, $Q = 2000$ pC, is at the origin of spherical coordinates. A concentric spherical distribution of charge at $r = 1$ m has a charge density $\rho_s = 40\pi$ pC/m^2. What surface charge density on a concentric shell at $r = 2$ m would result in $\mathbf{D} = 0$ for $r > 2$ m? *Ans.* -71.2 pC/m^2

3.42. Given a charge distribution with density $\rho = 5r$ (C/m^3) in spherical coordinates, use Gauss's law to find **D**. *Ans.* $(5r^2/4)\mathbf{a}_r$ (C/m^2)

3.43. A uniform charge density of 2 C/m^3 exists in the volume $2 \le x \le 4$ m (cartesian coordinates). Use Gauss's law to find **D** in all regions. *Ans.* $-2\mathbf{a}_x$ C/m^2, $2(x-3)\mathbf{a}_x$ (C/m^2), $2\mathbf{a}_x$ C/m^2

3.44. Use Gauss's law to find **D** and **E** in the region between the concentric conductors of a cylindrical capacitor. The inner cylinder is of radius a. Neglect fringing. *Ans.* $\rho_{sa}(a/r)$, $\rho_{sa}(a/\epsilon_0 r)$

3.45. A conductor of substantial thickness has a surface charge of density ρ_s. Assuming that $\Psi = 0$ within the conductor, show that $D = \pm\rho_s$ just outside the conductor, by constructing a small special gaussian surface.

Chapter 4

Divergence and the Divergence Theorem

4.1 DIVERGENCE

There are two characteristics of the manner in which a vector field changes from point to point throughout space. The first of these is *divergence*, which will be examined here. It is a scalar and bears a similarity to the derivative of a function. The second is *curl*, a vector which will be examined when magnetic fields are discussed in Chapter 9.

When the divergence of a vector field is nonzero, that region is said to contain *sources* or *sinks*, sources when the divergence is positive, sinks when negative. In static electric fields there is a correspondence between positive divergence, sources, and positive electric charge Q. Electric flux Ψ by definition originates on positive charge. Thus, a region which contains positive charges contains the *sources* of Ψ. The divergence of the electric flux density **D** will be positive in this region. A similar correspondence exists between negative divergence, sinks, and negative electric charge.

Divergence of the vector field **A** at the point P is defined by

$$\text{div}\,\mathbf{A} \equiv \lim_{\Delta v \to 0} \frac{\oint \mathbf{A} \cdot d\mathbf{S}}{\Delta v}$$

Here the integration is over the surface of an infinitesimal volume Δv that shrinks to point P.

4.2 DIVERGENCE IN CARTESIAN COORDINATES

The divergence can be expressed for any vector field in any coordinate system. For the development in cartesian coordinates a cube is selected with edges Δx, Δy, and Δz parallel to the x, y, and z axes, as shown in Fig. 4-1. Then the vector field **A** is defined at P, the corner of the cube with the lowest values of the coordinates x, y, and z.

$$\mathbf{A} = A_x \mathbf{a}_x + A_y \mathbf{a}_y + A_z \mathbf{a}_z$$

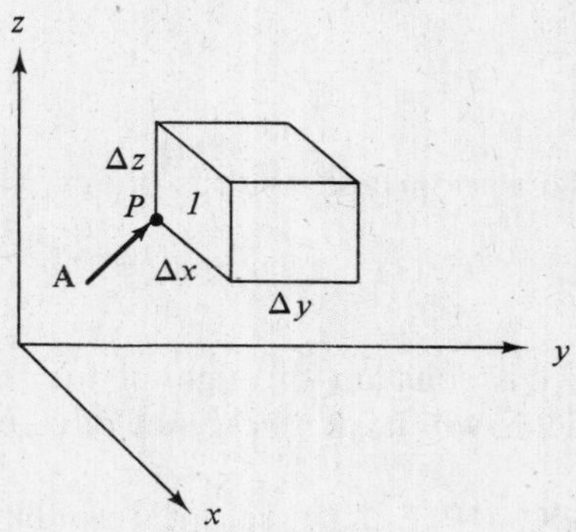

Fig. 4-1

In order to express $\oint \mathbf{A} \cdot d\mathbf{S}$ for the cube, all six faces must be covered. On each face, the direction of $d\mathbf{S}$ is outward. Since the faces are normal to the three axes, only one component of **A** will cross any two parallel faces.

In Fig. 4-2 the cube is turned such that face *1* is in full view; the x components of **A** over the faces to the left and right of *1* are indicated. Since the faces are small,

$$\int_{\substack{\text{left}\\ \text{face}}} \mathbf{A} \cdot d\mathbf{S} \approx -A_x(x)\,\Delta y\,\Delta z$$

$$\int_{\substack{\text{right}\\ \text{face}}} \mathbf{A} \cdot d\mathbf{S} \approx A_x(x + \Delta x)\,\Delta y\,\Delta z$$

$$\approx \left[A_x(x) + \frac{\partial A_x}{\partial x}\Delta x\right]\Delta y\,\Delta z$$

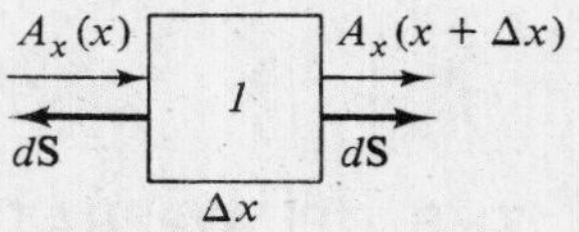

Fig. 4-2

so that the total for these two faces is

$$\frac{\partial A_x}{\partial x}\Delta x\,\Delta y\,\Delta z$$

The same procedure is applied to the remaining two pairs of faces and the results combined.

$$\oint \mathbf{A} \cdot d\mathbf{S} \approx \left(\frac{\partial A_x}{\partial x} + \frac{\partial A_y}{\partial y} + \frac{\partial A_z}{\partial z} \right) \Delta x \, \Delta y \, \Delta z$$

Dividing by $\Delta x \, \Delta y \, \Delta z = \Delta v$ and letting $\Delta v \to 0$, one obtains

$$\text{div } \mathbf{A} = \frac{\partial A_x}{\partial x} + \frac{\partial A_y}{\partial y} + \frac{\partial A_z}{\partial z} \qquad \text{(cartesian)}$$

The same approach may be used in cylindrical (Problem 4.1) and in spherical coordinates.

$$\text{div } \mathbf{A} = \frac{1}{r} \frac{\partial}{\partial r}(rA_r) + \frac{1}{r} \frac{\partial A_\phi}{\partial \phi} + \frac{\partial A_z}{\partial z} \qquad \text{(cylindrical)}$$

$$\text{div } \mathbf{A} = \frac{1}{r^2} \frac{\partial}{\partial r}(r^2 A_r) + \frac{1}{r \sin\theta} \frac{\partial}{\partial \theta}(A_\theta \sin\theta) + \frac{1}{r \sin\theta} \frac{\partial A_\phi}{\partial \phi} \qquad \text{(spherical)}$$

4.3 DIVERGENCE OF D

From Gauss's law (Section 3.3),

$$\frac{\oint \mathbf{D} \cdot d\mathbf{S}}{\Delta v} = \frac{Q_{\text{enc}}}{\Delta v}$$

In the limit,

$$\lim_{\Delta v \to 0} \frac{\oint \mathbf{D} \cdot d\mathbf{S}}{\Delta v} = \text{div } \mathbf{D} = \lim_{\Delta v \to 0} \frac{Q_{\text{enc}}}{\Delta v} = \rho$$

This important result is one of Maxwell's equations for static fields:

$$\text{div } \mathbf{D} = \rho \qquad \text{and} \qquad \text{div } \mathbf{E} = \frac{\rho}{\epsilon}$$

if ϵ is constant throughout the region under examination (if not, $\text{div } \epsilon\mathbf{E} = \rho$). Thus both **E** and **D** fields will have divergence of zero in any charge-free region.

EXAMPLE 1 In spherical coordinates the region $r \le a$ contains a uniform charge density ρ, while for $r > a$ the charge density is zero. From Problem 2.56, $\mathbf{E} = E_r \mathbf{a}_r$, where $E_r = (\rho r / 3\epsilon_0)$ for $r \le a$ and $E_r = (\rho a^3 / 3\epsilon_0 r^2)$ for $r > a$. Then, for $r \le a$,

$$\text{div } \mathbf{E} = \frac{1}{r^2} \frac{\partial}{\partial r} \left(r^2 \frac{\rho r}{3\epsilon_0} \right) = \frac{1}{r^2} \left(3r^2 \frac{\rho}{3\epsilon_0} \right) = \frac{\rho}{\epsilon_0}$$

and, for $r > a$,

$$\text{div } \mathbf{E} = \frac{1}{r^2} \frac{\partial}{\partial r} \left(r^2 \frac{\rho a^3}{3\epsilon_0 r^2} \right) = 0$$

4.4 THE DEL OPERATOR

Vector analysis has its own shorthand, which the reader must note with care. At this point a vector operator, symbolized ∇, is defined *in cartesian coordinates* by

$$\nabla \equiv \frac{\partial(\;)}{\partial x} \mathbf{a}_x + \frac{\partial(\;)}{\partial y} \mathbf{a}_y + \frac{\partial(\;)}{\partial z} \mathbf{a}_z$$

In the calculus a differential operator D is sometimes used to represent d/dx. The symbols $\sqrt{\ }$ and $\int$ are also operators. Standing alone, without any indication of what they are to operate on, they look strange. And so ∇, standing alone, simply suggests the taking of certain partial derivatives, each followed by a unit vector. However, when ∇ is dotted with a vector **A**, the result is the divergence of **A**.

$$\nabla \cdot \mathbf{A} = \left(\frac{\partial}{\partial x}\mathbf{a}_x + \frac{\partial}{\partial y}\mathbf{a}_y + \frac{\partial}{\partial z}\mathbf{a}_z\right) \cdot (A_x\mathbf{a}_x + A_y\mathbf{a}_y + A_z\mathbf{a}_z) = \frac{\partial A_x}{\partial x} + \frac{\partial A_y}{\partial y} + \frac{\partial A_z}{\partial z} = \text{div } \mathbf{A}$$

Hereafter, the divergence of a vector field will be written $\nabla \cdot \mathbf{A}$.

Warning! The del operator is defined only in cartesian coordinates. When $\nabla \cdot \mathbf{A}$ is written for the divergence of **A** in other coordinate systems, it does not mean that a del operator can be defined for these systems. For example, the divergence in cylindrical coordinates will be written as

$$\nabla \cdot \mathbf{A} = \frac{1}{r}\frac{\partial}{\partial r}(rA_r) + \frac{1}{r}\frac{\partial A_\phi}{\partial \phi} + \frac{\partial A_z}{\partial z}$$

(see Section 4.2). This *does not imply that*

$$\nabla = \frac{1}{r}\frac{\partial}{\partial r}(r\quad)\mathbf{a}_r + \frac{1}{r}\frac{\partial(\quad)}{\partial \phi}\mathbf{a}_\phi + \frac{\partial(\quad)}{\partial z}\mathbf{a}_z$$

in cylindrical coordinates. In fact, the expression would give *false results* when used in ∇V (the gradient, Chapter 5) or $\nabla \times \mathbf{A}$ (the curl, Chapter 9).

4.5 THE DIVERGENCE THEOREM

Gauss's law states that the closed surface integral of $\mathbf{D} \cdot d\mathbf{S}$ is equal to the charge enclosed. If the charge density function ρ is known throughout the volume, then the charge enclosed may be obtained from an integration of ρ throughout the volume. Thus,

$$\oint \mathbf{D} \cdot d\mathbf{S} = \int \rho \, dv = Q_{enc}$$

But $\rho = \nabla \cdot \mathbf{D}$, and so

$$\oint \mathbf{D} \cdot d\mathbf{S} = \int (\nabla \cdot \mathbf{D}) \, dv$$

This is the *divergence theorem*, also known as *Gauss's divergence theorem*. It is a three-dimensional analog of Green's theorem for the plane. While it was arrived at from known relationships among **D**, Q, and ρ, the theorem is applicable to any vector field.

divergence theorem $$\oint_S \mathbf{A} \cdot d\mathbf{S} = \int_v (\nabla \cdot \mathbf{A}) \, dv$$

Of course, the volume v is that which is enclosed by the surface S.

EXAMPLE 2 The region $r \leq a$ in spherical coordinates has an electric field intensity

$$\mathbf{E} = \frac{\rho r}{3\epsilon}\mathbf{a}_r$$

Examine both sides of the divergence theorem for this vector field.

For S, choose the spherical surface $r = b \leq a$.

$\oint \mathbf{E}\cdot d\mathbf{S}$		$\int (\nabla\cdot\mathbf{E})\,dv$
$\iint \left(\frac{\rho b}{3\epsilon}\mathbf{a}_r\right)\cdot(b^2 \sin\theta\, d\theta\, d\phi\, \mathbf{a}_r)$		$\nabla\cdot\mathbf{E} = \frac{1}{r^2}\frac{\partial}{\partial r}\left(r^2 \frac{\rho r}{3\epsilon}\right) = \frac{\rho}{\epsilon}$
$= \int_0^{2\pi}\int_0^{\pi} \frac{\rho b^3}{3\epsilon}\sin\theta\, d\theta\, d\phi$	and	$\int_0^{2\pi}\int_0^{\pi}\int_0^{b} \frac{\rho}{\epsilon} r^2 \sin\theta\, dr\, d\theta\, d\phi$
$= \frac{4\pi\rho b^3}{3\epsilon}$		$= \frac{4\pi\rho b^3}{3\epsilon}$

The divergence theorem applies to time-varying as well as static fields in any coordinate system. The theorem is used most often in derivations where it becomes necessary to change from a closed surface integration to a volume integration. But then it may also be used to convert the volume integral of a function that can be expressed as the divergence of a vector field into a closed surface integral.

Solved Problems

4.1. Develop the expression for divergence in cylindrical coordinates.

A delta volume is shown in Fig. 4-3 with edges Δr, $r\,\Delta\phi$, and Δz. The vector field $\mathbf{A}$ is defined at P, the corner with the lowest values of the coordinates r, ϕ, and z, as

$$\mathbf{A} = A_r \mathbf{a}_r + A_\phi \mathbf{a}_\phi + A_z \mathbf{a}_z$$

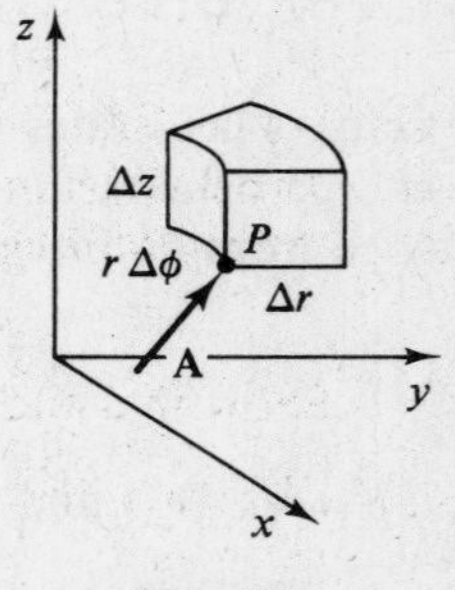

Fig. 4-3

By definition,

$$\operatorname{div}\mathbf{A} = \lim_{\Delta v \to 0} \frac{\oint \mathbf{A}\cdot d\mathbf{S}}{\Delta v}$$

To express $\oint \mathbf{A}\cdot d\mathbf{S}$ all six faces of the volume must be covered. For the radial component of $\mathbf{A}$ refer to Fig. 4-4.

Over the left face,

$$\int \mathbf{A}\cdot d\mathbf{S} \approx -A_r\, r\,\Delta\phi\,\Delta z$$

and over the right face,

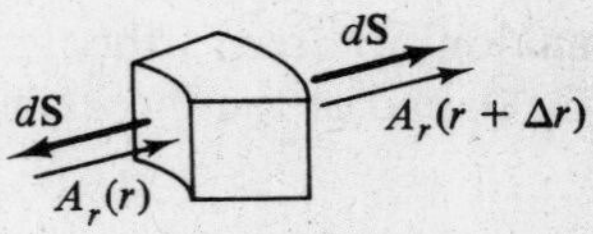

Fig. 4-4

$$\int \mathbf{A}\cdot d\mathbf{S} \approx A_r(r+\Delta r)(r+\Delta r)\Delta\phi\,\Delta z$$

$$\approx \left(A_r + \frac{\partial A_r}{\partial r}\Delta r\right)(r+\Delta r)\Delta\phi\,\Delta z$$

$$\approx A_r\, r\,\Delta\phi\,\Delta z + \left(A_r + r\frac{\partial A_r}{\partial r}\right)\Delta r\,\Delta\phi\,\Delta z$$

where the term in $(\Delta r)^2$ has been neglected. The net contribution of this pair of faces is then

$$\left(A_r + r\frac{\partial A_r}{\partial r}\right)\Delta r\,\Delta\phi\,\Delta z = \frac{\partial}{\partial r}(rA_r)\Delta r\,\Delta\phi\,\Delta z = \frac{1}{r}\frac{\partial}{\partial r}(rA_r)\Delta v \qquad (1)$$

since $\Delta v = r\,\Delta r\,\Delta\phi\,\Delta z$.

Similarly, the faces normal to $\mathbf{a}_\phi$ yield

$$A_\phi \, \Delta r \, \Delta z \qquad \text{and} \qquad \left(A_\phi + \frac{\partial A_\phi}{\partial \phi} \Delta \phi\right) \Delta r \, \Delta z$$

for a net contribution of

$$\frac{1}{r} \frac{\partial A_\phi}{\partial \phi} \Delta v \tag{2}$$

and the faces normal to $\mathbf{a}_z$ yield

$$A_z \, r \, \Delta r \, \Delta \phi \qquad \text{and} \qquad \left(A_z + \frac{\partial A_z}{\partial z} \Delta z\right) r \, \Delta r \, \Delta \phi$$

for a net contribution of

$$\frac{\partial A_z}{\partial z} \Delta v \tag{3}$$

When (*1*), (*2*) and (*3*) are combined to give $\oint \mathbf{A} \cdot d\mathbf{S}$, the definition of divergence gives

$$\text{div } \mathbf{A} = \frac{1}{r} \frac{\partial (r A_r)}{\partial r} + \frac{1}{r} \frac{\partial A_\phi}{\partial \phi} + \frac{\partial A_z}{\partial z}$$

4.2. Show that $\nabla \cdot \mathbf{E}$ is zero for the field of a uniform line charge.

For a line charge, in cylindrical coordinates,

$$\mathbf{E} = \frac{\rho_\ell}{2\pi\epsilon_0 r} \mathbf{a}_r$$

Then

$$\nabla \cdot \mathbf{E} = \frac{1}{r} \frac{\partial}{\partial r} \left(r \frac{\rho_\ell}{2\pi\epsilon_0 r} \right) = 0$$

The divergence of $\mathbf{E}$ for this charge configuration is zero everywhere except at $r = 0$, where the expression is indeterminate.

4.3. Show that the $\mathbf{D}$ field due to a point charge has a divergence of zero.

For a point charge, in spherical coordinates,

$$\mathbf{D} = \frac{Q}{4\pi r^2} \mathbf{a}_r$$

Then, for $r > 0$,

$$\nabla \cdot \mathbf{D} = \frac{1}{r^2} \frac{\partial}{\partial r} \left(r^2 \frac{Q}{4\pi r^2} \right) = 0$$

4.4. Given $\mathbf{A} = e^{-y}(\cos x \, \mathbf{a}_x - \sin x \, \mathbf{a}_y)$, find $\nabla \cdot \mathbf{A}$.

$$\nabla \cdot \mathbf{A} = \frac{\partial}{\partial x} (e^{-y} \cos x) + \frac{\partial}{\partial y} (-e^{-y} \sin x) = e^{-y}(-\sin x) + e^{-y}(\sin x) = 0$$

4.5. Given $\mathbf{A} = x^2 \mathbf{a}_x + yz \mathbf{a}_y + xy \mathbf{a}_z$, find $\nabla \cdot \mathbf{A}$.

$$\nabla \cdot \mathbf{A} = \frac{\partial}{\partial x} (x^2) + \frac{\partial}{\partial y} (yz) + \frac{\partial}{\partial z} (xy) = 2x + z$$

4.6. Given $\mathbf{A} = 5x^2\left(\sin\frac{\pi x}{2}\right)\mathbf{a}_x$, find $\nabla \cdot \mathbf{A}$ at $x = 1$.

$$\nabla \cdot \mathbf{A} = \frac{\partial}{\partial x}\left(5x^2 \sin\frac{\pi x}{2}\right)$$

$$= 5x^2\left(\cos\frac{\pi x}{2}\right)\frac{\pi}{2} + 10x\sin\frac{\pi x}{2} = \frac{5}{2}\pi x^2 \cos\frac{\pi x}{2} + 10x\sin\frac{\pi x}{2}$$

and $\nabla \cdot \mathbf{A}\Big|_{x=1} = 10.$

4.7. Given $\mathbf{A} = (x^2 + y^2)^{-1/2}\mathbf{a}_x$, find $\nabla \cdot \mathbf{A}$ at $(2, 2, 0)$.

$$\nabla \cdot \mathbf{A} = -\frac{1}{2}(x^2 + y^2)^{-3/2}(2x) \qquad \text{and} \qquad \nabla \cdot \mathbf{A}\Big|_{(2,2,0)} = -8.84 \times 10^{-2}$$

4.8. Given $\mathbf{A} = r\sin\phi\,\mathbf{a}_r + 2r\cos\phi\,\mathbf{a}_\phi + 2z^2\mathbf{a}_z$, find $\nabla \cdot \mathbf{A}$.

$$\nabla \cdot \mathbf{A} = \frac{1}{r}\frac{\partial}{\partial r}(r^2 \sin\phi) + \frac{1}{r}\frac{\partial}{\partial \phi}(2r\cos\phi) + \frac{\partial}{\partial z}(2z^2)$$

$$= 2\sin\phi - 2\sin\phi + 4z = 4z$$

4.9. Given $\mathbf{A} = r\sin\phi\,\mathbf{a}_r + r^2\cos\phi\,\mathbf{a}_\phi + 2re^{-5z}\mathbf{a}_z$, find $\nabla \cdot \mathbf{A}$ at $(1/2, \pi/2, 0)$.

$$\nabla \cdot \mathbf{A} = \frac{1}{r}\frac{\partial}{\partial r}(r^2\sin\phi) + \frac{1}{r}\frac{\partial}{\partial \phi}(r^2\cos\phi) + \frac{\partial}{\partial z}(2re^{-5z}) = 2\sin\phi - r\sin\phi - 10re^{-5z}$$

and

$$\nabla \cdot \mathbf{A}\Big|_{(1/2,\pi/2,0)} = 2\sin\frac{\pi}{2} - \frac{1}{2}\sin\frac{\pi}{2} - 10\left(\frac{1}{2}\right)e^0 = -\frac{7}{2}$$

4.10. Given $\mathbf{A} = 10\sin^2\phi\,\mathbf{a}_r + r\mathbf{a}_\phi + [(z^2/r)\cos^2\phi]\mathbf{a}_z$, find $\nabla \cdot \mathbf{A}$ at $(2, \phi, 5)$.

$$\nabla \cdot \mathbf{A} = \frac{10\sin^2\phi + 2z\cos^2\phi}{r} \qquad \text{and} \qquad \nabla \cdot \mathbf{A}\Big|_{(2,\phi,5)} = 5$$

4.11. Given $\mathbf{A} = (5/r^2)\sin\theta\,\mathbf{a}_r + r\cot\theta\,\mathbf{a}_\theta + r\sin\theta\cos\phi\,\mathbf{a}_\phi$, find $\nabla \cdot \mathbf{A}$.

$$\nabla \cdot \mathbf{A} = \frac{1}{r^2}\frac{\partial}{\partial r}(5\sin\theta) + \frac{1}{r\sin\theta}\frac{\partial}{\partial\theta}(r\sin\theta\cot\theta) + \frac{1}{r\sin\theta}\frac{\partial}{\partial\phi}(r\sin\theta\cos\phi) = -1 - \sin\phi$$

4.12. Given $\mathbf{A} = (5/r^2)\mathbf{a}_r + (10/\sin\theta)\mathbf{a}_\theta - r^2\phi\sin\theta\,\mathbf{a}_\phi$, find $\nabla \cdot \mathbf{A}$.

$$\nabla \cdot \mathbf{A} = \frac{1}{r^2}\frac{\partial}{\partial r}(5) + \frac{1}{r\sin\theta}\frac{\partial}{\partial\theta}(10) + \frac{1}{r\sin\theta}\frac{\partial}{\partial\phi}(-r^2\phi\sin\theta) = -r$$

4.13. Given $\mathbf{A} = 5\sin\theta\,\mathbf{a}_\theta + 5\sin\phi\,\mathbf{a}_\phi$, find $\nabla \cdot \mathbf{A}$ at $(0.5, \pi/4, \pi/4)$.

$$\nabla \cdot \mathbf{A} = \frac{1}{r\sin\theta}\frac{\partial}{\partial\theta}(5\sin^2\theta) + \frac{1}{r\sin\theta}\frac{\partial}{\partial\phi}(5\sin\phi) = 10\frac{\cos\theta}{r} + 5\frac{\cos\phi}{r\sin\theta}$$

and

$$\nabla \cdot \mathbf{A}\Big|_{(0.5,\pi/4,\pi/4)} = 24.14$$

4.14. Given that $\mathbf{D} = \rho_0 z\mathbf{a}_z$ in the region $-1 \leq z \leq 1$ in cartesian coordinates and $\mathbf{D} = (\rho_0 z/|z|)\mathbf{a}_z$ elsewhere, find the charge density.

$$\nabla \cdot \mathbf{D} = \rho$$

For $-1 \leq z \leq 1$,

$$\rho = \frac{\partial}{\partial z}(\rho_0 z) = \rho_0$$

and for $z < -1$ or $z > 1$,

$$\rho = \frac{\partial}{\partial z}(\mp \rho_0) = 0$$

The charge distribution is shown in Fig. 4-5.

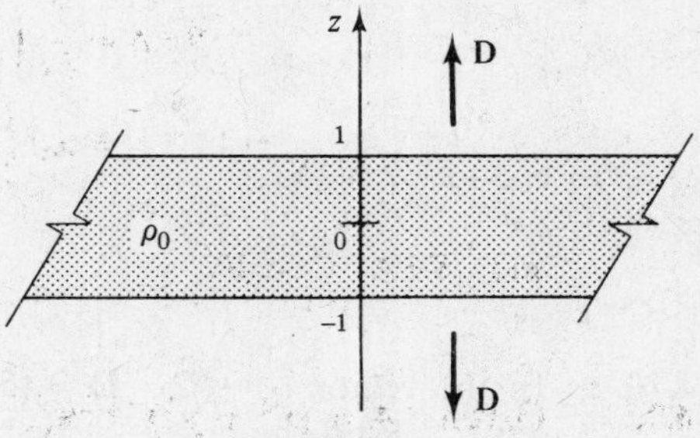

Fig. 4-5

4.15. Given that

$$\mathbf{D} = b(r^2 + z^2)^{-3/2}(r\mathbf{a}_r + z\mathbf{a}_z)$$

in cylindrical coordinates, find the charge density.

$$\begin{aligned}\rho &= \frac{1}{r}\frac{\partial}{\partial r}[b(r^2+z^2)^{-3/2}r^2] + \frac{\partial}{\partial z}[b(r^2+z^2)^{-3/2}z] \\ &= \frac{b}{r}\left[-\frac{3}{2}(r^2+z^2)^{-5/2}(2r^3) + (r^2+z^2)^{-3/2}(2r)\right] + b\left[-\frac{3}{2}(r^2+z^2)^{-5/2}(2z^2) + (r^2+z^2)^{-3/2}\right] \\ &= b(r^2+z^2)^{-5/2}[-3r^2 + (r^2+z^2)(2) - 3z^2 + (r^2+z^2)] = 0\end{aligned}$$

unless $r = z = 0$. (The given **D** field corresponds to a point charge at the origin.)

4.16. Given that $\mathbf{D} = (10r^3/4)\mathbf{a}_r$ (C/m^2) in the region $0 < r \leq 3$ m in cylindrical coordinates and $\mathbf{D} = (810/4r)\mathbf{a}_r$ (C/m^2) elsewhere, find the charge density.

For $0 < r \leq 3$ m,

$$\rho = \frac{1}{r}\frac{\partial}{\partial r}(10r^4/4) = 10r^2 \quad (\text{C/m}^3)$$

and for $r > 3$ m,

$$\rho = \frac{1}{r}\frac{\partial}{\partial r}(810/4) = 0$$

4.17. Given that

$$\mathbf{D} = \frac{Q}{\pi r^2}(1 - \cos 3r)\mathbf{a}_r$$

in spherical coordinates, find the charge density.

$$\rho = \frac{1}{r^2}\frac{\partial}{\partial r}\left[r^2 \frac{Q}{\pi r^2}(1 - \cos 3r)\right] = \frac{3Q}{\pi r^2}\sin 3r$$

4.18. Given that $\mathbf{D} = 7r^2\,\mathbf{a}_r + 28\sin\theta\,\mathbf{a}_\theta$ in spherical coordinates, find the charge density.

$$\rho = \frac{1}{r^2}\frac{\partial}{\partial r}(7r^4) + \frac{1}{r\sin\theta}\frac{\partial}{\partial\theta}(28\sin^2\theta) = 28r + \frac{56\cos\theta}{r}$$

4.19. In the region $0 < r \le 1$ m, $\mathbf{D} = (-2 \times 10^{-4}/r)\mathbf{a}_r$ (C/m²) and for $r > 1$ m, $\mathbf{D} = (-4 \times 10^{-4}/r^2)\mathbf{a}_r$ (C/m²), in spherical coordinates. Find the charge density in both regions.

For $0 < r \le 1$ m,

$$\rho = \frac{1}{r^2}\frac{\partial}{\partial r}(-2 \times 10^{-4} r) = \frac{-2 \times 10^{-4}}{r^2} \quad (\text{C/m}^3)$$

and for $r > 1$ m,

$$\rho = \frac{1}{r^2}\frac{\partial}{\partial r}(-4 \times 10^{-4}) = 0$$

4.20. In the region $r \le 2$, $\mathbf{D} = (5r^2/4)\mathbf{a}_r$ and for $r > 2$, $\mathbf{D} = (20/r^2)\mathbf{a}_r$, in spherical coordinates. Find the charge density.

For $r \le 2$,

$$\rho = \frac{1}{r^2}\frac{\partial}{\partial r}(5r^4/4) = 5r$$

and for $r > 2$,

$$\rho = \frac{1}{r^2}\frac{\partial}{\partial r}(20) = 0$$

4.21. Given that $\mathbf{D} = (10x^3/3)\mathbf{a}_x$ (C/m²), evaluate both sides of the divergence theorem for the volume of a cube, 2 m on an edge, centered at the origin and with edges parallel to the axes.

$$\oint \mathbf{D} \cdot d\mathbf{S} = \int_{\text{vol}} (\nabla \cdot \mathbf{D})\, dv$$

Since $\mathbf{D}$ has only an x component, $\mathbf{D} \cdot d\mathbf{S}$ is zero on all but the faces at $x = 1$ m and $x = -1$ m (see Fig. 4-6).

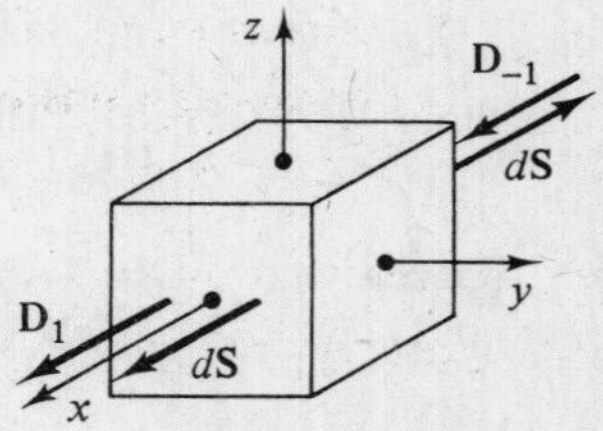

Fig. 4-6

$$\oint \mathbf{D} \cdot d\mathbf{S} = \int_{-1}^{1}\int_{-1}^{1} \frac{10(1)}{3}\mathbf{a}_x \cdot dy\,dz\,\mathbf{a}_x + \int_{-1}^{1}\int_{-1}^{1} \frac{10(-1)}{3}\mathbf{a}_x \cdot dy\,dz\,(-\mathbf{a}_x)$$

$$= \frac{40}{3} + \frac{40}{3} = \frac{80}{3}\ \text{C}$$

Now for the right side of the divergence theorem. Since $\nabla \cdot \mathbf{D} = 10x^2$,

$$\int_{\text{vol}} (\nabla \cdot \mathbf{D})\, dv = \int_{-1}^{1}\int_{-1}^{1}\int_{-1}^{1} (10x^2)\, dx\, dy\, dz = \int_{-1}^{1}\int_{-1}^{1} \left[10\frac{x^3}{3}\right]_{-1}^{1} dy\, dz = \frac{80}{3}\ \text{C}$$

4.22. Given that $\mathbf{A} = 30e^{-r}\mathbf{a}_r - 2z\mathbf{a}_z$ in cylindrical coordinates, evaluate both sides of the divergence theorem for the volume enclosed by $r = 2$, $z = 0$, and $z = 5$ (Fig. 4-7).

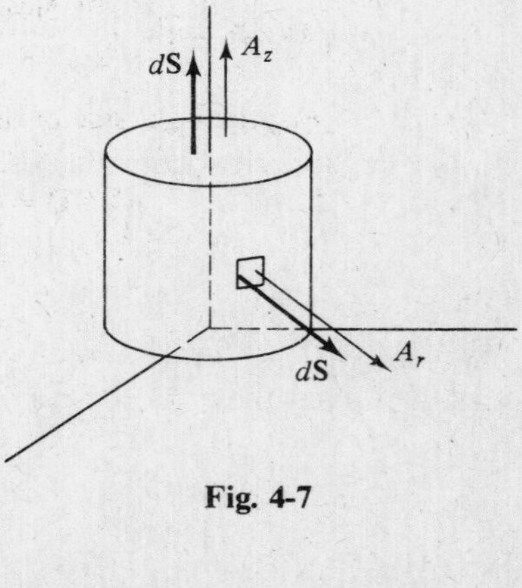

Fig. 4-7

$$\oint \mathbf{A} \cdot d\mathbf{S} = \int (\nabla \cdot \mathbf{A})\, dv$$

It is noted that $A_z = 0$ for $z = 0$ and hence $\mathbf{A} \cdot d\mathbf{S}$ is zero over that part of the surface.

$$\oint \mathbf{A} \cdot d\mathbf{S} = \int_0^5\int_0^{2\pi} 30e^{-2}\mathbf{a}_r \cdot 2\,d\phi\,dz\,\mathbf{a}_r + \int_0^{2\pi}\int_0^2 -2(5)\mathbf{a}_z \cdot r\,dr\,d\phi\,\mathbf{a}_z$$

$$= 60e^{-2}(2\pi)(5) - 10(2\pi)(2) = 129.4$$

For the right side of the divergence theorem:

$$\nabla \cdot \mathbf{A} = \frac{1}{r}\frac{\partial}{\partial r}(30re^{-r}) + \frac{\partial}{\partial z}(-2z) = \frac{30e^{-r}}{r} - 30e^{-r} - 2$$

and

$$\int (\nabla \cdot \mathbf{A})\,dv = \int_0^5 \int_0^{2\pi} \int_0^2 \left(\frac{30e^{-r}}{r} - 30e^{-r} - 2\right) r\,dr\,d\phi\,dz = 129.4$$

4.23. Given that $\mathbf{D} = (10r^3/4)\mathbf{a}_r$ (C/m²) in cylindrical coordinates, evaluate both sides of the divergence theorem for the volume enclosed by $r = 1$ m, $r = 2$ m, $z = 0$ and $z = 10$ m (see Fig. 4-8).

$$\oint \mathbf{D} \cdot d\mathbf{S} = \int (\nabla \cdot \mathbf{D})\,dv$$

Since $\mathbf{D}$ has no z component, $\mathbf{D} \cdot d\mathbf{S}$ is zero for the top and bottom. On the inner cylindrical surface $d\mathbf{S}$ is in the direction $-\mathbf{a}_r$.

$$\oint \mathbf{D} \cdot d\mathbf{S} = \int_0^{10} \int_0^{2\pi} \frac{10}{4}(1)^3\mathbf{a}_r \cdot (1)\,d\phi\,dz\,(-\mathbf{a}_r)$$

$$+ \int_0^{10} \int_0^{2\pi} \frac{10}{4}(2)^3\mathbf{a}_r \cdot (2)\,d\phi\,dz\,\mathbf{a}_r$$

$$= \frac{-200\pi}{4} + 16\frac{200\pi}{4} = 750\pi \text{ C}$$

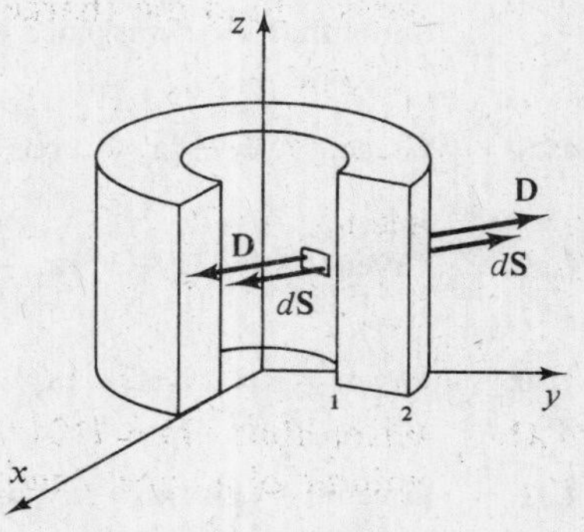

Fig. 4-8

For the right side of the divergence theorem:

$$\nabla \cdot \mathbf{D} = \frac{1}{r}\frac{\partial}{\partial r}(10r^4/4) = 10r^2$$

and

$$\int (\nabla \cdot \mathbf{D})\,dv = \int_0^{10} \int_0^{2\pi} \int_1^2 (10r^2)\,r\,dr\,d\phi\,dz = 750\pi \text{ C}$$

4.24. Given that $\mathbf{D} = (5r^2/4)\mathbf{a}_r$ (C/m²) in spherical coordinates, evaluate both sides of the divergence theorem for the volume enclosed by $r = 4$ m and $\theta = \pi/4$ (see Fig. 4-9).

$$\oint \mathbf{D} \cdot d\mathbf{S} = \int (\nabla \cdot \mathbf{D})\,dv$$

Since $\mathbf{D}$ has only a radial component, $\mathbf{D} \cdot d\mathbf{S}$ has a nonzero value only on the surface $r = 4$ m.

$$\oint \mathbf{D} \cdot d\mathbf{S} = \int_0^{2\pi} \int_0^{\pi/4} \frac{5(4)^2}{4}\mathbf{a}_r \cdot (4)^2 \sin\theta\,d\theta\,d\phi\,\mathbf{a}_r = 589.1 \text{ C}$$

For the right side of the divergence theorem:

$$\nabla \cdot \mathbf{D} = \frac{1}{r^2}\frac{\partial}{\partial r}(5r^4/4) = 5r$$

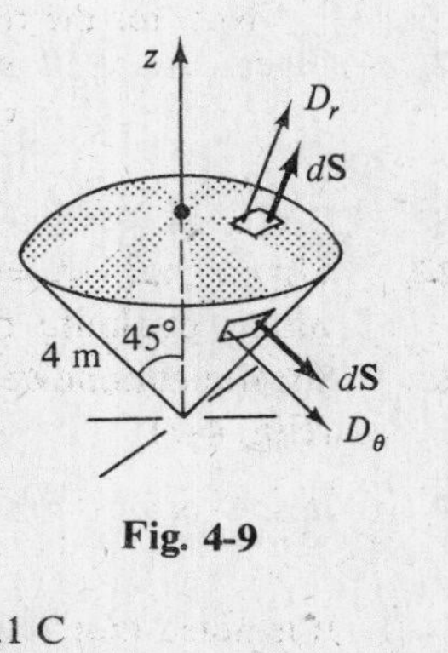

Fig. 4-9

and

$$\int (\nabla \cdot \mathbf{D})\,dv = \int_0^{2\pi} \int_0^{\pi/4} \int_0^4 (5r)\,r^2 \sin\theta\,dr\,d\theta\,d\phi = 589.1 \text{ C}$$

Supplementary Problems

4.25. Develop the divergence in spherical coordinates. Use the delta volume with edges Δr, $r\,\Delta\theta$ and $r\sin\theta\,\Delta\phi$.

4.26. Show that $\nabla\cdot\mathbf{E}$ is zero for the field of a uniform sheet charge.

4.27. The field of an electric dipole with the charges at $\pm d/2$ on the z axis is

$$\mathbf{E} = \frac{Qd}{4\pi\epsilon_0 r^3}(2\cos\theta\,\mathbf{a}_r + \sin\theta\,\mathbf{a}_\theta)$$

Show that the divergence of this field is zero.

4.28. Given $\mathbf{A} = e^{5x}\,\mathbf{a}_x + 2\cos y\,\mathbf{a}_y + 2\sin z\,\mathbf{a}_z$, find $\nabla\cdot\mathbf{A}$ at the origin. *Ans.* 7.0

4.29. Given $\mathbf{A} = (3x + y^2)\,\mathbf{a}_x + (x - y^2)\,\mathbf{a}_y$, find $\nabla\cdot\mathbf{A}$. *Ans.* $3 - 2y$

4.30. Given $\mathbf{A} = 2xy\mathbf{a}_x + z\mathbf{a}_y + yz^2\mathbf{a}_z$, find $\nabla\cdot\mathbf{A}$ at $(2, -1, 3)$. *Ans.* -8.0

4.31. Given $\mathbf{A} = 4xy\mathbf{a}_x - xy^2\mathbf{a}_y + 5\sin z\,\mathbf{a}_z$, find $\nabla\cdot\mathbf{A}$ at $(2, 2, 0)$. *Ans.* 5.0

4.32. Given $\mathbf{A} = 2r\cos^2\phi\,\mathbf{a}_r + 3r^2\sin z\,\mathbf{a}_\phi + 4z\sin^2\phi\,\mathbf{a}_z$, find $\nabla\cdot\mathbf{A}$. *Ans.* 4.0

4.33. Given $\mathbf{A} = (10/r^2)\mathbf{a}_r + 5e^{-2z}\,\mathbf{a}_z$, find $\nabla\cdot\mathbf{A}$ at $(2, \phi, 1)$. *Ans.* -2.60

4.34. Given $\mathbf{A} = 5\cos r\,\mathbf{a}_r + (3ze^{-2r}/r)\mathbf{a}_z$, find $\nabla\cdot\mathbf{A}$ at (π, ϕ, z). *Ans.* -1.59

4.35. Given $\mathbf{A} = 10\mathbf{a}_r + 5\sin\theta\,\mathbf{a}_\theta$, find $\nabla\cdot\mathbf{A}$. *Ans.* $(2 + \cos\theta)(10/r)$

4.36. Given $\mathbf{A} = r\mathbf{a}_r - r^2\cot\theta\,\mathbf{a}_\theta$, find $\nabla\cdot\mathbf{A}$. *Ans.* $3 - r$

4.37. Given $\mathbf{A} = [(10\sin^2\theta)/r]\mathbf{a}_r$, find $\nabla\cdot\mathbf{A}$ at $(2, \pi/4, \phi)$. *Ans.* 1.25

4.38. Given $\mathbf{A} = r^2\sin\theta\,\mathbf{a}_r + 13\phi\,\mathbf{a}_\theta + 2r\mathbf{a}_\phi$, find $\nabla\cdot\mathbf{A}$. *Ans.* $4r\sin\theta + \left(\frac{13\phi}{r}\right)\cot\theta$

4.39. Show that the divergence of $\mathbf{E}$ is zero if $\mathbf{E} = (100/r)\mathbf{a}_\phi + 40\mathbf{a}_z$.

4.40. In the region $a \le r \le b$ (cylindrical coordinates),

$$\mathbf{D} = \rho_0\left(\frac{r^2 - a^2}{2r}\right)\mathbf{a}_r$$

and for $r > b$,

$$\mathbf{D} = \rho_0\left(\frac{b^2 - a^2}{2r}\right)\mathbf{a}_r$$

For $r < a$, $\mathbf{D} = 0$. Find ρ in all three regions. *Ans.* $0, \rho_0, 0$

4.41. In the region $0 < r \leq 2$ (cylindrical coordinates), $\mathbf{D} = (4r^{-1} + 2e^{-0.5r} + 4r^{-1}e^{-0.5r})\mathbf{a}_r$, and for $r > 2$, $\mathbf{D} = (2.057/r)\mathbf{a}_r$. Find ρ in both regions. *Ans.* $-e^{-0.5r}$, 0

4.42. In the region $r \leq 2$ (cylindrical coordinates), $\mathbf{D} = [10r + (r^2/3)]\mathbf{a}_r$, and for $r > 2$, $\mathbf{D} = [3/(128r)]\mathbf{a}_r$. Find ρ in both regions. *Ans.* $20 + r$, 0

4.43. Given $\mathbf{D} = 10 \sin\theta\, \mathbf{a}_r + 2\cos\theta\, \mathbf{a}_\theta$, find the charge density.

Ans. $\dfrac{\sin\theta}{r}(18 + 2\cot^2\theta)$

4.44. Given

$$\mathbf{D} = \frac{3r}{r^2+1}\mathbf{a}_r$$

in spherical coordinates, find the charge density. *Ans.* $3(r^2+3)/(r^2+1)^2$

4.45. Given

$$\mathbf{D} = \frac{10}{r^2}[1 - e^{-2r}(1 + 2r + 2r^2)]\mathbf{a}_r$$

in spherical coordinates, find the charge density. *Ans.* $40e^{-2r}$

4.46. In the region $r \leq 1$ (spherical coordinates),

$$\mathbf{D} = \left(\frac{4r}{3} - \frac{r^3}{5}\right)\mathbf{a}_r$$

and for $r > 1$, $\mathbf{D} = [5/(63r^2)]\mathbf{a}_r$. Find the charge density in both regions. *Ans.* $4 - r^2$, 0

4.47. The region $r \leq 2$ m (spherical coordinates) has a field $\mathbf{E} = (5r \times 10^{-5}/\epsilon_0)\mathbf{a}_r$ (V/m). Find the net charge enclosed by the shell $r = 2$ m. *Ans.* 5.03×10^{-3} C

4.48. Given that $\mathbf{D} = (5r^2/4)\mathbf{a}_r$ in spherical coordinates, evaluate both sides of the divergence theorem for the volume enclosed between $r = 1$ and $r = 2$. *Ans.* 75π

4.49. Given that $\mathbf{D} = (10r^3/4)\mathbf{a}_r$ in cylindrical coordinates, evaluate both sides of the divergence theorem for the volume enclosed by $r = 2$, $z = 0$, and $z = 10$. *Ans.* 800π

4.50. Given that $\mathbf{D} = 10\sin\theta\, \mathbf{a}_r + 2\cos\theta\, \mathbf{a}_\theta$, evaluate both sides of the divergence theorem for the volume enclosed by the shell $r = 2$. *Ans.* $40\pi^2$

Chapter 5

Energy and Electric Potential of Charge Systems

5.1 WORK DONE IN MOVING A POINT CHARGE

In an electric field $\mathbf{E}$ a point charge Q experiences a force given by

$$\mathbf{F} = Q\mathbf{E}$$

This unbalanced force will result in an acceleration of the charged particle, and its motion will be in the direction of the field if Q is positive. See Fig. 5-1.

To put the charge in equilibrium an *applied force* is required which is equal in magnitude and opposite in direction to the force from the field:

$$\mathbf{F}_a = -Q\mathbf{E}$$

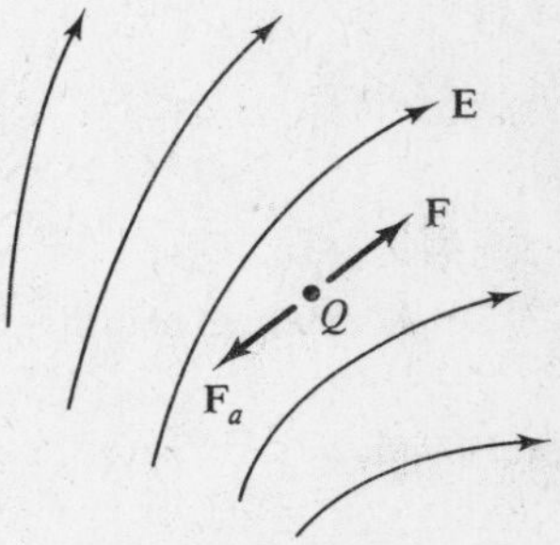

Fig. 5-1

Work is defined as a force acting over a distance. Therefore a differential amount of work dW is done by the applied force when the charged particle moves (at constant speed) through a differential distance $d\ell$. Now, work may be positive or negative, depending on the direction of $d\mathbf{l}$, the vector displacement, with respect to the applied force, $\mathbf{F}_a$. And when $d\mathbf{l}$ and $\mathbf{F}_a$ are not in the same direction, the component of the force in the direction of $d\mathbf{l}$ must be used. All this is simply expressed as

$$dW = F_a\,d\ell\cos\theta = \mathbf{F}_a \cdot d\mathbf{l}$$

Thus, in an electric field the differential work done by an external agency is

$$dW = -Q\mathbf{E}\cdot d\mathbf{l}$$

With this as the defining expression for work in moving a charged particle in an electric field, a positive value will mean that work had to be done by the external agent in order to bring about the change in position; a negative result will mean that work was done by the field.

In the three coordinate systems the expressions for $d\mathbf{l}$ are:

$$d\mathbf{l} = dx\,\mathbf{a}_x + dy\,\mathbf{a}_y + dz\,\mathbf{a}_z \qquad \text{(cartesian)}$$

$$d\mathbf{l} = dr\,\mathbf{a}_r + r\,d\phi\,\mathbf{a}_\phi + dz\,\mathbf{a}_z \qquad \text{(cylindrical)}$$

$$d\mathbf{l} = dr\,\mathbf{a}_r + r\,d\theta\,\mathbf{a}_\theta + r\sin\theta\,d\phi\,\mathbf{a}_\phi \qquad \text{(spherical)}$$

EXAMPLE 1 Find the work done in moving a charge of +2 C from (2,0,0) m to (0,2,0) m along the straight-line path joining the two points, if the electric field is

$$\mathbf{E} = 2x\mathbf{a}_x - 4y\mathbf{a}_y \quad \text{(V/m)}$$

The differential work is

$$dW = -2(2x\mathbf{a}_x - 4y\mathbf{a}_y)\cdot(dx\,\mathbf{a}_x + dy\,\mathbf{a}_y + dz\,\mathbf{a}_z)$$
$$= -4x\,dx + 8y\,dy$$

The equation of the path is $x + y = 2$, from which $dy = -dx$ along the path. Hence,

$$dW = -4x\,dx + 8(2-x)(-dx) = (4x - 16)\,dx$$

and

$$W = \int_2^0 (4x - 16)\,dx = 24\text{ J}$$

(Remember that 1 V/m = 1 N/C = 1 J/C · m.)

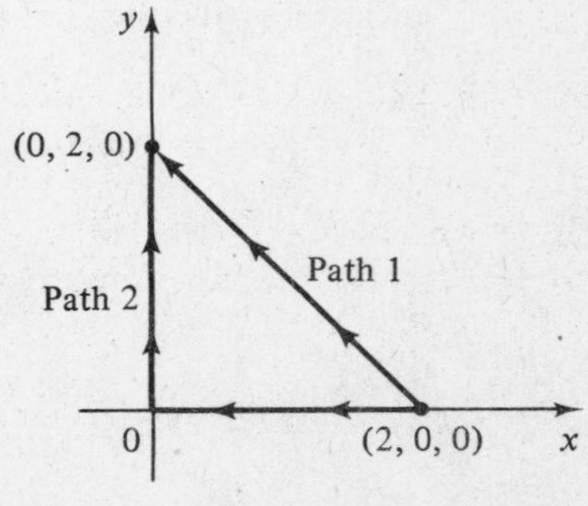

Fig. 5-2

The work done in moving a point charge Q from point B to point A in a static electric field is the same for any path chosen. Equivalently, the work done in moving the charge around any closed loop is zero:

$$\oint \mathbf{E} \cdot d\mathbf{l} = 0 \qquad \text{(static fields)}$$

Such a vector field is called a *conservative* field.

EXAMPLE 2 Find the work done in the field of Example 1 when the 2 C charge is moved from (2,0,0) m to (0,0,0) along the x axis and then from (0,0,0) to (0,2,0) m along the y axis.

The path is shown in Fig. 5-2. On the first segment, $y = dy = dz = 0$, so that

$$dW = -2(2x\mathbf{a}_x - 0\mathbf{a}_y) \cdot (dx\,\mathbf{a}_x + 0\mathbf{a}_y + 0\mathbf{a}_z) = -4x\,dx$$

On the second segment, $x = dx = dz = 0$, so that

$$dW = -2(0\mathbf{a}_x - 4y\mathbf{a}_y) \cdot (0\mathbf{a}_x + dy\,\mathbf{a}_y + 0\mathbf{a}_z) = 8y\,dy$$

Consequently,

$$W = -4\int_2^0 x\,dx + 8\int_0^2 y\,dy = 24 \text{ J}$$

the same value as for the path of Example 1.

5.2 ELECTRIC POTENTIAL BETWEEN TWO POINTS

The *potential* of point A with respect to point B is defined as the work done in moving a unit positive charge, Q_u, from B to A.

$$V_{AB} = \frac{W}{Q_u} = -\int_B^A \mathbf{E} \cdot d\mathbf{l} \quad \text{(J/C or V)}$$

It should be observed that the initial, or reference, point is the lower limit of the line integral. Then, too, the minus sign must not be omitted. This sign came into the expression by way of the force $\mathbf{F}_a = -Q\mathbf{E}$, which had to be applied to put the charge in equilibrium.

Because $\mathbf{E}$ is a conservative field,

$$V_{AB} = V_{AC} - V_{BC}$$

whence V_{AB} may be considered as the *potential difference* between points A and B. When V_{AB} is positive, work must be done to move the unit positive charge from B to A, and point A is said to be at a higher potential than point B. In Example 1, if point B is taken to be (2,0,0) m and A is (0,2,0) m, then

$$V_{AB} = \frac{24 \text{ J}}{2 \text{ C}} = 12 \text{ V}$$

Point A is at a higher potential than B, 12 V higher. Also, the potential V_{BA} must be -12 V, since V_{BA} differs from V_{AB} only by an interchange of the lower and upper limits in the defining integral, which simply changes the sign of the result.

EXAMPLE 3 Find the potential of A, $(1, \phi, z)$, with respect to B, $(3, \phi', z')$, in cylindrical coordinates, where the electric field due to a line charge on the z axis is given by $\mathbf{E} = (50/r)\mathbf{a}_r$ (V/m).

It is noted first that $d\mathbf{l}$ has components in the directions $\mathbf{a}_r$, $\mathbf{a}_\phi$, and $\mathbf{a}_z$ but that $\mathbf{E}$ is completely in the radial direction. Then $\mathbf{E} \cdot d\mathbf{l} = E_r\,dr$, and so

$$V_{AB} = -\int_B^A \mathbf{E} \cdot d\mathbf{l} = -\int_3^1 \frac{50}{r}\,dr = -50 \ln\frac{1}{3} = 54.9 \text{ V}$$

Point A is at a higher potential than point B.

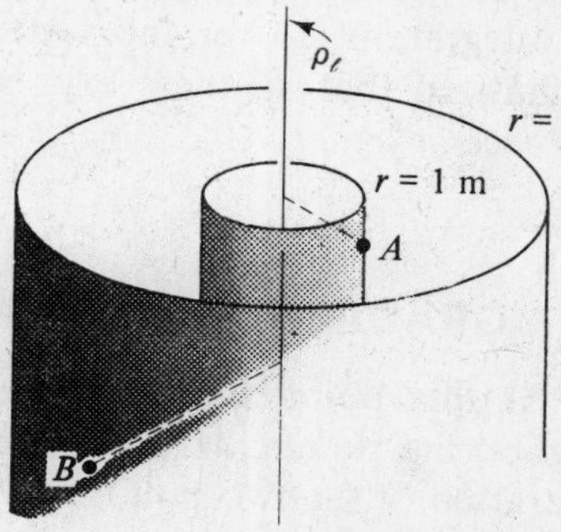

Fig. 5-3

Because no work is done in motions along $\mathbf{a}_\phi$ or $\mathbf{a}_z$, all points on the cylinder $r =$ const. must be at the same potential. In other words, for a uniform line charge, concentric right circular cylinders are *equipotential surfaces.*

5.3 POTENTIAL OF A POINT CHARGE

Since the electric field due to a point charge Q is completely in the radial direction,

$$V_{AB} = -\int_B^A \mathbf{E} \cdot d\mathbf{l} = -\int_{r_B}^{r_A} E_r\, dr = -\frac{Q}{4\pi\epsilon_0}\int_{r_B}^{r_A}\frac{dr}{r^2} = \frac{Q}{4\pi\epsilon_0}\left(\frac{1}{r_A} - \frac{1}{r_B}\right)$$

For a positive charge Q, point A is at a higher potential than point B when r_A is smaller than r_B. The equipotential surfaces are concentric spherical shells.

If the reference point B is now allowed to move out to infinity,

$$V_{A\infty} = \frac{Q}{4\pi\epsilon_0}\left(\frac{1}{r_A} - \frac{1}{\infty}\right)$$

or

$$V = \frac{Q}{4\pi\epsilon_0 r}$$

Considerable use will be made of this equation in the materials that follow. The greatest danger lies in forgetting where the reference is and attempting to apply the equation to charge distributions which themselves extend to infinity.

5.4 POTENTIAL OF A CHARGE DISTRIBUTION

If charge is distributed throughout some finite volume with a known charge density ρ (C/m^3), then the potential at some external point can be determined. To do so, a differential charge at a general point within the volume is identified, as shown in Fig. 5-4. Then at P,

$$dV = \frac{dQ}{4\pi\epsilon_0 R}$$

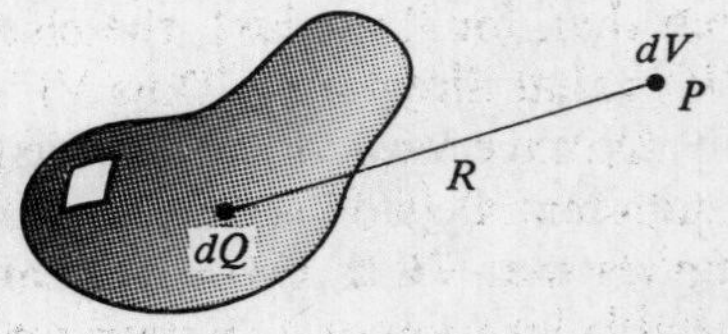

Fig. 5-4

Integration over the volume gives the total potential at P:

$$V = \int_{\text{vol}} \frac{\rho\, dv}{4\pi\epsilon_0 R}$$

where dQ is replaced by $\rho\, dv$. Now R must not be confused with r of the spherical coordinate system. And R is not a vector but the distance from dQ to the fixed point P. Finally, R almost always varies from place to place throughout the volume and so cannot be removed from the integrand.

If charge is distributed over a surface or a line, the above expression for V holds, provided that the integration is over the surface or the line and that ρ_s or ρ_ℓ is used in place of ρ. It must be emphasized that all these expressions for the potential at an external point are based upon a *zero reference at infinity.*

5.5 GRADIENT

At this point another operation of vector analysis is introduced. Figure 5-5(a) shows two neighboring points, M and N, of the region in which a scalar function V is defined. The vector separation of the two points is

$$d\mathbf{r} = dx\,\mathbf{a}_x + dy\,\mathbf{a}_y + dz\,\mathbf{a}_z$$

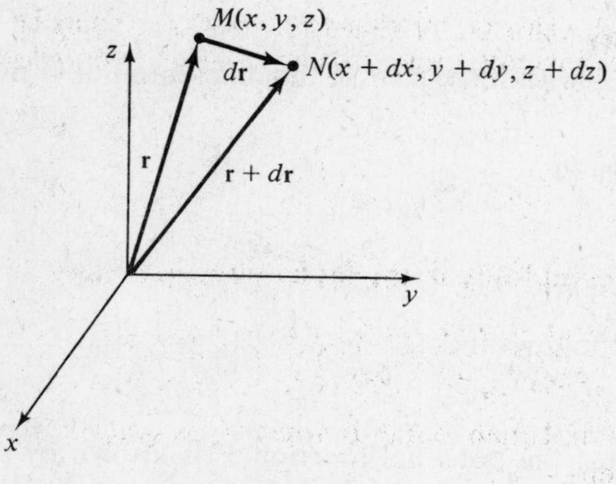

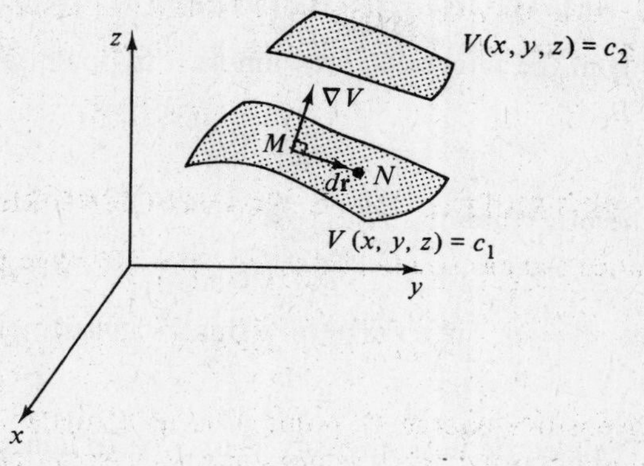

Fig. 5-5

From the calculus, the change in V from M to N is given by

$$dV = \frac{\partial V}{\partial x}dx + \frac{\partial V}{\partial y}dy + \frac{\partial V}{\partial z}dz$$

Now, the del operator, introduced in Section 4.4, operating on V gives

$$\nabla V = \frac{\partial V}{\partial x}\mathbf{a}_x + \frac{\partial V}{\partial y}\mathbf{a}_y + \frac{\partial V}{\partial z}\mathbf{a}_z$$

It follows that

$$dV = \nabla V \cdot d\mathbf{r}$$

The vector field ∇V (also written grad V) is called the *gradient* of the scalar function V. It is seen that, for fixed $|d\mathbf{r}|$, the change in V in a given direction $d\mathbf{r}$ is proportional to the projection of ∇V in that direction. Thus ∇V *lies in the direction of maximum increase of the function* V.

Another view of the gradient is obtained by allowing the points M and N to lie on the same equipotential (if V is a potential) surface, $V(x, y, z) = c_1$ [see Fig. 5-5(*b*)]. Then $dV = 0$, which implies that ∇V is perpendicular to $d\mathbf{r}$. But $d\mathbf{r}$ is tangent to the equipotential surface; indeed, for a suitable location of N, it represents *any* tangent through M. Therefore, ∇V must be along the surface normal at M. Since ∇V is in the direction of increasing V, it points from $V(x, y, z) = c_1$ to $V(x, y, z) = c_2$, where $c_2 > c_1$. *The gradient of a potential function is a vector field that is everywhere normal to the equipotential surfaces.*

The gradient in the cylindrical and spherical coordinate systems follows directly from that in the cartesian system. It is noted that each term contains the partial derivative of V with respect to distance in the direction of that particular unit vector.

$$\nabla V = \frac{\partial V}{\partial x}\mathbf{a}_x + \frac{\partial V}{\partial y}\mathbf{a}_y + \frac{\partial V}{\partial z}\mathbf{a}_z \qquad \text{(cartesian)}$$

$$\nabla V = \frac{\partial V}{\partial r}\mathbf{a}_r + \frac{\partial V}{r\,\partial \phi}\mathbf{a}_\phi + \frac{\partial V}{\partial z}\mathbf{a}_z \qquad \text{(cylindrical)}$$

$$\nabla V = \frac{\partial V}{\partial r}\mathbf{a}_r + \frac{\partial V}{r\,\partial \theta}\mathbf{a}_\theta + \frac{\partial V}{r\sin\theta\,\partial \phi}\mathbf{a}_\phi \qquad \text{(spherical)}$$

While ∇V is written for grad V in any coordinate system, it must be remembered that the del operator is defined only in cartesian coordinates.

5.6 RELATIONSHIP BETWEEN E AND V

From the integral expression for the potential of A with respect to B, the differential of V may be written

$$dV = -\mathbf{E} \cdot d\mathbf{l}$$

On the other hand,

$$dV = \nabla V \cdot d\mathbf{r}$$

Since $d\mathbf{l} = d\mathbf{r}$ is an arbitrary small displacement, it follows that

$$\mathbf{E} = -\nabla V$$

The electric field intensity **E** may be obtained when the potential function V is known by simply taking the negative of the gradient of V. The gradient was found to be a vector normal to the equipotential surfaces, directed to a positive change in V. With the negative sign here, the **E** field is found to be directed from higher to lower levels of potential V.

5.7 ENERGY IN STATIC ELECTRIC FIELDS

Consider the work required to assemble, charge by charge, a distribution of $n = 3$ point charges. The region is assumed initially to be charge-free and with $\mathbf{E} = 0$ throughout.

Referring to Fig. 5-6, the work required to place the first charge, Q_1, into position 1 is zero. Then, when Q_2 is moved toward the region, work equal to the product of this charge and the potential due to Q_1 is required. The total work to position the three charges is

$$\begin{aligned} W_E &= W_1 + W_2 + W_3 \\ &= 0 + (Q_2 V_{2,1}) + (Q_3 V_{3,1} + Q_3 V_{3,2}) \end{aligned}$$

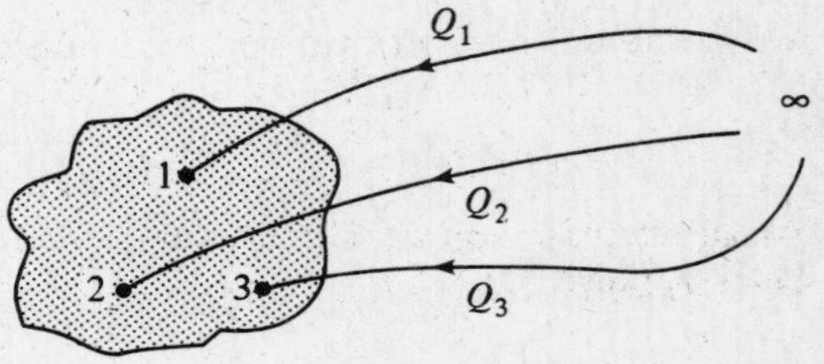

Fig. 5-6

The potential $V_{2,1}$ must be read "the potential at point 2 due to charge Q_1 at position 1." (This rather unusual notation will not appear again in this book.) The work W_E is the energy stored in the electric field of the charge distribution. (See Problem 5.20 for a comment on this identification.)

Now if the three charges were brought into place in reverse order, the total work would be

$$\begin{aligned} W_E &= W_3 + W_2 + W_1 \\ &= 0 + (Q_2 V_{2,3}) + (Q_1 V_{1,3} + Q_1 V_{1,2}) \end{aligned}$$

When the two expressions above are added, the result is twice the stored energy:

$$2W_E = Q_1(V_{1,2} + V_{1,3}) + Q_2(V_{2,1} + V_{2,3}) + Q_3(V_{3,1} + V_{3,2})$$

The term $Q_1(V_{1,2} + V_{1,3})$ was the work done against the fields of Q_2 and Q_3, the only other charges in the region. Hence, $V_{1,2} + V_{1,3} = V_1$, the potential at position 1. Then

$$2W_E = Q_1 V_1 + Q_2 V_2 + Q_3 V_3$$

and

$$W_E = \frac{1}{2} \sum_{m=1}^{n} Q_m V_m$$

for a region containing n point charges. For a region with a charge density ρ (C/m^3) the summation becomes an integration,

$$W_E = \frac{1}{2} \int \rho V \, dv$$

Other forms (see Problem 5.15) of the expression for stored energy are

$$W_E = \frac{1}{2}\int \mathbf{D}\cdot\mathbf{E}\,dv \qquad W_E = \frac{1}{2}\int \epsilon E^2\,dv \qquad W_E = \frac{1}{2}\int \frac{D^2}{\epsilon}\,dv$$

In an electric circuit, the energy stored in the field of a capacitor is given by

$$W_E = \frac{1}{2}QV = \frac{1}{2}CV^2$$

where C is the capacitance (in farads), V is the voltage difference between the two conductors making up the capacitor, and Q is the magnitude of the total charge on one of the conductors.

EXAMPLE 4 A parallel-plate capacitor, for which $C = \epsilon A/d$, has a constant voltage V applied across the plates (Fig. 5-7). Find the stored energy in the electric field.

With fringing neglected, the field is $\mathbf{E} = (V/d)\mathbf{a}_n$ between the plates and $\mathbf{E} = 0$ elsewhere.

$$W_E = \frac{1}{2}\int \epsilon E^2\,dv$$

$$= \frac{\epsilon}{2}\left(\frac{V}{d}\right)^2 \int dv$$

$$= \frac{\epsilon A V^2}{2d}$$

$$= \frac{1}{2}CV^2$$

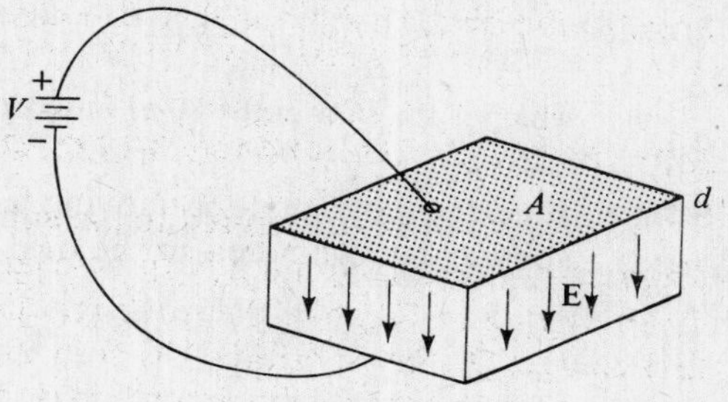

Fig. 5-7

As an alternate approach, the total charge on one conductor may be found from $\mathbf{D}$ at the surface via Gauss's law (Section 3.3).

$$\mathbf{D} = \frac{\epsilon V}{d}\mathbf{a}_n$$

$$Q = |\mathbf{D}|\,A = \frac{\epsilon V A}{d}$$

Then

$$W = \frac{1}{2}QV = \frac{1}{2}\left(\frac{\epsilon A V^2}{d}\right) = \frac{1}{2}CV^2$$

Solved Problems

5.1. Find the work done in moving a point charge $Q = -20\ \mu\text{C}$ from the origin to $(4, 0, 0)$ m in the field

$$\mathbf{E} = \left(\frac{x}{2} + 2y\right)\mathbf{a}_x + 2x\mathbf{a}_y \quad \text{(V/m)}$$

For a path along the x axis, $d\mathbf{l} = dx\,\mathbf{a}_x$.

$$dW = -Q\mathbf{E}\cdot d\mathbf{l}$$

$$= (20 \times 10^{-6})\left(\frac{x}{2} + 2y\right)dx$$

and

$$W = (20 \times 10^{-6})\int_0^4 \left(\frac{x}{2} + 2y\right)dx$$

$$= 80\ \mu\text{J}$$

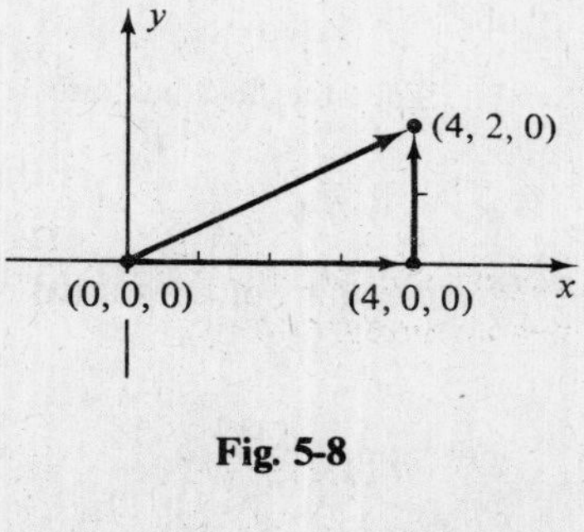

Fig. 5-8

5.2. In the field of Problem 5.1, move the charge from (4,0,0) m to (4,2,0) m and determine the work done.

Now (see Fig. 5-8) $d\mathbf{l} = dy\,\mathbf{a}_y$, and so

$$W = (20 \times 10^{-6}) \int_0^2 2x\,dy = (20 \times 10^{-6})(2)(4) \int_0^2 dy = 320\ \mu\text{J}$$

5.3. In the **E** field of Problem 5.1, find the work done in moving the charge from the origin to (4,2,0) m along the straight line connecting the points.

The equation of the line is $x = 2y$, from which $dx = 2\,dy$, $dz = 0$. Then

$$W = (20 \times 10^{-6}) \int \left[\left(\frac{x}{2} + 2y\right)\mathbf{a}_x + 2x\mathbf{a}_y\right] \cdot (dx\,\mathbf{a}_x + dy\,\mathbf{a}_y) = (20 \times 10^{-6}) \int \left(\frac{x}{2} + 2y\right) dx + 2x\,dy$$

To integrate with respect to x, y and dy are changed to $x/2$ and $dx/2$.

$$W = (20 \times 10^{-6}) \int_0^4 \frac{5}{2} x\,dx = 400\ \mu\text{J}$$

which is the sum of the 80 μJ and 320 μJ found in Problems 5.1 and 5.2.

5.4. Find the work done in moving a point charge $Q = 5\ \mu\text{C}$ from the origin to $(2\text{ m}, \pi/4, \pi/2)$, spherical coordinates, in the field

$$\mathbf{E} = 5e^{-r/4}\mathbf{a}_r + \frac{10}{r \sin\theta}\mathbf{a}_\phi \quad (\text{V/m})$$

In spherical coordinates,

$$d\mathbf{l} = dr\,\mathbf{a}_r + r\,d\theta\,\mathbf{a}_\theta + r\sin\theta\,d\phi\,\mathbf{a}_\phi$$

Choose the path shown in Fig. 5-9. Along segment I, $d\theta = d\phi = 0$, and

$$dW = -Q\mathbf{E} \cdot d\mathbf{l} = (-5 \times 10^{-6})(5e^{-r/4}dr)$$

Along segment II, $dr = d\theta = 0$, and

$$dW = -Q\mathbf{E} \cdot d\mathbf{l} = (-5 \times 10^{-6})(10\,d\phi)$$

Along segment III, $dr = d\phi = 0$, and

$$dW = -Q\mathbf{E} \cdot d\mathbf{l} = 0$$

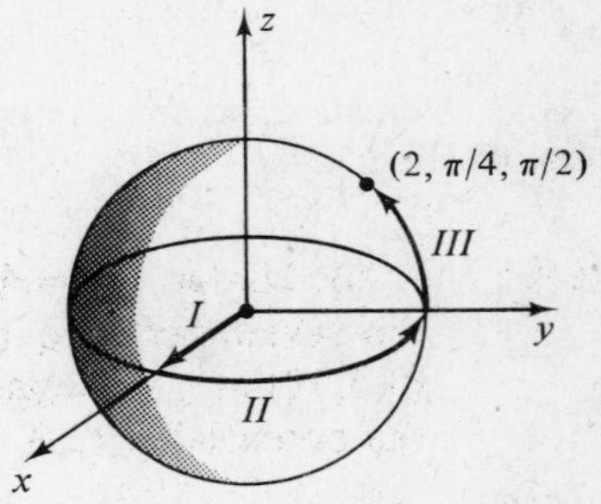

Fig. 5-9

Therefore,

$$W = (-25 \times 10^{-6}) \int_0^2 e^{-r/4}dr + (-50 \times 10^{-6}) \int_0^{\pi/2} d\phi = -117.9\ \mu\text{J}$$

In this case, the field does 117.9 μJ of work on the moving charge.

5.5. Given the field $\mathbf{E} = (k/r)\mathbf{a}_r$ in cylindrical coordinates, show that the work needed to move a point charge Q from any radial distance r to a point at twice that radial distance is independent of r.

Since the field has only a radial component,

$$dW = -Q\mathbf{E} \cdot d\mathbf{l} = -QE_r\,dr = \frac{-kQ}{r}\,dr$$

For the limits of integration use r_1 and $2r_1$.

$$W = -kQ \int_{r_1}^{2r_1} \frac{dr}{r} = -kQ \ln 2$$

independent of r_1.

5.6. For a line charge $\rho_\ell = (10^{-9}/2)$ C/m on the z axis, find V_{AB}, where A is $(2 \text{ m}, \pi/2, 0)$ and B is $(4 \text{ m}, \pi, 5 \text{ m})$.

$$V_{AB} = -\int_B^A \mathbf{E} \cdot d\mathbf{l} \qquad \text{where} \qquad \mathbf{E} = \frac{\rho_\ell}{2\pi\epsilon_0 r}\,\mathbf{a}_r$$

Since the field due to the line charge is completely in the radial direction, the dot product with $d\mathbf{l}$ results in $E_r\,dr$.

$$V_{AB} = -\int_B^A \frac{10^{-9}}{2(2\pi\epsilon_0 r)}\,dr = -9[\ln r]_4^2 = 6.24 \text{ V}$$

5.7. In the field of Problem 5.6, find V_{BC}, where $r_B = 4$ m and $r_C = 10$ m. Then find V_{AC} and compare with the sum of V_{AB} and V_{BC}.

$$V_{BC} = -9[\ln r]_{r_C}^{r_B} = -9(\ln 4 - \ln 10) = 8.25 \text{ V}$$

$$V_{AC} = -9[\ln r]_{r_C}^{r_A} = -9(\ln 2 - \ln 10) = 14.49 \text{ V}$$

$$V_{AB} + V_{BC} = 6.24 \text{ V} + 8.25 \text{ V} = 14.49 \text{ V} = V_{AC}$$

5.8. Given the field $\mathbf{E} = (-16/r^2)\mathbf{a}_r$ (V/m) in spherical coordinates, find the potential of point $(2 \text{ m}, \pi, \pi/2)$ with respect to $(4 \text{ m}, 0, \pi)$.

The equipotential surfaces are concentric spherical shells. Let $r = 2$ m be A and $r = 4$ m, B. Then

$$V_{AB} = -\int_4^2 \left(\frac{-16}{r^2}\right) dr = -4 \text{ V}$$

5.9. A line charge $\rho_\ell = 400$ pC/m lies along the x axis and the surface of zero potential passes through the point $(0, 5, 12)$ m in cartesian coordinates (see Fig. 5-10). Find the potential at $(2, 3, -4)$ m.

With the line charge along the x axis, the x coordinates of the two points may be ignored.

$$r_A = \sqrt{9 + 16} = 5 \text{ m} \qquad r_B = \sqrt{25 + 144} = 13 \text{ m}$$

Then

$$V_{AB} = -\int_{r_B}^{r_A} \frac{\rho_\ell}{2\pi\epsilon_0 r}\,dr = -\frac{\rho_\ell}{2\pi\epsilon_0} \ln \frac{r_A}{r_B} = 6.88 \text{ V}$$

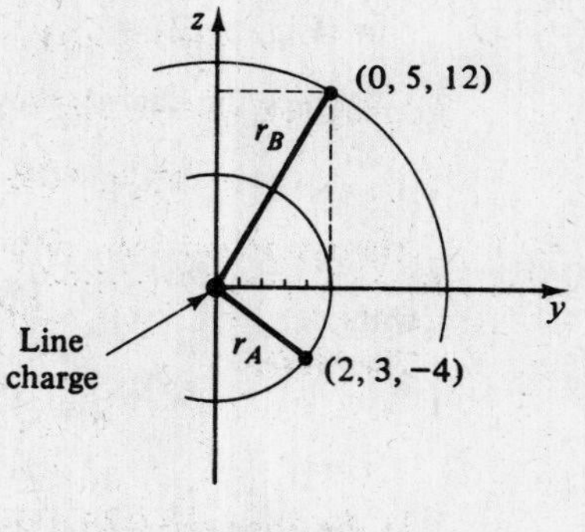

Fig. 5-10

5.10. Find the potential at $r_A = 5$ m with respect to $r_B = 15$ m due to a point charge $Q = 500$ pC at the origin and zero reference at infinity.

Due to a point charge,

$$V_{AB} = \frac{Q}{4\pi\epsilon_0}\left(\frac{1}{r_A} - \frac{1}{r_B}\right)$$

To find the potential difference, the zero reference is not needed.

$$V_{AB} = \frac{500 \times 10^{-12}}{4\pi(10^{-9}/36\pi)}\left(\frac{1}{5} - \frac{1}{15}\right) = 0.60 \text{ V}$$

The zero reference at infinity may be used to find V_5 and V_{15}.

$$V_5 = \frac{Q}{4\pi\epsilon_0}\left(\frac{1}{5}\right) = 0.90 \text{ V} \qquad V_{15} = \frac{Q}{4\pi\epsilon_0}\left(\frac{1}{15}\right) = 0.30 \text{ V}$$

Then $$V_{AB} = V_5 - V_{15} = 0.60 \text{ V}$$

5.11. A total charge of (40/3) nC is uniformly distributed around a circular ring of radius 2 m. Find the potential at a point on the axis 5 m from the plane of the ring. Compare with the result where all the charge is at the origin in the form of a point charge.

With the charge in a line,

$$V = \int \frac{\rho_\ell \, d\ell}{4\pi\epsilon_0 R}$$

Here $$\rho_\ell = \frac{(40/3) \times 10^{-9}}{2\pi(2)} = \frac{10^{-8}}{3\pi} \text{ C/m}$$

and (see Fig. 5-11) $R = \sqrt{29}$ m, $d\ell = (2 \text{ m})\,d\phi$.

$$V = \int_0^{2\pi} \frac{(10^{-8}/3\pi)(2)\,d\phi}{4\pi(10^{-9}/36\pi)\sqrt{29}} = 22.3 \text{ V}$$

If the charge is concentrated at the origin,

$$V = \frac{(40/3) \times 10^{-9}}{4\pi\epsilon_0(5)} = 24.0 \text{ V}$$

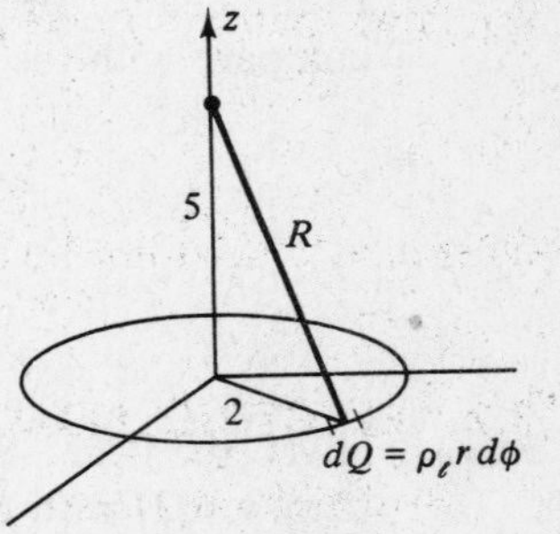

Fig. 5-11

5.12. Repeat Problem 5.11 with the total charge distributed uniformly over a circular disk of radius 2 m (Fig. 5-12).

Since the charge is over a surface,

$$V = \int \frac{\rho_s \, dS}{4\pi\epsilon_0 R}$$

with $$\rho_s = \frac{(40/3) \times 10^{-9}}{\pi(2)^2} = \frac{10^{-8}}{3\pi} \text{ C/m}^2$$

$$R = \sqrt{25 + r^2} \quad \text{(m)}$$

$$V = \frac{10^{-8}/3\pi}{4\pi(10^{-9}/36\pi)} \int_0^{2\pi} \int_0^2 \frac{r\,dr\,d\phi}{\sqrt{25 + r^2}} = 23.1 \text{ V}$$

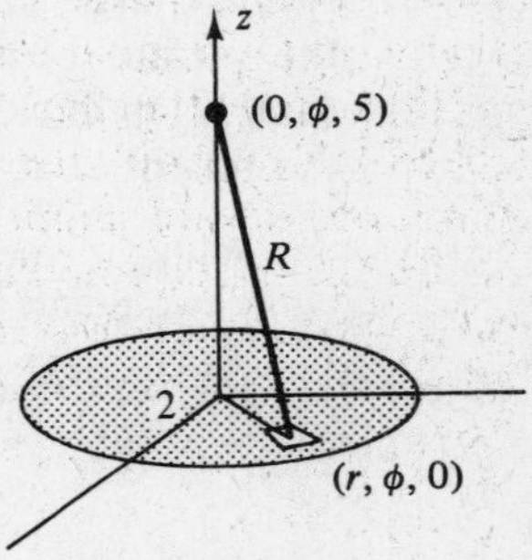

Fig. 5-12

5.13. Five equal point charges, $Q = 20$ nC, are located at $x = 2, 3, 4, 5$ and 6 m. Find the potential at the origin.

$$V = \frac{1}{4\pi\epsilon_0} \sum_{m=1}^{n} \frac{Q_m}{R_m} = \frac{20 \times 10^{-9}}{4\pi\epsilon_0}\left(\frac{1}{2} + \frac{1}{3} + \frac{1}{4} + \frac{1}{5} + \frac{1}{6}\right) = 261 \text{ V}$$

5.14. Charge is distributed uniformly along a straight line of finite length $2L$ (Fig. 5-13). Show that for two external points near the midpoint, such that r_1 and r_2 are small compared to the length, the potential V_{12} is the same as for an infinite line charge.

The potential at point 1 with zero reference at infinity is

$$V_1 = 2\int_0^L \frac{\rho_\ell\,dz}{4\pi\epsilon_0(z^2+r_1^2)^{1/2}}$$

$$= \frac{2\rho_\ell}{4\pi\epsilon_0}\left[\ln\left(z+\sqrt{z^2+r_1^2}\right)\right]_0^L$$

$$= \frac{\rho_\ell}{2\pi\epsilon_0}\left[\ln\left(L+\sqrt{L^2+r_1^2}\right)-\ln r_1\right]$$

Similarly, the potential at point 2 is

$$V_2 = \frac{\rho_\ell}{2\pi\epsilon_0}\left[\ln\left(L+\sqrt{L^2+r_2^2}\right)-\ln r_2\right]$$

Now if $L \gg r_1$ and $L \gg r_2$,

$$V_1 \approx \frac{\rho_\ell}{2\pi\epsilon_0}(\ln 2L - \ln r_1)$$

$$V_2 \approx \frac{\rho_\ell}{2\pi\epsilon_0}(\ln 2L - \ln r_2)$$

Then

$$V_{12} = V_1 - V_2 \approx \frac{\rho_\ell}{2\pi\epsilon_0}\ln\frac{r_2}{r_1}$$

which agrees with the expression found in Problem 5.9 for the infinite line.

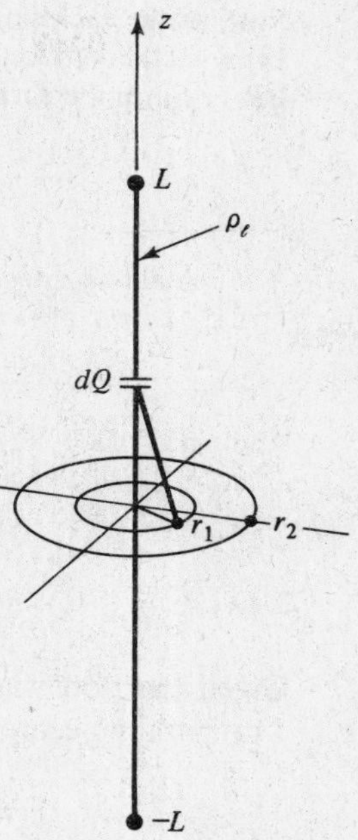

Fig. 5-13

5.15. Charge distributed throughout a volume v with density ρ gives rise to an electric field with energy content

$$W_E = \frac{1}{2}\int_v \rho V\,dv$$

Show that an equivalent expression for the stored energy is

$$W_E = \frac{1}{2}\int \epsilon E^2\,dv$$

Figure 5-14 shows the charge-containing volume v enclosed within a large sphere of radius R. Since ρ vanishes outside v,

$$W_E = \frac{1}{2}\int_v \rho V\,dv = \frac{1}{2}\int_{\substack{\text{spherical}\\\text{volume}}} \rho V\,dv = \frac{1}{2}\int_{\substack{\text{spherical}\\\text{volume}}} (\nabla\cdot\mathbf{D})V\,dv$$

The vector identity $\nabla\cdot V\mathbf{A} = \mathbf{A}\cdot\nabla V + V(\nabla\cdot\mathbf{A})$, applied to the integrand, gives

$$W_E = \frac{1}{2}\int_{\substack{\text{spherical}\\\text{volume}}} (\nabla\cdot V\mathbf{D})\,dv - \frac{1}{2}\int_{\substack{\text{spherical}\\\text{volume}}} (\mathbf{D}\cdot\nabla V)\,dv$$

This expression holds for an arbitrarily large radius R; the plan is to let $R\to\infty$.

The first integral on the right equals, by the divergence theorem,

$$\frac{1}{2}\oint_{\substack{\text{spherical}\\\text{surface}}} V\mathbf{D}\cdot d\mathbf{S}$$

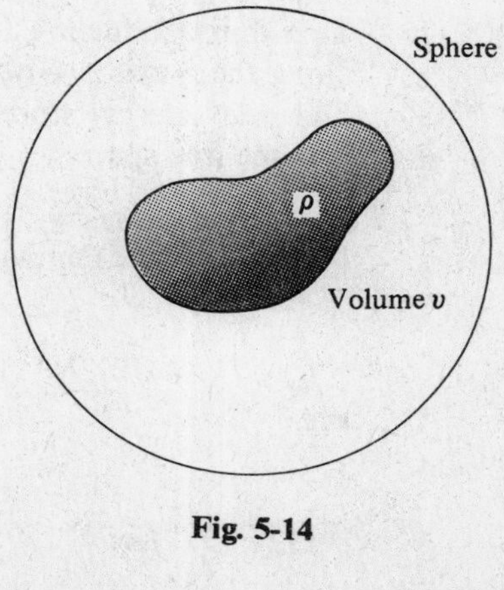

Fig. 5-14

Now, as the enclosing sphere becomes very large, the enclosed volume charge looks like a point charge. Thus, at the surface, D appears as k_1/R^2 and V appears as k_2/R. So the integrand is decreasing as $1/R^3$. Since the surface area increases only as R^2, it follows that

$$\lim_{R\to\infty} \oint_{\substack{\text{spherical}\\ \text{surface}}} V\mathbf{D}\cdot d\mathbf{S} = 0$$

The remaining integral gives, in the limit,

$$W_E = -\frac{1}{2}\int (\mathbf{D}\cdot\nabla V)\,dv = \frac{1}{2}\int (\mathbf{D}\cdot\mathbf{E})\,dv$$

And since $\mathbf{D} = \epsilon\mathbf{E}$, the stored energy is also given by

$$W_E = \frac{1}{2}\int \epsilon E^2\,dv \qquad \text{or} \qquad W_E = \frac{1}{2}\int \frac{D^2}{\epsilon}\,dv$$

5.16. Given the potential function $V = 2x + 4y$ (V) in free space, find the stored energy in a 1 m^3 volume centered at the origin. Examine other 1 m^3 volumes.

$$\mathbf{E} = -\nabla V = -\left(\frac{\partial V}{\partial x}\mathbf{a}_x + \frac{\partial V}{\partial y}\mathbf{a}_y + \frac{\partial V}{\partial z}\mathbf{a}_z\right) = -2\mathbf{a}_x - 4\mathbf{a}_y \quad \text{(V/m)}$$

This field is constant in magnitude ($E = \sqrt{20}$ V/m) and direction over all space, and so the total stored energy is infinite. (The field could be that within an infinite parallel-plate capacitor. It would take an infinite amount of work to charge such a capacitor.)

Nevertheless, it is possible to speak of an *energy density* for this and other fields. The expression

$$W_E = \frac{1}{2}\int \epsilon E^2\,dv$$

suggests that each tiny volume dv be assigned the energy content $w\,dv$, where

$$w = \frac{1}{2}\epsilon E^2$$

For the present field, the energy density is constant:

$$w = \frac{1}{2}\epsilon_0(20) = \frac{10^{-8}}{36\pi}\ \text{J/m}^3$$

and so every 1 m^3 volume contains $(10^{-8}/36\pi)$ J of energy.

5.17. Two thin conducting half-planes, at $\phi = 0$ and $\phi = \pi/6$, are insulated from each other along the z axis. Given that the potential function for $0 \le \phi \le \pi/6$ is $V = (-60\phi/\pi)$ V, find the energy stored between the half-planes for $0.1 \le r \le 0.6$ m and $0 \le z \le 1$ m. Assume free space.

To find the energy, W'_E, stored in a limited region of space, one must integrate the energy density (see Problem 5.16) through the region. Between the half-planes,

$$\mathbf{E} = -\nabla V = -\frac{1}{r}\frac{\partial}{\partial\phi}\left(\frac{-60\phi}{\pi}\right)\mathbf{a}_\phi = \frac{60}{\pi r}\mathbf{a}_\phi \quad \text{(V/m)}$$

and so

$$W'_E = \frac{\epsilon_0}{2}\int_0^1\int_0^{\pi/6}\int_{0.1}^{0.6}\left(\frac{60}{\pi r}\right)^2 r\,dr\,d\phi\,dz = \frac{300\epsilon_0}{\pi}\ln 6 = 1.51\ \text{nJ}$$

5.18. The electric field between two concentric cylindrical conductors at $r = 0.01$ m and $r = 0.05$ m is given by $\mathbf{E} = (10^5/r)\mathbf{a}_r$ (V/m), fringing neglected. Find the energy stored in a 0.5 m length. Assume free space.

$$W'_E = \frac{1}{2}\int \epsilon_0 E^2\,dv = \frac{\epsilon_0}{2}\int_h^{h+0.5}\int_0^{2\pi}\int_{0.01}^{0.05}\left(\frac{10^5}{r}\right)^2 r\,dr\,d\phi\,dz = 0.224 \text{ J}$$

5.19. Find the stored energy in a system of four identical point charges, $Q = 4$ nC, at the corners of a square 1 m on a side. What is the stored energy in the system when only two charges at opposite corners are in place?

$$2W_E = Q_1V_1 + Q_2V_2 + Q_3V_3 + Q_4V_4 = 4Q_1V_1$$

where the last equality follows from the symmetry of the system.

$$V_1 = \frac{Q_2}{4\pi\epsilon_0 R_{12}} + \frac{Q_3}{4\pi\epsilon_0 R_{13}} + \frac{Q_4}{4\pi\epsilon_0 R_{14}} = \frac{4\times 10^{-9}}{4\pi\epsilon_0}\left(\frac{1}{1}+\frac{1}{1}+\frac{1}{\sqrt{2}}\right) = 97.5 \text{ V}$$

Then
$$W_E = 2Q_1V_1 = 2(4\times 10^{-9})(97.5) = 780 \text{ nJ}$$

For only two charges in place,

$$2W_E = Q_1V_1 + Q_2V_2 = 2Q_1V_1$$

or
$$W_E = Q_1V_1 = (4\times 10^{-9})\left(\frac{4\times 10^{-9}}{4\pi\epsilon_0\sqrt{2}}\right) = 102 \text{ nJ}$$

5.20. What energy is stored in the system of two point charges, $Q_1 = 3$ nC and $Q_2 = -3$ nC, separated by a distance of $d = 0.2$ m?

$$2W_E = Q_1V_1 + Q_2V_2 = Q_1\left(\frac{Q_2}{4\pi\epsilon_0 d}\right) + Q_2\left(\frac{Q_1}{4\pi\epsilon_0 d}\right)$$

whence
$$W_E = \frac{Q_1Q_2}{4\pi\epsilon_0 d} = -\frac{(3\times 10^{-9})^2}{4\pi(10^{-9}/36\pi)(0.2)} = -405 \text{ nJ}$$

It may seem paradoxical that the stored energy turns out to be negative here, whereas $\frac{1}{2}\epsilon E^2$, and hence

$$W_E = \frac{1}{2}\int_{\text{all space}} \epsilon E^2\,dv$$

is necessarily positive. The reason for the discrepancy is that in equating the work done in assembling a system of point charges to the energy stored in the field, one neglects the infinite energy already in the field when the charges were at infinity. (It took an infinite amount of work to create the separate charges at infinity.) Thus, the above result, $W_E = -405$ nJ, may be taken to mean that the energy is 405 nJ below the (infinite) reference level at infinity. Since only energy *differences* have physical significance, the reference level may properly be disregarded.

5.21. A spherical conducting shell of radius a, centered at the origin, has a potential field

$$V = \begin{cases} V_0 & r \le a \\ V_0\,a/r & r > a \end{cases}$$

with the zero reference at infinity. Find an expression for the stored energy that this potential represents.

$$\mathbf{E} = -\nabla V = \begin{cases} 0 & r < a \\ (V_0\,a/r^2)\mathbf{a}_r & r > a \end{cases}$$

$$W_E = \frac{1}{2}\int \epsilon_0 E^2\,dv = 0 + \frac{\epsilon_0}{2}\int_0^{2\pi}\int_0^{\pi}\int_a^{\infty}\left(\frac{V_0\,a}{r^2}\right)^2 r^2\sin\theta\,dr\,d\theta\,d\phi = 2\pi\epsilon_0 V_0^2\,a$$

Note that the total charge on the shell is, from Gauss's law,

$$Q = DA = \left(\frac{\epsilon_0 V_0 a}{a^2}\right)(4\pi a^2) = 4\pi\epsilon_0 V_0 a$$

while the potential at the shell is $V = V_0$. Thus, $W_E = \frac{1}{2}QV$, the familiar result for the energy stored in a capacitor (in this case, a spherical capacitor with the other plate of infinite radius).

Supplementary Problems

5.22. Find the work done in moving a point charge $Q = -20\ \mu C$ from the origin to (4, 2, 0) m in the field

$$\mathbf{E} = 2(x + 4y)\mathbf{a}_x + 8x\mathbf{a}_y \quad \text{(V/m)}$$

along the path $x^2 = 8y$. *Ans.* 1.60 mJ

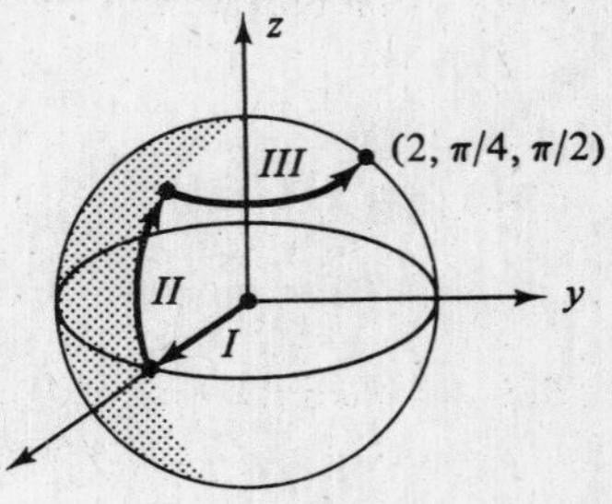

Fig. 5-15

5.23. Repeat Problem 5.4 using the direct radial path. *Ans.* $-39.35\ \mu J$ (the nature of the singularity along the z axis makes the field nonconservative)

5.24. Repeat Problem 5.4 using the path shown in Fig. 5-15. *Ans.* $-117.9\ \mu J$

5.25. Find the work done in moving a point charge $Q = 3\ \mu C$ from (4 m, π, 0) to (2 m, $\pi/2$, 2 m), cylindrical coordinates, in the field $\mathbf{E} = (10^5/r)\mathbf{a}_r + 10^5 z\mathbf{a}_z$ (V/m). *Ans.* -0.392 J

5.26. Find the difference in the amounts of work required to bring a point charge $Q = 2$ nC from infinity to $r = 2$ m and from infinity to $r = 4$ m, in the field $\mathbf{E} = (10^5/r)\mathbf{a}_r$ (V/m). *Ans.* 1.39×10^{-4} J

5.27. A total charge of (40/3) nC is uniformly distributed in the form of a circular disk of radius 2 m. Find the potential due to this charge at a point on the axis, 2 m from the disk. Compare this potential with that which results if all of the charge is at the center of the disk. *Ans.* 49.7 V, 60 V

5.28. A uniform line charge of density $\rho_\ell = 1$ nC/m is arranged in the form of a square 6 m on a side, as shown in Fig. 5-16. Find the potential at (0, 0, 5) m. *Ans.* 35.6 V

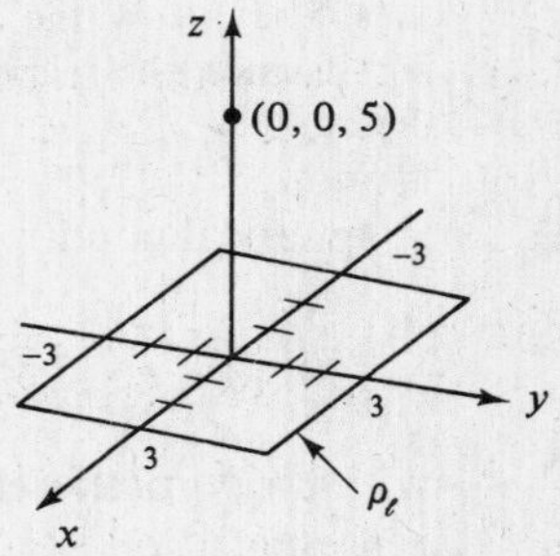

Fig. 5-16

5.29. Develop an expression for the potential at a point d meters radially outward from the midpoint of a finite line charge L meters long and of uniform density ρ_ℓ (C/m). Apply this result to Problem 5.28 as a check.

Ans. $$\frac{\rho_\ell}{2\pi\epsilon_0} \ln \frac{L/2 + \sqrt{d^2 + L^2/4}}{d} \quad \text{(V)}$$

5.30. Show that the potential at the origin due to a uniform surface charge density ρ_s over the ring $z = 0$, $R \le r \le R + 1$ is independent of R.

5.31. A total charge of 160 nC is first separated into four equal point charges spaced at 90° intervals around a circle of 3 m radius. Find the potential at a point on the axis, 5 m from the plane of the circle. Separate the total charge in eight equal parts and repeat with the charges at 45° intervals. What would be the answer in the limit $\rho_\ell = (160/6\pi)$ nC/m? *Ans.* 247 V

5.32. In spherical coordinates, point A is at a radius 2 m while B is at 4 m. Given the field $\mathbf{E} = (-16/r^2)\mathbf{a}_r$ (V/m), find the potential of point A, zero reference at infinity. Repeat for point B. Now express the potential difference $V_A - V_B$ and compare the result with Problem 5.8.
Ans. $V_A = 2V_B = -8$ V

5.33. If the zero potential reference is at $r = 10$ m and a point charge $Q = 0.5$ nC is at the origin, find the potentials at $r = 5$ m and $r = 15$ m. At what radius is the potential the same in magnitude as that at $r = 5$ m but opposite in sign? *Ans.* 0.45 V, -0.15 V, ∞

5.34. A point charge $Q = 0.4$ nC is located at $(2, 3, 3)$ m in cartesian coordinates. Find the potential difference V_{AB}, where point A is $(2, 2, 3)$ m and B is $(-2, 3, 3)$ m. *Ans.* 2.70 V

5.35. Find the potential in spherical coordinates due to two equal but opposite point charges on the y axis at $y = \pm d/2$. Assume $r \gg d$. *Ans.* $(Qd \sin\theta)/(4\pi\epsilon_0 r^2)$

5.36. Repeat Problem 5.35 with the charges on the z axis. *Ans.* $(Qd \cos\theta)/(4\pi\epsilon_0 r^2)$

5.37. Find the charge densities on the conductors in Problem 5.17.

Ans. $\dfrac{+60\epsilon_0}{\pi r}$ (C/m²) on $\phi = 0$, $\dfrac{-60\epsilon_0}{\pi r}$ (C/m²) on $\phi = \pi/6$

5.38. A uniform line charge $\rho_\ell = 2$ nC/m lies in the $z = 0$ plane parallel to the x axis at $y = 3$ m. Find the potential difference V_{AB} for the points $A(2\text{ m}, 0, 4\text{ m})$ and $B(0, 0, 0)$. *Ans.* -18.4 V

5.39. A uniform sheet of charge, $\rho_s = (1/6\pi)$ nC/m², is at $x = 0$ and a second sheet, $\rho_s = (-1/6\pi)$ nC/m², is at $x = 10$ m. Find V_{AB}, V_{BC} and V_{AC} for $A(10\text{ m}, 0, 0)$, $B(4\text{ m}, 0, 0)$ and $C(0, 0, 0)$.
Ans. -36 V, -24 V, -60 V

5.40. Given the cylindrical coordinate electric fields $\mathbf{E} = (5/r)\mathbf{a}_r$ (V/m) for $0 < r \le 2$ m and $\mathbf{E} = 2.5\,\mathbf{a}_r$ V/m for $r > 2$ m, find the potential difference V_{AB} for $A(1\text{ m}, 0, 0)$ and $B(4\text{ m}, 0, 0)$.
Ans. 8.47 V

5.41. A parallel-plate capacitor 0.5 m by 1.0 m, has a separation distance of 2 cm and a voltage difference of 10 V. Find the stored energy, assuming that $\epsilon = \epsilon_0$. *Ans.* 11.1 nJ

5.42. The capacitor described in Problem 5.41 has an applied voltage of 200 V.
(*a*) Find the stored energy.
(*b*) Hold d_1 (Fig. 5-17) at 2 cm and the voltage difference at 200 V, while increasing d_2 to 2.2 cm. Find the final stored energy. (*Hint*: $\Delta W_E = \frac{1}{2}(\Delta C)V^2$)
Ans. (*a*) 4.4 μJ; (*b*) 4.2 μJ

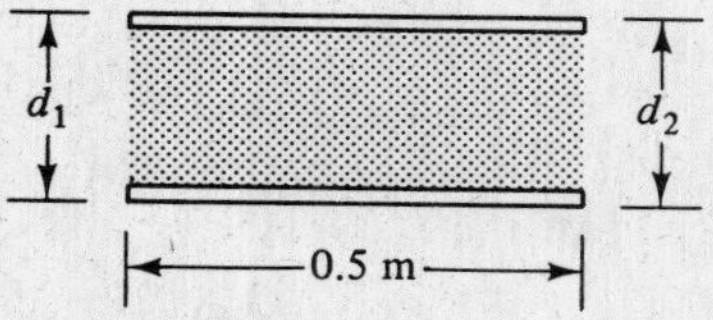

Fig. 5-17

5.43. Find the energy stored in a system of three equal point charges, $Q = 2$ nC, arranged in a line with 0.5 m separation between them. *Ans.* 180 nJ

5.44. Repeat Problem 5.43 if the charge in the center is -2 nC. *Ans.* -108 nJ

5.45. Four equal point charges, $Q = 2$ nC, are to be placed at the corners of a square (1/3) m on a side, one at a time. Find the energy in the system after each charge is positioned.
Ans. 0, 108 nJ, 292 nJ, 585 nJ

5.46. Given the electric field $\mathbf{E} = -5e^{-r/a}\mathbf{a}_r$ in cylindrical coordinates, find the energy stored in the volume described by $r \leq 2a$ and $0 \leq z \leq 5a$. *Ans.* $7.89 \times 10^{-10}a^3$

5.47. Given a potential $V = 3x^2 + 4y^2$ (V), find the energy stored in the volume described by $0 \leq x \leq 1$ m, $0 \leq y \leq 1$ m, and $0 \leq z \leq 1$ m. *Ans.* 147 pJ

Chapter 6

Current, Current Density and Conductors

6.1 INTRODUCTION

Electric current is the rate of transport of electric charge past a specified point or across a specified surface. The symbol I is generally used for constant currents and i for time-variable currents. The unit of current is the *ampere* ($1\ \text{A} = 1\ \text{C/s}$; in the SI, the ampere is the basic unit and the coulomb is the derived unit).

Ohm's law relates current to voltage and resistance. For simple dc circuits, $I = V/R$. However, when charges are suspended in a liquid or a gas, or where both positive and negative charge carriers are present with different characteristics, the simple form of Ohm's law is insufficient. Consequently, the current density $\mathbf{J}$ (A/m^2) receives more attention in electromagnetics than does current I.

6.2 CHARGES IN MOTION

Consider the force on a positively charged particle in an electric field in vacuum, as shown in Fig. 6-1(*a*). This force, $\mathbf{F} = +Q\mathbf{E}$, is unopposed and results in constant acceleration. Thus the charge moves in the direction of $\mathbf{E}$ with a velocity $\mathbf{U}$ that increases as long as the particle is in the $\mathbf{E}$ field. When the charge is in a liquid or gas, as shown in Fig. 6-1(*b*), it collides repeatedly with particles in the medium, resulting in random changes in direction. But, for constant $\mathbf{E}$ and a homogeneous medium, the random velocity components cancel out, leaving a constant average velocity, known as the *drift velocity* $\mathbf{U}$, along the direction of $\mathbf{E}$. Conduction in metals takes place by movement of the electrons in the outermost shells of the atoms making up the crystalline structure. According to the *electron-gas theory*, these electrons reach an average drift velocity in much the same way as a charged particle moving through a liquid or gas. The drift velocity is directly proportional to the electric field intensity,

$$\mathbf{U} = \mu \mathbf{E}$$

where μ, the *mobility*, has the units $\text{m}^2/\text{V}\cdot\text{s}$. Each cubic meter of a conductor contains on the order of 10^{28} atoms. Good conductors have one or two electrons from each atom free to move upon application of the field. The mobility μ varies with temperature and the crystalline structure of the solid. The particles in the solid have a vibratory motion which increases with temperature. This makes it more difficult for the charges to move. Thus, at higher temperatures the mobility μ is reduced, resulting in a smaller drift velocity (or current) for a given $\mathbf{E}$. In circuit analysis this phenomenon is accounted for by stating a *resistivity* for each material and specifying an increase in this resistivity with increasing temperature.

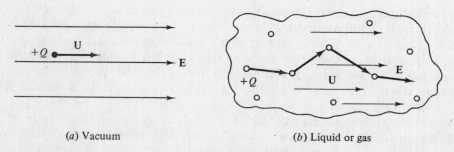

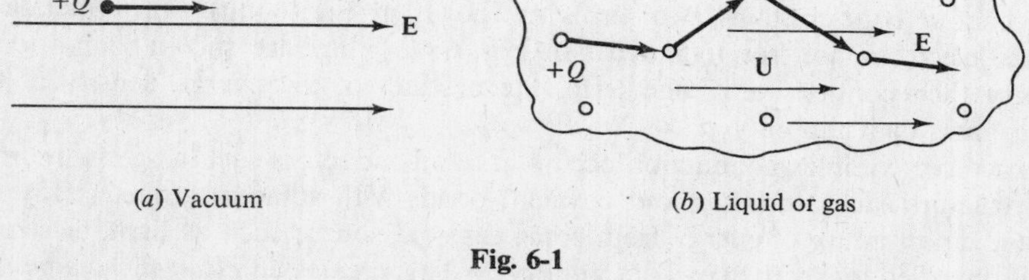

(*a*) Vacuum (*b*) Liquid or gas

Fig. 6-1

6.3 CONVECTION CURRENT DENSITY J

A set of charged particles giving rise to a charge density ρ in a volume v is shown in Fig. 6-2 to have a velocity $\mathbf{U}$ to the right. The particles are assumed to maintain their relative positions within the volume. As this charge configuration passes a surface S it constitutes a *convection current*, with density

$$\mathbf{J} = \rho\mathbf{U} \quad (\mathrm{A/m^2})$$

Of course, if the cross section of v varies or if the density ρ is not constant throughout v, then $\mathbf{J}$ will not be constant with time. Further, $\mathbf{J}$ will be zero when the last portion of the volume crosses S. Nevertheless, the concept of a current density caused by a cloud of charged particles in motion is at times useful in the study of electromagnetic field theory.

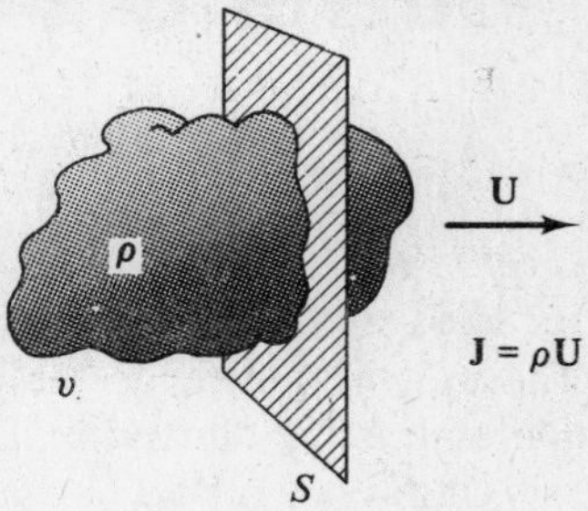

Fig. 6-2

6.4 CONDUCTION CURRENT DENSITY J

Of more interest is the *conduction current* that occurs in the presence of an electric field within a conductor of fixed cross section. The current density is again given by

$$\mathbf{J} = \rho\mathbf{U} \quad (\mathrm{A/m^2})$$

which, in view of the relation $\mathbf{U} = \mu\mathbf{E}$, can be written

$$\mathbf{J} = \sigma\mathbf{E}$$

where $\sigma = \rho\mu$ is the *conductivity* of the material, in *siemens per meter* (S/m). In metallic conductors the charge carriers are electrons, which drift in a direction opposite to that of the electric field (Fig. 6-3). Hence, for electrons, both ρ and μ are negative, which results in a positive conductivity σ, just as in the case of positive charge carriers. It follows that $\mathbf{J}$ and $\mathbf{E}$ have the same direction regardless of the sign of the charge carriers. It is conventional to treat electrons moving to the left as positive charges moving to the right, and always to report ρ and μ as positive.

The relation $\mathbf{J} = \sigma\mathbf{E}$ is often referred to as the *point form of Ohm's law*. The factor σ takes into account the density of the electrons free to move (ρ) and the relative ease with which they move through the crystalline structure (μ). As might be expected, σ is a function of temperature.

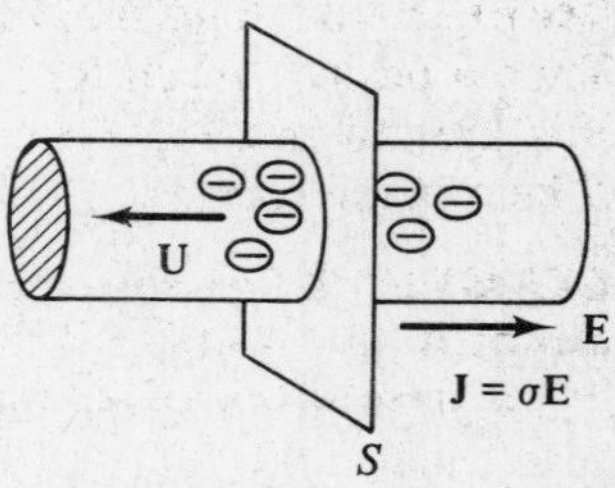

Fig. 6-3

6.5 CONDUCTIVITY σ

In a liquid or gas there are generally present both positive and negative ions, some singly charged and others doubly charged, and possibly of different masses. A conductivity expression would include all such factors. However, if it is assumed that all the negative ions are alike and so too the positive ions, then the conductivity contains two terms as shown in Fig. 6-4(a). In a metallic conductor, only the valence electrons are free to move. In Fig. 6-4(b) they are shown in motion to the left. The conductivity then contains only one term, the product of the charge density of the electrons free to move, ρ_e, and their mobility, μ_e.

A somewhat more complex conduction occurs in semiconductors such as germanium and silicon. In the crystal structure each atom has four covalent bonds with adjacent atoms. However, at room temperature, and upon influx of energy from some external source such as light, electrons can move out of the position called for by the covalent bonding. This creates an *electron–hole pair* available for

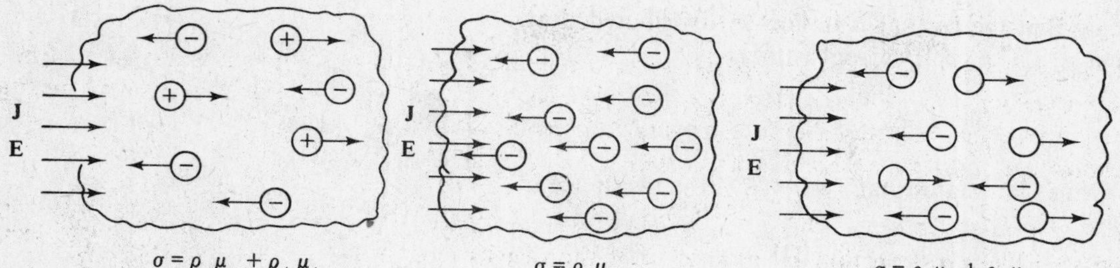

Fig. 6-4

conduction. Such materials are called *intrinsic* semiconductors. Electron–hole pairs have a short lifetime, disappearing by recombination. However, others are constantly being formed and at all times some are available for conduction. As shown in Fig. 6-4(*c*), the conductivity σ consists of two terms, one for the electrons and another for the holes. In practice, impurities, in the form of valence-three or valence-five elements, are added to create *p-type* and *n-type* semiconductor materials. The intrinsic behavior just described continues, but is far overshadowed by the presence of extra electrons in *n*-type, or holes in *p*-type, materials. Then, in the conductivity σ, one of the densities, ρ_e or ρ_h, will far exceed the other.

6.6 CURRENT I

The total current I (in A) crossing a surface S is given by

$$I = \int_S \mathbf{J} \cdot d\mathbf{S}$$

(see Fig. 6-5). A normal vector must be chosen for the differential surface $d\mathbf{S}$. Then a positive result for I indicates a current crossing S in the direction of this normal vector. Of course, $\mathbf{J}$ need not be uniform over S, nor need S be a plane surface.

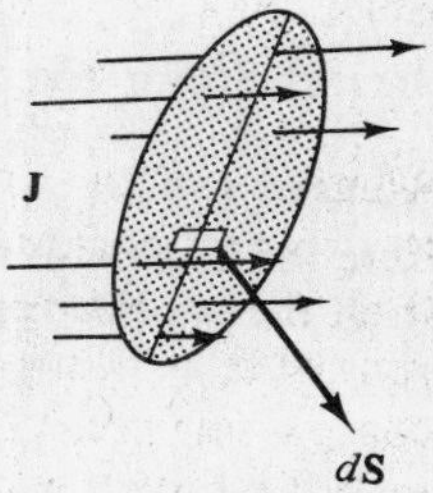

Fig. 6-5

EXAMPLE 1 Find the current in the circular wire shown in Fig. 6-6 if the current density is $\mathbf{J} = 15(1 - e^{-1000r})\mathbf{a}_z$ (A/m^2). The radius of the wire is 2 mm.

A cross section of the wire is chosen for S. Then

$$dI = \mathbf{J} \cdot d\mathbf{S}$$
$$= 15(1 - e^{-1000r})\mathbf{a}_z \cdot r\,dr\,d\phi\,\mathbf{a}_z$$

and

$$I = \int_0^{2\pi} \int_0^{0.002} 15(1 - e^{-1000r})r\,dr\,d\phi$$
$$= 1.33 \times 10^{-4}\ \text{A} = 0.133\ \text{mA}$$

Any surface S which has a perimeter that meets the outer surface of the conductor all the way around will have the same total current, $I = 0.133$ mA, crossing it.

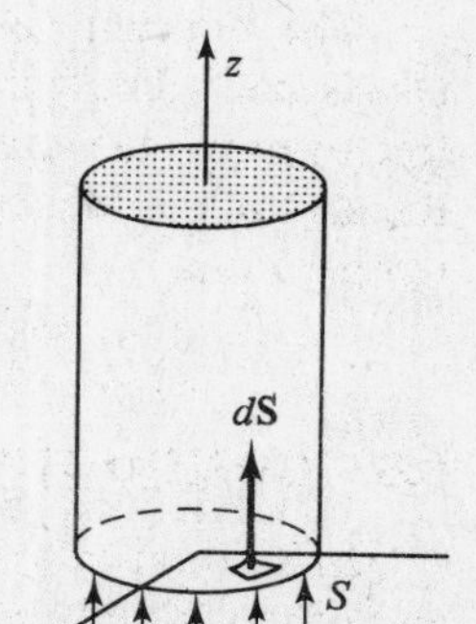

Fig. 6-6

6.7 RESISTANCE R

If a conductor of uniform cross-sectional area A and length ℓ, as shown in Fig. 6-7, has a voltage difference V between its ends, then

$$E = \frac{V}{\ell} \quad \text{and} \quad J = \frac{\sigma V}{\ell}$$

assuming that the current is uniformly distributed over the area A. The total current is then

$$I = JA = \frac{\sigma A V}{\ell}$$

Since Ohm's law states that $V = IR$, the resistance is

$$R = \frac{\ell}{\sigma A} \quad (\Omega)$$

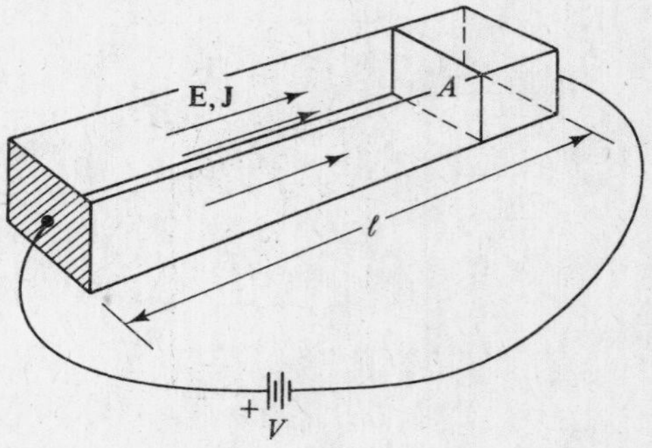

Fig. 6-7

(Note that $1\ \mathrm{S}^{-1} = 1\ \Omega$; the siemens was formerly known as the *mho*.) This expression for resistance is generally applied to all conductors where the cross section remains constant over the length ℓ. However, if the current density is greater along the surface area of the conductor than in the center, then the expression is not valid. For such nonuniform current distributions the resistance is given by

$$R = \frac{V}{\int \mathbf{J} \cdot d\mathbf{S}} = \frac{V}{\int \sigma \mathbf{E} \cdot d\mathbf{S}}$$

If **E** is known rather than the voltage difference between the two faces, the resistance is given by

$$R = \frac{\int \mathbf{E} \cdot d\mathbf{l}}{\int \sigma \mathbf{E} \cdot d\mathbf{S}}$$

The numerator gives the voltage drop across the sample, while the denominator gives the total current I.

EXAMPLE 2 Find the resistance between the inner and outer curved surfaces of the block shown in Fig. 6-8, where the material is silver for which $\sigma = 6.17 \times 10^7$ S/m.

If the same current I crosses both the inner and outer curved surfaces,

$$\mathbf{J} = \frac{k}{r}\,\mathbf{a}_r \qquad \text{and} \qquad \mathbf{E} = \frac{k}{\sigma r}\,\mathbf{a}_r$$

Then ($5° = 0.0873$ rad),

$$R = \frac{\displaystyle\int_{0.2}^{3.0} \frac{k}{\sigma r}\,\mathbf{a}_r \cdot dr\,\mathbf{a}_r}{\displaystyle\int_0^{0.05}\int_0^{0.0873} \frac{k}{r}\,\mathbf{a}_r \cdot r\,d\phi\,dz\,\mathbf{a}_r}$$

$$= \frac{\ln 15}{\sigma(0.05)(0.0873)} = 1.01 \times 10^{-5}\ \Omega = 10.1\ \mu\Omega$$

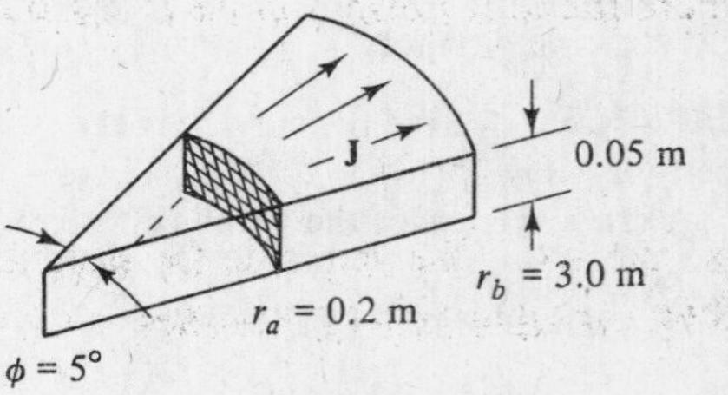

Fig. 6-8

6.8 CURRENT SHEET DENSITY K

At times current is confined to the surface of a conductor, such as the inside walls of a waveguide. For such a *current sheet* it is helpful to define the density vector **K** (in A/m), which gives the rate of charge transport per unit transverse length. (Some books use the notation $\mathbf{J}_s$.) Figure 6-9 shows a total current of I, in the form of a cylindrical sheet of radius r, flowing in the positive z direction. In this case,

$$\mathbf{K} = \frac{I}{2\pi r}\,\mathbf{a}_z$$

at each point of the sheet. For other sheets, **K** might vary from point to point (see Problem 6.19). In general, the current flowing through a curve C within a current sheet is obtained by integrating

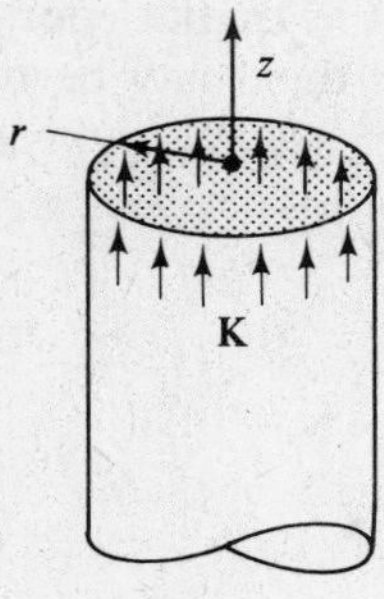

Fig. 6-9

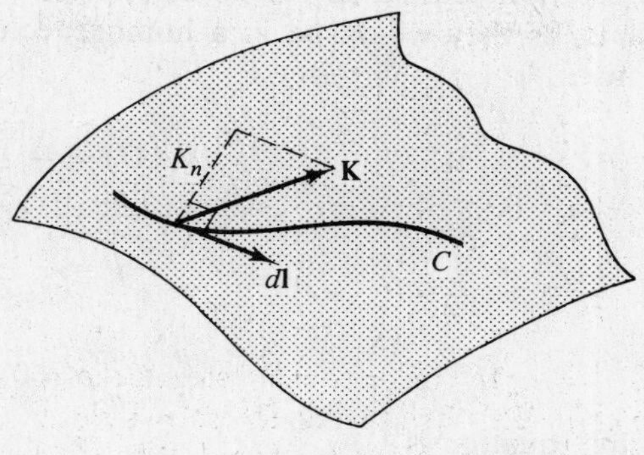

Fig. 6-10

the *normal* component of **K** along the curve (see Fig. 6-10). Thus

$$I = \int_C K_n \, d\ell$$

6.9 CONTINUITY OF CURRENT

Current I crossing a general surface S has been examined where **J** at the surface was known. Now, if the surface is *closed*, in order for net current to come out there must be a decrease of positive charge within:

$$\oint \mathbf{J} \cdot d\mathbf{S} = I = -\frac{dQ}{dt} = -\frac{\partial}{\partial t}\int \rho \, dv$$

where the unit normal in $d\mathbf{S}$ is the outward-directed normal. Dividing by Δv,

$$\frac{\oint \mathbf{J} \cdot d\mathbf{S}}{\Delta v} = -\frac{\partial}{\partial t}\frac{\int \rho \, dv}{\Delta v}$$

As $\Delta v \to 0$, the left side by definition approaches $\nabla \cdot \mathbf{J}$, the divergence of the current density, while the right side approaches $-\partial\rho/\partial t$. Thus

$$\nabla \cdot \mathbf{J} = -\frac{\partial \rho}{\partial t}$$

This is the *continuity of current* equation. In it ρ stands for the *net charge* density, not just the density of mobile charge. As will be shown below, $\partial\rho/\partial t$ can be nonzero within a conductor only transiently. Then the continuity equation, $\nabla \cdot \mathbf{J} = 0$, becomes the field equivalent of Kirchhoff's current law, which states that the net current leaving a junction of several conductors is zero.

In the process of conduction, valence electrons are free to move upon the application of an electric field. So, to the extent that these electrons are in motion, static conditions no longer exist. However, these electrons should not be confused with *net charge*, for each conduction electron is balanced by a proton in the nucleus such that there is zero net charge in every Δv of the material. Suppose, however, that through a temporary imbalance a region within a solid conductor has a *net* charge density ρ_0 at time $t = 0$. Then, since $\mathbf{J} = \sigma\mathbf{E} = (\sigma/\epsilon)\mathbf{D}$,

$$\nabla \cdot \frac{\sigma}{\epsilon}\mathbf{D} = -\frac{\partial \rho}{\partial t}$$

The divergence operation consists of partial derivatives with respect to the spatial coordinates. If σ and ϵ are constants, as they would be in a homogeneous sample, then they may be removed from the partial derivatives.

$$\frac{\sigma}{\epsilon}(\nabla \cdot \mathbf{D}) = -\frac{\partial \rho}{\partial t}$$

$$\frac{\sigma}{\epsilon}\rho = -\frac{\partial \rho}{\partial t}$$

or

$$\frac{\partial \rho}{\partial t} + \frac{\sigma}{\epsilon}\rho = 0$$

The solution to this equation is

$$\rho = \rho_0 e^{-(\sigma/\epsilon)t}$$

It is seen that ρ, and with it

$$\frac{\partial \rho}{\partial t} = -\frac{\sigma}{\epsilon}\rho$$

decays exponentially with a time constant ϵ/σ, also known as the *relaxation time* for the particular material. For silver, with $\sigma = 6.17 \times 10^7$ S/m and $\epsilon \approx \epsilon_0$, the relaxation time is 1.44×10^{-19} s. Thus, if a charge density of ρ_0 could somehow be set up in the interior of a block of silver, the charges would separate due to coulomb forces, and after 1.44×10^{-19} s the remaining density would be 36.8% of ρ_0. After five time constants, or 7.20×10^{-19} s, only 0.67% of ρ_0 remains. Thus, for static fields, it may be said that the *net charge within a conductor is zero.* If any net charge is present, it must reside on the outer surface.

6.10 CONDUCTOR–DIELECTRIC BOUNDARY CONDITIONS

Under static conditions all net charge will be on the outer surfaces of a conductor and both **E** and **D** are therefore zero within the conductor. Because the electric field is a conservative field, the closed line integral of $\mathbf{E} \cdot d\mathbf{l}$ is zero for any path. A rectangular path with corners *1, 2, 3, 4* is shown in Fig. 6-11.

$$\int_1^2 \mathbf{E} \cdot d\mathbf{l} + \int_2^3 \mathbf{E} \cdot d\mathbf{l} + \int_3^4 \mathbf{E} \cdot d\mathbf{l} + \int_4^1 \mathbf{E} \cdot d\mathbf{l} = 0$$

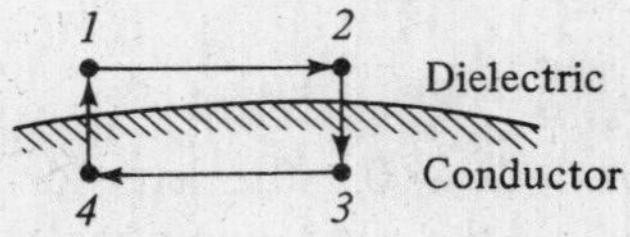

Fig. 6-11

If the path lengths *2* to *3* and *4* to *1* are now permitted to approach zero, keeping the interface between them, then the second and fourth integrals are zero. The path from *3* to *4* is within the conductor where **E** must be zero. This leaves

$$\int_1^2 \mathbf{E} \cdot d\mathbf{l} = \int_1^2 E_t \, d\ell = 0$$

where E_t is the tangential component of **E** at the surface of the dielectric. Since the interval *1* to *2* can be chosen arbitrarily,

$$E_t = D_t = 0$$

at each point of the surface.

To discover the conditions on the normal components, a small, closed, right circular cylinder is placed across the interface as shown in Fig. 6-12. Gauss's law applied to this surface gives

$$\oint \mathbf{D} \cdot d\mathbf{S} = Q_{\text{enc}}$$

or

$$\int_{\text{top}} \mathbf{D} \cdot d\mathbf{S} + \int_{\text{bottom}} \mathbf{D} \cdot d\mathbf{S} + \int_{\text{side}} \mathbf{D} \cdot d\mathbf{S} = \int_A \rho_s \, dS$$

The third integral is zero since, as just determined, $D_t = 0$ on either side of the interface. The second integral is also zero, since the bottom of the cylinder is within the conductor, where **D** and **E** are zero. Then,

$$\int_{\text{top}} \mathbf{D} \cdot d\mathbf{S} = \int_{\text{top}} D_n\, dS = \int_A \rho_s\, dS$$

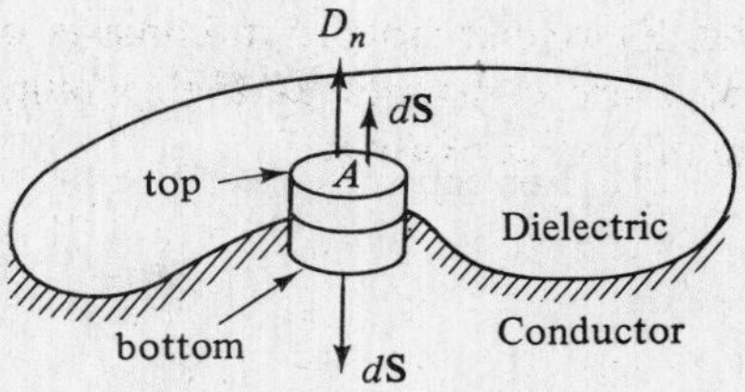

Fig. 6-12

which can hold only if

$$D_n = \rho_s \qquad \text{and} \qquad E_n = \frac{\rho_s}{\epsilon}$$

In short, under static conditions the field just outside a conductor is zero (both tangential and normal components) unless there exists a surface charge distribution. A surface charge does not imply a *net* charge in the conductor, however. To illustrate this, consider a positive charge at the origin of spherical coordinates. Now if this point charge is enclosed by an *uncharged* conducting spherical shell of finite thickness, as shown in Fig. 6-13(*a*), then the field is still given by

$$\mathbf{E} = \frac{+Q}{4\pi\epsilon r^2}\,\mathbf{a}_r$$

except within the conductor itself, where **E** must be zero. The coulomb forces caused by $+Q$ attract the conduction electrons to the inner surface, where they create a ρ_{s1} of negative sign. Then the deficiency of electrons on the outer surface constitutes a positive surface charge density ρ_{s2}. The electric flux lines Ψ, leaving the point charge $+Q$, terminate at the electrons on the inner surface of the conductor, as shown in Fig. 6-13(*b*). Then electric flux lines Ψ originate once again on the positive charges on the outer surface of the conductor. It should be noted that the flux does not pass through the conductor and the *net* charge on the conductor remains zero.

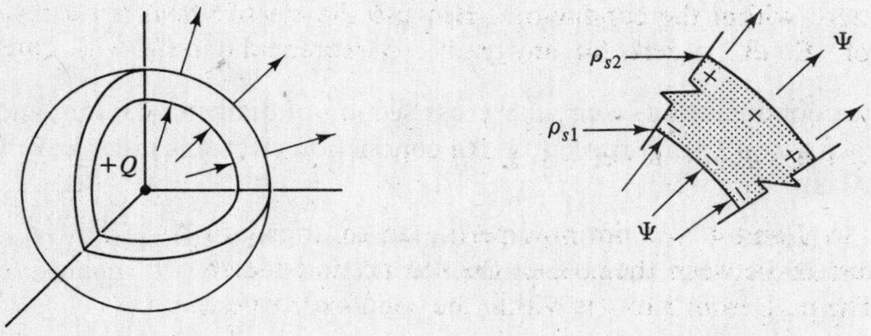

Fig. 6-13

Solved Problems

6.1. An AWG #12 copper conductor has an 80.8 mil diameter. A 50 foot length carries a current of 20 A. Find the electric field intensity *E*, drift velocity *U*, the voltage drop and the resistance for the 50 foot length.

Since a mil is 1/1000 inch, the cross-sectional area is

$$A = \pi\left[\left(\frac{0.0808\text{ in}}{2}\right)\left(\frac{2.54 \times 10^{-2}\text{ m}}{1\text{ in}}\right)\right]^2 = 3.31 \times 10^{-6}\text{ m}^2$$

Then $$J = \frac{I}{A} = \frac{20}{3.31 \times 10^{-6}} = 6.04 \times 10^{6}\ \text{A/m}^2$$

For copper, $\sigma = 5.8 \times 10^7$ S/m. Then

$$E = \frac{J}{\sigma} = \frac{6.04 \times 10^6}{5.8 \times 10^7} = 1.04 \times 10^{-1}\ \text{V/m}$$

$$V = E\ell = (1.04 \times 10^{-1})(50)(12)(0.0254) = 1.59\ \text{V}$$

$$R = \frac{V}{I} = \frac{1.59}{20} = 7.95 \times 10^{-2}\ \Omega$$

The electron mobility in copper is $\mu = 0.0032\ \text{m}^2/\text{V}\cdot\text{s}$, and since $\sigma = \rho\mu$, the charge density is

$$\rho = \frac{\sigma}{\mu} = \frac{5.8 \times 10^7}{0.0032} = 1.81 \times 10^{10}\ \text{C/m}^3$$

From $J = \rho U$ the drift velocity is now found as

$$U = \frac{J}{\rho} = \frac{6.05 \times 10^6}{1.81 \times 10^{10}} = 3.34 \times 10^{-4}\ \text{m/s}$$

With this drift velocity an electron takes approximately 30 seconds to move a distance of one centimeter in the #12 copper conductor.

6.2. What current density and electric field intensity correspond to a drift velocity of 5.3×10^{-4} m/s in aluminum?

For aluminum, the conductivity is $\sigma = 3.82 \times 10^7$ S/m and the mobility is $\mu = 0.0014\ \text{m}^2/\text{V}\cdot\text{s}$.

$$J = \rho U = \frac{\sigma}{\mu} U = \frac{3.82 \times 10^7}{0.0014}(5.3 \times 10^{-4}) = 1.45 \times 10^7\ \text{A/m}^2$$

$$E = \frac{J}{\sigma} = \frac{U}{\mu} = 3.79 \times 10^{-1}\ \text{V/m}$$

6.3. A long copper conductor has a circular cross section of diameter 3.0 mm and carries a current of 10 A. Each second, what percent of the conduction electrons must leave (to be replaced by others) a 100 mm length?

Avogadro's number is $N = 6.02 \times 10^{26}$ atoms/kmol. The specific gravity of copper is 8.96 and the atomic weight is 63.54. Assuming one conduction electron per atom, the number of electrons per unit volume is

$$N_e = \left(6.02 \times 10^{26}\ \frac{\text{atoms}}{\text{kmol}}\right)\left(\frac{1\ \text{kmol}}{63.54\ \text{kg}}\right)\left(8.96 \times 10^3\ \frac{\text{kg}}{\text{m}^3}\right)\left(1\ \frac{\text{electron}}{\text{atom}}\right)$$
$$= 8.49 \times 10^{28}\ \text{electrons/m}^3$$

The number of electrons in a 100 mm length is

$$N = \pi\left(\frac{3 \times 10^{-3}}{2}\right)^2(0.100)(8.49 \times 10^{28}) = 6.00 \times 10^{22}$$

A 10 A current requires that

$$\left(10\ \frac{\text{C}}{\text{s}}\right)\left(\frac{1}{1.6 \times 10^{-19}}\ \frac{\text{electron}}{\text{C}}\right) = 6.25 \times 10^{19}\ \text{electrons/s}$$

pass a fixed point. Then the percent leaving the 100 mm length per second is

$$\frac{6.25 \times 10^{19}}{6.00 \times 10^{22}}(100) = 0.104\%\ \text{per s}$$

6.4. What current would result if all the conduction electrons in a one centimeter cube of aluminum passed a specified point in 2.0 s? Assume one conduction electron per atom.

The density of aluminum is 2.70×10^3 kg/m³ and the atomic weight is 26.98 kg/kmol. Then

$$N_e = (6.02 \times 10^{26})\left(\frac{1}{26.98}\right)(2.70 \times 10^3) = 6.02 \times 10^{28} \text{ electrons/m}^3$$

and

$$I = \frac{\Delta Q}{\Delta t} = \frac{(6.02 \times 10^{28} \text{ electrons/m}^3)(10^{-2} \text{ m})^3(1.6 \times 10^{-19} \text{ C/electron})}{2 \text{ s}} = 4.82 \text{ kA}$$

6.5. What is the density of free electrons in a metal for a mobility of $0.0046 \text{ m}^2/\text{V}\cdot\text{s}$ and a conductivity of 29.1 MS/m?

Since $\sigma = \mu\rho$,

$$\rho = \frac{\sigma}{\mu} = \frac{29.1 \times 10^6}{0.0046} = 6.33 \times 10^9 \text{ C/m}^3$$

and

$$N_e = \frac{6.33 \times 10^9}{1.6 \times 10^{-19}} = 3.96 \times 10^{28} \text{ electrons/m}^3$$

6.6. Determine the conductivity of intrinsic germanium at room temperature.

At 300 K there are 2.5×10^{19} electron–hole pairs per cubic meter. The electron mobility is $\mu_e = 0.38 \text{ m}^2/\text{V}\cdot\text{s}$ and the hole mobility is $\mu_h = 0.18 \text{ m}^2/\text{V}\cdot\text{s}$. Since the material is not doped, the numbers of electrons and holes are equal.

$$\sigma = N_e e(\mu_e + \mu_h) = (2.5 \times 10^{19})(1.6 \times 10^{-19})(0.38 + 0.18) = 2.24 \text{ S/m}$$

6.7. Find the conductivity of *n*-type germanium at room temperature, assuming one donor atom in each 10^8 atoms. The density of germanium is 5.32×10^3 kg/m³ and the atomic weight is 72.6 kg/kmol.

There are

$$N = (6.02 \times 10^{26})\left(\frac{1}{72.6}\right)(5.32 \times 10^3) = 4.41 \times 10^{28} \text{ atoms/m}^3$$

and these yield

$$N_e = 10^{-8}(4.41 \times 10^{28}) = 4.41 \times 10^{20} \text{ electrons/m}^3$$

The intrinsic concentration n_i for germanium at 300 K is 2.5×10^{19} per m³. The *mass-action law*, $N_e N_h = n_i^2$, then gives the density of holes as

$$N_h = \frac{(2.5 \times 10^{19})^2}{4.41 \times 10^{20}} = 1.42 \times 10^{18} \text{ holes/m}^3$$

Then, using the mobilities from Problem 6.6,

$$\begin{aligned}\sigma &= N_e e\mu_e + N_h e\mu_h \\ &= (4.41 \times 10^{20})(1.6 \times 10^{-19})(0.38) + (1.42 \times 10^{18})(1.6 \times 10^{-19})(0.18) \\ &= 26.8 + 0.041 = 26.8 \text{ S/m}\end{aligned}$$

In this *n*-type germanium the number of electrons in a cubic meter is 4.41×10^{20}, as against 1.42×10^{18} holes. The conductivity then is controlled by the electrons provided by the valence-five doping agent.

6.8. A conductor of uniform cross section and 150 m long has a voltage drop of 1.3 V and a current density of 4.65×10^5 A/m^2. What is the conductivity of the material in the conductor?

Since $E = V/\ell$ and $J = \sigma E$,

$$4.65 \times 10^5 = \sigma\left(\frac{1.3}{150}\right) \qquad \text{or} \qquad \sigma = 5.37 \times 10^7 \text{ S/m}$$

6.9. A table of resistivities gives 10.4 ohm · circular mils per foot for annealed copper. What is the corresponding conductivity in siemens per meter?

A *circular mil* is the area of a circle with a diameter of one mil (10^{-3} in).

$$1 \text{ cir mil} = \pi\left[\left(\frac{10^{-3}\text{ in}}{2}\right)\left(0.0254\,\frac{\text{m}}{\text{in}}\right)\right]^2 = 5.07 \times 10^{-10}\text{ m}^2$$

The conductivity is the reciprocal of the resistivity.

$$\sigma = \left(\frac{1}{10.4\ \Omega\cdot\text{cir mil}}\,\frac{\text{ft}}{}\right)\left(12\,\frac{\text{in}}{\text{ft}}\right)\left(0.0254\,\frac{\text{m}}{\text{in}}\right)\left(\frac{1\text{ cir mil}}{5.07 \times 10^{-10}\text{ m}^2}\right) = 5.78 \times 10^7\text{ S/m}$$

6.10. An AWG #20 aluminum wire has a resistance of 16.7 ohms per 1000 feet. What conductivity does this imply for aluminum?

From wire tables, a #20 wire has a diameter of 32 mils.

$$A = \pi\left[\frac{32 \times 10^{-3}}{2}(0.0254)\right]^2 = 5.19 \times 10^{-7}\text{ m}^2$$

$$\ell = (1000\text{ ft})(12\text{ in/ft})(0.0254\text{ m/in}) = 3.05 \times 10^2\text{ m}$$

Then from $R = \ell/\sigma A$,

$$\sigma = \frac{3.05 \times 10^2}{(16.7)(5.19 \times 10^{-7})} = 35.2\text{ MS/m}$$

6.11. In a cylindrical conductor of radius 2 mm, the current density varies with the distance from the axis according to

$$J = 10^3 e^{-400r}\quad (\text{A/m}^2)$$

Find the total current I.

$$I = \int \mathbf{J}\cdot d\mathbf{S} = \int J\,dS = \int_0^{2\pi}\int_0^{0.002} 10^3 e^{-400r}\,r\,dr\,d\phi$$

$$= 2\pi(10^3)\left[\frac{e^{-400r}}{(-400)^2}(-400r - 1)\right]_0^{0.002} = 7.51\text{ mA}$$

6.12. Find the current crossing the portion of the $y = 0$ plane defined by $-0.1 \le x \le 0.1$ m and $-0.002 \le z \le 0.002$ m if

$$\mathbf{J} = 10^2|x|\mathbf{a}_y\quad (\text{A/m}^2)$$

$$I = \int \mathbf{J}\cdot d\mathbf{S} = \int_{-0.002}^{0.002}\int_{-0.1}^{0.1} 10^2|x|\mathbf{a}_y\cdot dx\,dz\,\mathbf{a}_y = 4\text{ mA}$$

6.13. Find the current crossing the portion of the $x = 0$ plane defined by $-\pi/4 \le y \le \pi/4$ m and $-0.01 \le z \le 0.01$ m if

$$\mathbf{J} = 100 \cos 2y \, \mathbf{a}_x \quad (\mathrm{A/m^2})$$

$$I = \int \mathbf{J} \cdot d\mathbf{S} = \int_{-0.01}^{0.01} \int_{-\pi/4}^{\pi/4} 100 \cos 2y \, \mathbf{a}_x \cdot dy \, dz \, \mathbf{a}_x = 2.0 \text{ A}$$

6.14. Given $\mathbf{J} = 10^3 \sin\theta \, \mathbf{a}_r$ A/m² in spherical coordinates, find the current crossing the spherical shell $r = 0.02$ m.

Since **J** and

$$d\mathbf{S} = r^2 \sin\theta \, d\theta \, d\phi \, \mathbf{a}_r$$

are radial,

$$I = \int_0^{2\pi} \int_0^{\pi} 10^3 (0.02)^2 \sin^2\theta \, d\theta \, d\phi = 3.95 \text{ A}$$

6.15. Show that the resistance of any conductor of constant cross-sectional area A and length ℓ is given by $R = \ell/\sigma A$, assuming uniform current distribution.

A constant cross section along the length ℓ results in constant E, and the voltage drop is

$$V = \int \mathbf{E} \cdot d\mathbf{l} = E\ell$$

If the current is uniformly distributed over the area A,

$$I = \int \mathbf{J} \cdot d\mathbf{S} = JA = \sigma E A$$

where σ is the conductivity. Then, since $R = V/I$,

$$R = \frac{\ell}{\sigma A}$$

6.16. Determine the resistance of the insulation in a length ℓ of coaxial cable, as shown in Fig. 6-14.

Assume a total current I from the inner conductor to the outer conductor. Then, at a radial distance r,

$$J = \frac{I}{A} = \frac{I}{2\pi r \ell}$$

and so

$$E = \frac{I}{2\pi\sigma r \ell}$$

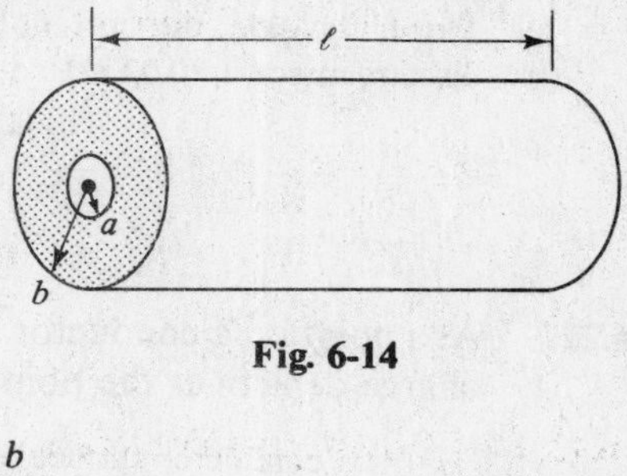

Fig. 6-14

The voltage difference between the conductors is then

$$V_{ab} = -\int_b^a \frac{I}{2\pi\sigma r\ell} \, dr = \frac{I}{2\pi\sigma\ell} \ln\frac{b}{a}$$

and the resistance is

$$R = \frac{V}{I} = \frac{1}{2\pi\sigma\ell} \ln\frac{b}{a}$$

6.17. A current sheet of width 4 m lies in the $z = 0$ plane and contains a total current of 10 A in a direction from the origin to $(1, 3, 0)$ m. Find an expression for **K**.

At each point of the sheet, the direction of **K** is the unit vector

$$\frac{\mathbf{a}_x + 3\mathbf{a}_y}{\sqrt{10}}$$

and the magnitude of **K** is (10/4) A/m. Thus

$$\mathbf{K} = \frac{10}{4}\left(\frac{\mathbf{a}_x + 3\mathbf{a}_y}{\sqrt{10}}\right) \text{ A/m}$$

6.18. As shown in Fig. 6-15, a current I_T follows a filament down the z axis and enters a thin conducting sheet at $z = 0$. Express **K** for this sheet.

Consider a circle in the $z = 0$ plane. The current I_T on the sheet spreads out uniformly over the circumference $2\pi r$. The direction of **K** is $\mathbf{a}_r$. Then

$$\mathbf{K} = \frac{I_T}{2\pi r}\mathbf{a}_r$$

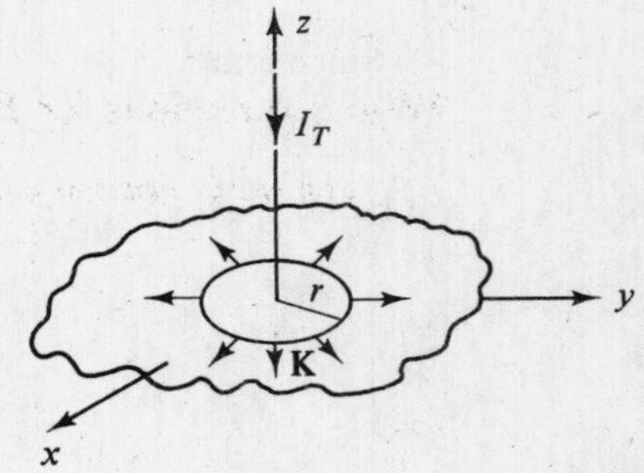

Fig. 6-15

6.19. For the current sheet of Problem 6.18 find the current in a 30° section of the plane (Fig. 6-16).

$$I = \int K_n \, d\ell = \int_0^{\pi/6} \frac{I_T}{2\pi r} r \, d\phi = \frac{I_T}{12}$$

However, integration is not necessary, since for uniformly distributed current a 30° segment will contain 30°/360° or 1/12 of the total.

Fig. 6-16

6.20. A current I (A) enters a thin right circular cylinder at the top, as shown in Fig. 6-17. Express **K** if the radius of the cylinder is 2 cm.

On the top, the current is uniformly distributed over any circumference $2\pi r$, so that

$$\mathbf{K} = \frac{I}{2\pi r}\mathbf{a}_r \quad \text{(A/m)}$$

Down the side, the current is uniformly distributed over the circumference $2\pi(0.02 \text{ m})$, so that

$$\mathbf{K} = \frac{I}{0.04\pi}(-\mathbf{a}_z) \quad \text{(A/m)}$$

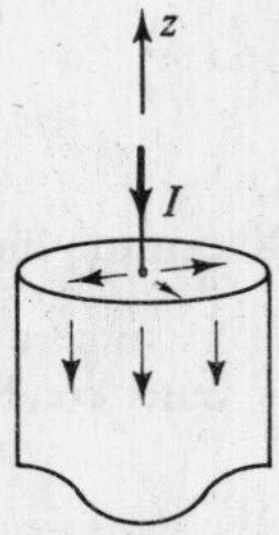

Fig. 6-17

6.21. At a point on a conductor surface, $\mathbf{E} = 0.70\,\mathbf{a}_x - 0.35\,\mathbf{a}_y - 1.00\,\mathbf{a}_z$ V/m. What is the surface charge density at the point?

At a conductor surface under static conditions, the tangential component E_t is zero. Therefore, the given vector must be normal to the conductor. Assuming free space at the surface,

$$\rho_s = D_n = \epsilon_0 E_n = \pm\epsilon_0 |\mathbf{E}| = \pm\frac{10^{-9}}{36\pi}\sqrt{(0.70)^2 + (0.35)^2 + (1.00)^2} = \pm 11.2 \text{ pC/m}^2$$

The plus sign would be chosen if **E** were known to point out of the surface.

6.22. A cylindrical conductor of radius 0.05 m with its axis along the z axis has a surface charge density $\rho_s = (\rho_0/z)$ (C/m^2). Write an expression for **E** at the surface.

Since $D_n = \rho_s$, $E_n = \rho_s/\epsilon_0$. At $(0.05, \phi, z)$,

$$\mathbf{E} = E_n \mathbf{a}_r = \frac{\rho_0}{\epsilon_0 z} \mathbf{a}_r$$

6.23. A conductor occupying the region $x \geq 5$ has a surface charge density

$$\rho_s = \frac{\rho_0}{\sqrt{y^2 + z^2}}$$

Write expressions for **E** and **D** just outside the conductor.

The outer normal is $-\mathbf{a}_x$. Then, just outside the conductor,

$$\mathbf{D} = D_n(-\mathbf{a}_x) = \rho_s(-\mathbf{a}_x) = \frac{\rho_0}{\sqrt{y^2 + z^2}}(-\mathbf{a}_x)$$

and

$$\mathbf{E} = \frac{\rho_0}{\epsilon_0\sqrt{y^2 + z^2}}(-\mathbf{a}_x)$$

6.24. Two concentric cylindrical conductors, $r_a = 0.01$ m and $r_b = 0.08$ m, have charge densities $\rho_{sa} = 40$ pC/m^2 and ρ_{sb}, such that **D** and **E** fields exist between the two cylinders but are zero elsewhere. See Fig. 6-18. Find ρ_{sb} and write expressions for **D** and **E** between the cylinders.

By symmetry, the field between the cylinders must be radial and a function of r only. Then, for $r_a < r < r_b$,

$$\nabla \cdot \mathbf{D} = \frac{1}{r}\frac{d}{dr}(rD_r) = 0 \qquad \text{or} \qquad rD_r = c$$

To evaluate the constant c, use the fact that $D_n = D_r = \rho_{sa}$ at $r = r_a + 0$.

$$c = (0.01)(40 \times 10^{-12}) = 4 \times 10^{-13} \text{ C/m}$$

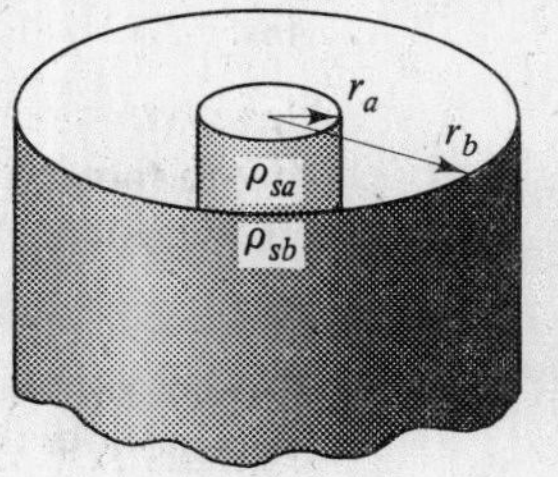

Fig. 6-18

and so

$$\mathbf{D} = \frac{4 \times 10^{-13}}{r}\mathbf{a}_r \ (\text{C/m}^2) \qquad \text{and} \qquad \mathbf{E} = \frac{\mathbf{D}}{\epsilon_0} = \frac{4.52 \times 10^{-2}}{r}\mathbf{a}_r \ (\text{V/m})$$

The density ρ_{sb} is now found from

$$\rho_{sb} = D_n\Big|_{r=r_b-0} = -D_r\Big|_{r=r_b-0} = -\frac{4 \times 10^{-13}}{0.08} = -5 \text{ pC/m}^2$$

Supplementary Problems

6.25. Find the mobility of the conduction electrons in aluminum, given a conductivity 38.2 MS/m and conduction electron density 1.70×10^{29} m^{-3}. *Ans.* 1.40×10^{-3} m^2/V · s

6.26. Repeat Problem 6.25 (*a*) for copper, where $\sigma = 58.0$ MS/m and $N_e = 1.13 \times 10^{29}$ m^{-3}; (*b*) for silver, where $\sigma = 61.7$ MS/m and $N_e = 7.44 \times 10^{28}$ m^{-3}.
Ans. (*a*) 3.21×10^{-3} m^2/V · s; (*b*) 5.18×10^{-3} m^2/V · s

6.27. Find the concentration of holes, N_h, in p-type germanium, where $\sigma = 10^4$ S/m and the hole mobility is $\mu_h = 0.18\ \text{m}^2/\text{V}\cdot\text{s}$. *Ans.* $3.47 \times 10^{23}\ \text{m}^{-3}$

6.28. Using the data of Problem 6.27, find the concentration of electrons, N_e, if the intrinsic concentration is $n_i = 2.5 \times 10^{19}\ \text{m}^{-3}$. *Ans.* $1.80 \times 10^{15}\ \text{m}^{-3}$

6.29. Find the electron and hole concentrations in n-type silicon for which $\sigma = 10.0$ S/m, $\mu_e = 0.13\ \text{m}^2/\text{V}\cdot\text{s}$ and $n_i = 1.5 \times 10^{16}\ \text{m}^{-3}$. *Ans.* $4.81 \times 10^{20}\ \text{m}^{-3}$, $4.68 \times 10^{11}\ \text{m}^{-3}$

6.30. Determine the number of conduction electrons in a one meter cube of tungsten, of which the density is $18.8 \times 10^3\ \text{kg/m}^3$ and the atomic weight is 184.0. Assume two conduction electrons per atom. *Ans.* 1.23×10^{29}

6.31. Find the number of conduction electrons in a one meter cube of copper if $\sigma = 58$ MS/m and $\mu = 3.2 \times 10^{-3}\ \text{m}^2/\text{V}\cdot\text{s}$. On the average, how many electrons is this per atom? The atomic weight is 63.54 and the density is $8.96 \times 10^3\ \text{kg/m}^3$. *Ans.* 1.13×10^{29}, 1.33

6.32. A copper bar of rectangular cross section 0.02 by 0.08 m and length 2.0 m has a voltage drop of 50 mV. Find the resistance, current, current density, electric field intensity and conduction electron drift velocity. *Ans.* 21.6 μΩ, 2.32 kA, 1.45 MA/m², 25 mV/m, 0.08 mm/s

6.33. An aluminum bus bar 0.01 by 0.07 m in cross section and of length 3 m carries a current of 300 A. Find the electric field intensity, current density and conduction electron drift velocity. *Ans.* 1.12×10^{-2} V/m, $4.28 \times 10^5\ \text{A/m}^2$, 1.57×10^{-5} m/s

6.34. A wire table gives for AWG #20 copper wire at 20 °C the resistance 33.31 Ω/km. What conductivity (in S/m) does this imply for copper? The diameter of AWG #20 is 32 mils. *Ans.* 5.8×10^7 S/m

6.35. A wire table gives for AWG #18 platinum wire the resistance 1.21×10^{-3} Ω/cm. What conductivity (in S/m) does this imply for platinum? The diameter of AWG #18 is 40 mils. *Ans.* 1.00×10^7 S/m

6.36. What is the conductivity of AWG #32 tungsten wire with a resistance of 0.0172 Ω/cm? The diameter of AWG #32 is 8.0 mils. *Ans.* 17.9 MS/m

6.37. Determine the resistance per meter of a hollow cylindrical aluminum conductor with an outer diameter of 32 mm and wall thickness 6 mm. *Ans.* 53.4 μΩ/m

6.38. Find the resistance of an aluminum foil 1.0 mil thick and 5.0 cm square (*a*) between opposite edges on a square face, (*b*) between the two square faces. *Ans.* (*a*) 1.03 mΩ; (*b*) 266 pΩ

6.39. Find the resistance of 100 ft of AWG #4/0 conductor in both copper and aluminum. An AWG #4/0 has a diameter of 460 mils. *Ans.* 4.91 mΩ, 7.46 mΩ

6.40. Determine the resistance of a copper conductor 2 m long with a circular cross section and a radius of 1 mm at one end increasing linearly to a radius of 5 mm at the other. *Ans.* 2.20 mΩ

6.41. Determine the resistance of a copper conductor 1 m long with a square cross section and a side 1 mm at one end increasing linearly to 3 mm at the other. *Ans.* 5.75 mΩ

6.42. Develop an expression for the resistance of a conductor of length ℓ if the cross section retains the same shape and the area increases linearly from A to kA over ℓ.
Ans. $\dfrac{\ell}{\sigma A}\left(\dfrac{\ln k}{k-1}\right)$

6.43. Find the current density in an AWG #12 conductor when it is carrying its rated current of 30 A. A #12 wire has a diameter of 81 mils. *Ans.* $9.09 \times 10^6\ \text{A/m}^2$

6.44. Find the total current in a circular conductor of radius 2 mm if the current density varies with r according to $J = 10^3/r\ (\text{A/m}^2)$. *Ans.* 4π A

6.45. In cylindrical coordinates, $\mathbf{J} = 10e^{-100r}\mathbf{a}_\phi\ (\text{A/m}^2)$ for the region $0.01 \le r \le 0.02$ m, $0 < z \le 1$ m. Find the total current crossing the intersection of this region with the plane $\phi = \text{const.}$
Ans. 2.33×10^{-2} A

6.46. Given a current density

$$\mathbf{J} = \left(\frac{10^3}{r^2}\cos\theta\right)\mathbf{a}_\theta \quad (\text{A/m}^2)$$

in spherical coordinates, find the current crossing the conical strip $\theta = \pi/4$, $0.001 \le r \le 0.080$ m.
Ans. 1.38×10^4 A

6.47. Find the total current outward directed from a one meter cube with one corner at the origin and edges parallel to the coordinate axes if $\mathbf{J} = 2x^2\mathbf{a}_x + 2xy^3\mathbf{a}_y + 2xy\mathbf{a}_z\ (\text{A/m}^2)$.
Ans. 3.0 A

6.48. As shown in Fig. 6-19, a current of 50 A passes down the z axis, enters a thin spherical shell of radius 0.03 m, and at $\theta = \pi/2$ enters a plane sheet. Write expressions for the current sheet densities $\mathbf{K}$ in the spherical shell and in the plane.
Ans. $\dfrac{265}{\sin\theta}\mathbf{a}_\theta$ (A/m), $\dfrac{7.96}{r}\mathbf{a}_r$ (A/m)

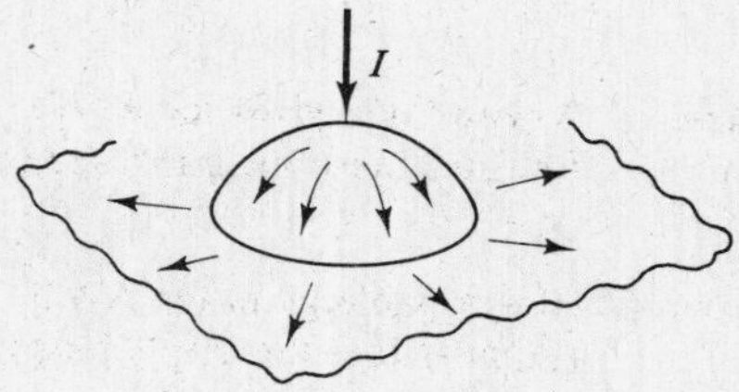

Fig. 6-19

6.49. A filamentary current of I (A) passes down the z axis to $z = 5 \times 10^{-2}$ m where it enters the portion $0 \le \phi \le \pi/4$ of a spherical shell of radius 5×10^{-2} m. Find $\mathbf{K}$ for this current sheet.
Ans. $\dfrac{80I}{\pi\sin\theta}\mathbf{a}_\theta$ (A/m)

6.50. A current sheet of density $\mathbf{K} = 20\mathbf{a}_z$ A/m lies in the plane $x = 0$ and a current density $\mathbf{J} = 10\mathbf{a}_z\ \text{A/m}^2$ also exists throughout space. (*a*) Find the current crossing the area enclosed by a circle of radius 0.5 m centered at the origin in the $z = 0$ plane. (*b*) Find the current crossing the square $|x| \le 0.25$ m, $|y| \le 0.25$ m, $z = 0$. *Ans.* (*a*) 27.9 A; (*b*) 12.5 A

6.51. A hollow, thin-walled, rectangular conductor 0.01 by 0.02 m carries a current of 10 A in the positive x direction. Express $\mathbf{K}$. *Ans.* $167\mathbf{a}_x$ A/m

6.52. A solid conductor has a surface described by $x + y = 3$ m and extends toward the origin. At the surface the electric field intensity is 0.35 V/m. Express $\mathbf{E}$ and $\mathbf{D}$ at the surface and find ρ_s.
Ans. $\pm 0.247(\mathbf{a}_x + \mathbf{a}_y)$ V/m, $\pm 2.19 \times 10^{-12}(\mathbf{a}_x + \mathbf{a}_y)\ \text{C/m}^2$, $\pm 3.10 \times 10^{-12}\ \text{C/m}^2$

6.53. A conductor that extends into the region $z < 0$ has one face in the plane $z = 0$, over which there is a surface charge density

$$\rho_s = 5 \times 10^{-10} e^{-10r} \sin^2 \phi \quad (\text{C/m}^2)$$

in cylindrical coordinates. Find the electric field intensity at $(0.15 \text{ m}, \pi/3, 0)$. *Ans.* $9.45\,\mathbf{a}_z$ V/m

6.54. A spherical conductor centered at the origin and of radius 3 has a surface charge density $\rho_s = \rho_0 \cos^2 \theta$. Find **E** at the surface.

Ans. $\dfrac{\rho_0}{\epsilon_0} \cos^2 \theta \, \mathbf{a}_r$

6.55. The electric field intensity at a point on a conductor surface is given by $\mathbf{E} = 0.2\,\mathbf{a}_x - 0.3\,\mathbf{a}_y - 0.2\,\mathbf{a}_z$ V/m. What is the surface charge density at the point? *Ans.* ± 3.65 pC/m²

6.56. A spherical conductor centered at the origin has an electric field intensity at its surface $\mathbf{E} = 0.53\,(\sin^2 \phi)\,\mathbf{a}_r$ V/m in spherical coordinates. What is the charge density where the sphere meets the y axis? *Ans.* 4.69 pC/m²

Chapter 7

Capacitance and Dielectric Materials

7.1 POLARIZATION P AND RELATIVE PERMITTIVITY ϵ_r

Dielectric materials become *polarized* in an electric field, with the result that the electric flux density **D** is greater than it would be under free-space conditions with the same field intensity. A simplified but satisfactory theory of polarization can be obtained by treating an atom of the dielectric as two superimposed positive and negative charge regions, as shown in Fig. 7-1(*a*). Upon application of an **E** field the positive charge region moves in the direction of the applied field and the negative charge region moves in the opposite direction. This displacement can be represented by an *electric dipole moment*, $\mathbf{p} = Q\mathbf{d}$, as shown in Fig. 7-1(*c*).

For most materials, the charge regions will return to their original superimposed positions when the applied field is removed. As with a spring obeying Hooke's law, the work done in the distortion is recoverable when the system is permitted to go back to its original state. Energy storage takes place in this distortion in the same manner as with the spring.

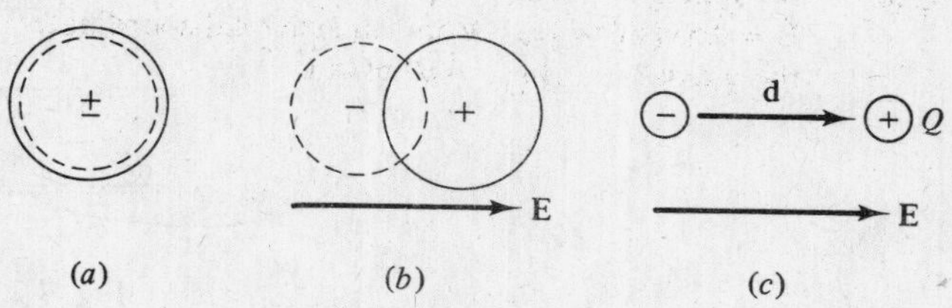

Fig. 7-1

A region Δv of a polarized dielectric will contain N dipole moments **p**. Polarization **P** is defined as the dipole moment per unit volume:

$$\mathbf{P} = \lim_{\Delta v \to 0} \frac{N\mathbf{p}}{\Delta v} \quad (\text{C/m}^2)$$

This suggests a smooth and continuous distribution of electric dipole moments throughout the volume, which, of course, is not the case. In the macroscopic view, however, polarization **P** can account for the increase in the electric flux density, the equation being

$$\mathbf{D} = \epsilon_0 \mathbf{E} + \mathbf{P}$$

This equation permits **E** and **P** to have different directions, as they do in certain crystalline dielectrics. In an isotropic, linear material **E** and **P** are parallel at each point, which is expressed by

$$\mathbf{P} = \chi_e \epsilon_0 \mathbf{E} \qquad \text{(isotropic material)}$$

where the *electric susceptibility* χ_e is a dimensionless constant. Then,

$$\mathbf{D} = \epsilon_0(1 + \chi_e)\mathbf{E} = \epsilon_0 \epsilon_r \mathbf{E} \qquad \text{(isotropic material)}$$

where $\epsilon_r \equiv 1 + \chi_e$ is also a pure number. Since $\mathbf{D} = \epsilon \mathbf{E}$ (Section 3.4),

$$\epsilon_r = \frac{\epsilon}{\epsilon_0}$$

whence ϵ_r is called the *relative permittivity*. (Compare Section 2.1.)

7.2 FIXED-VOLTAGE D AND E

A parallel-plate capacitor with free space between the plates and a constant applied voltage V, as shown in Fig. 7-2, has a constant electric field intensity **E**. With fringing neglected,

$$\mathbf{E} = \frac{V}{d}\,\mathbf{a}_n \qquad \mathbf{D} = \epsilon_0\,\mathbf{E} = \frac{\epsilon_0\,V}{d}\,\mathbf{a}_n \qquad D_n = \rho_s = \frac{Q}{A}$$

Now, when a dielectric with relative permittivity ϵ_r fills the space between the plates,

$$\mathbf{D} = \epsilon_0\,\mathbf{E} + \mathbf{P} = \epsilon_0\,\mathbf{E} + \epsilon_0\,\chi_e\,\mathbf{E}$$

and the equations are:

$$\mathbf{E} = \frac{V}{d}\,\mathbf{a}_n \qquad \text{(as in free space)}$$

$$\mathbf{D} = \epsilon_0\,\epsilon_r\,\mathbf{E}$$

Since $D_n = \rho_s = Q/A$, the charge and charge density increase by the factor ϵ_r over their free-space values. This charge increase is supplied by the source voltage V.

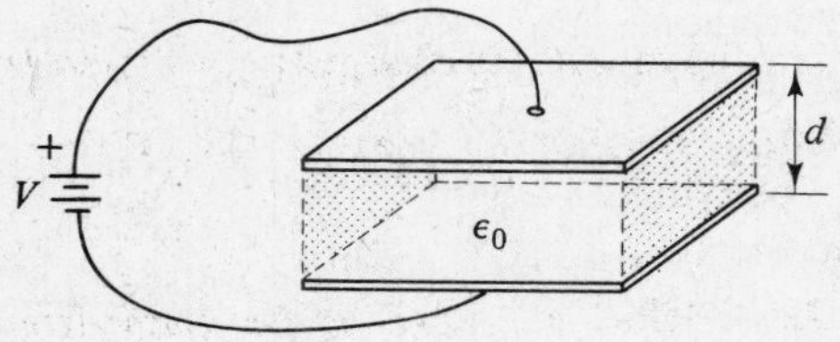

Fig. 7-2

7.3 FIXED-CHARGE D AND E

The parallel-plate capacitor in Fig. 7-3 has a charge $+Q$ on the upper plate and $-Q$ on the lower plate. This charge could have resulted from the connection of a voltage source V which was subsequently removed. With free space between the plates and fringing neglected,

$$D_n = \rho_s = \frac{Q}{A}$$

$$\mathbf{E} = \frac{\mathbf{D}}{\epsilon_0} = \frac{\rho_s}{\epsilon_0}\,\mathbf{a}_n$$

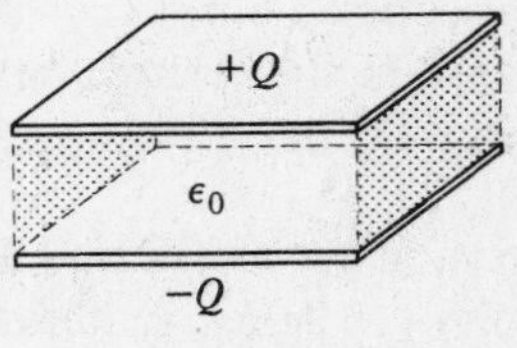

Fig. 7-3

In this arrangement there is no way for the charge to increase or decrease, since there is no conducting path to the plates. Now when a dielectric material is assumed to fill the space between the plates, the equations are as follows:

$$D_n = \rho_s = \frac{Q}{A} \qquad \text{(as in free space)}$$

$$\mathbf{E} = \frac{\mathbf{D}}{\epsilon_0\,\epsilon_r}$$

With Q and ρ_s constant, $\mathbf{D}$ must be the same as under free-space conditions, whereas the magnitude of $\mathbf{E}$ decreases by the factor $1/\epsilon_r$. The decrease in $\epsilon_0\,\mathbf{E}$ is made up for by the polarization $\mathbf{P}$ in the relation $\mathbf{D} = \epsilon_0\,\mathbf{E} + \mathbf{P}$. More generally, in a homogeneous dielectric of relative permittivity ϵ_r, the coulomb force between charges is reduced to $1/\epsilon_r$ of its free-space value:

$$\mathbf{F} = \frac{Q_1 Q_2}{4\pi\epsilon d^2}\,\mathbf{a} = \frac{1}{\epsilon_r}\left(\frac{Q_1 Q_2}{4\pi\epsilon_0\,d^2}\,\mathbf{a}\right)$$

7.4 BOUNDARY CONDITIONS AT THE INTERFACE OF TWO DIELECTRICS

If the conductor in Figs. 6-11 and 6-12 is replaced by a second, different, dielectric, then the same argument as was made in Section 6.10 establishes the following two boundary conditions:

(1) *The tangential component of* **E** *is continuous across a dielectric interface.* In symbols,

$$E_{t1} = E_{t2} \qquad \text{and} \qquad \frac{D_{t1}}{\epsilon_{r1}} = \frac{D_{t2}}{\epsilon_{r2}}$$

(2) *The normal component of* **D** *has a discontinuity of magnitude* $|\rho_s|$ *across a dielectric interface.* If the unit normal vector is chosen to point into dielectric 2, then this condition can be written

$$D_{n1} - D_{n2} = -\rho_s \qquad \text{and} \qquad \epsilon_{r1}E_{n1} - \epsilon_{r2}E_{n2} = -\frac{\rho_s}{\epsilon_0}$$

Generally the interface will have no free charges ($\rho_s = 0$), so that

$$D_{n1} = D_{n2} \qquad \text{and} \qquad \epsilon_{r1}E_{n1} = \epsilon_{r2}E_{n2}$$

EXAMPLE 1 Given that $\mathbf{E}_1 = 2\mathbf{a}_x - 3\mathbf{a}_y + 5\mathbf{a}_z$ V/m at the charge-free dielectric interface of Fig. 7-4, find $\mathbf{D}_2$ and the angles θ_1 and θ_2.

The interface is a $z =$ const. plane. The x and y components are tangential and the z components are normal. By continuity of the tangential component of **E** and the normal component of **D**:

$$\begin{aligned}
\mathbf{E}_1 &= 2\mathbf{a}_x - 3\mathbf{a}_y + 5\mathbf{a}_z \\
\mathbf{E}_2 &= 2\mathbf{a}_x - 3\mathbf{a}_y + E_{z2}\,\mathbf{a}_z \\
\mathbf{D}_1 = \epsilon_0\epsilon_{r1}\mathbf{E}_1 &= 4\epsilon_0\,\mathbf{a}_x - 6\epsilon_0\,\mathbf{a}_y + 10\epsilon_0\,\mathbf{a}_z \\
\mathbf{D}_2 &= D_{x2}\,\mathbf{a}_x + D_{y2}\,\mathbf{a}_y + 10\epsilon_0\,\mathbf{a}_z
\end{aligned}$$

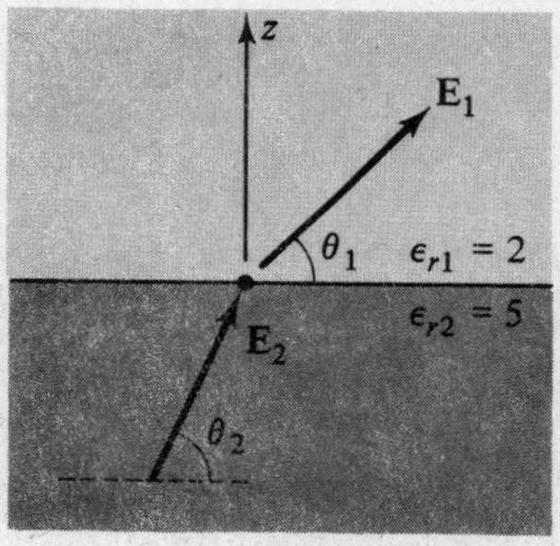

Fig. 7-4

The unknown components are now found from the relation $\mathbf{D}_2 = \epsilon_0\epsilon_{r2}\mathbf{E}_2$.

$$D_{x2}\,\mathbf{a}_x + D_{y2}\,\mathbf{a}_y + 10\epsilon_0\,\mathbf{a}_z = 2\epsilon_0\epsilon_{r2}\,\mathbf{a}_x - 3\epsilon_0\epsilon_{r2}\,\mathbf{a}_y + \epsilon_0\epsilon_{r2}E_{z2}\,\mathbf{a}_z$$

from which

$$D_{x2} = 2\epsilon_0\epsilon_{r2} = 10\epsilon_2 \qquad D_{y2} = -3\epsilon_0\epsilon_{r2} = -15\epsilon_0 \qquad E_{z2} = \frac{10}{\epsilon_{r2}} = 2$$

The angles made with the plane of the interface are easiest found from

$$\begin{aligned}
\mathbf{E}_1\cdot\mathbf{a}_z &= |\mathbf{E}_1|\cos(90^\circ - \theta_1) & \mathbf{E}_2\cdot\mathbf{a}_z &= |\mathbf{E}_2|\cos(90^\circ - \theta_2) \\
5 &= \sqrt{38}\sin\theta_1 & 2 &= \sqrt{17}\sin\theta_2 \\
\theta_1 &= 54.2^\circ & \theta_2 &= 29.0^\circ
\end{aligned}$$

A useful relation can be obtained from

$$\tan\theta_1 = \frac{E_{z1}}{\sqrt{E_{x1}^2 + E_{y1}^2}} = \frac{D_{z1}/\epsilon_0\epsilon_{r1}}{\sqrt{E_{x1}^2 + E_{y1}^2}}$$

$$\tan\theta_2 = \frac{E_{z2}}{\sqrt{E_{x2}^2 + E_{y2}^2}} = \frac{D_{z2}/\epsilon_0\epsilon_{r2}}{\sqrt{E_{x2}^2 + E_{y2}^2}}$$

In view of the continuity relations, division of these two equations gives

$$\frac{\tan\theta_1}{\tan\theta_2} = \frac{\epsilon_{r2}}{\epsilon_{r1}}$$

7.5 CAPACITANCE

Any two conducting bodies separated by free space or a dielectric material have a *capacitance* between them. A voltage difference applied results in a charge $+Q$ on one conductor and $-Q$ on the other. The ratio of the absolute value of the charge to the absolute value of the voltage difference is defined as the capacitance of the system:

$$C = \frac{Q}{V} \quad \text{(F)}$$

where 1 farad (F) = 1 C/V.

The capacitance depends only on the geometry of the system and the properties of the dielectric(s) involved. In Fig. 7-5, charge $+Q$ placed on conductor 1 and $-Q$ on conductor 2 creates a flux field as shown. The **D** and **E** fields are therefore also established. To double the charges would simply double **D** and **E**, and therefore double the voltage difference. Hence the ratio Q/V would remain fixed.

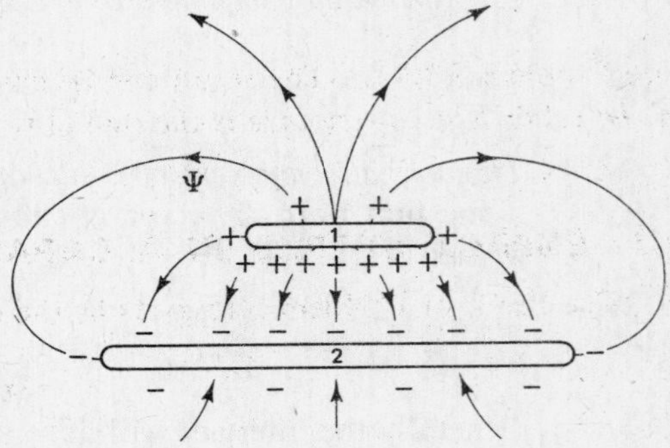

Fig. 7-5

EXAMPLE 2 Find the capacitance of the parallel plates in Fig. 7-6, neglecting fringing.

With $+Q$ on the upper plate and $-Q$ on the lower,

$$\rho_s = \frac{Q}{A} \qquad D_n = \rho_s = \frac{Q}{A}$$

Because the **D** field is uniform between the plates,

$$\mathbf{D} = \frac{Q}{A}(-\mathbf{a}_z) \qquad \mathbf{E} = \frac{Q}{\epsilon_0 \epsilon_r A}(-\mathbf{a}_z)$$

The voltage of the plate at $z = d$ with respect to the lower plate is

$$V = -\int_0^d \frac{Q}{\epsilon_0 \epsilon_r A}(-\mathbf{a}_z) \cdot dz\,\mathbf{a}_z = \frac{Qd}{\epsilon_0 \epsilon_r A}$$

Thus $$C = \frac{Q}{V} = \frac{\epsilon_0 \epsilon_r A}{d}$$

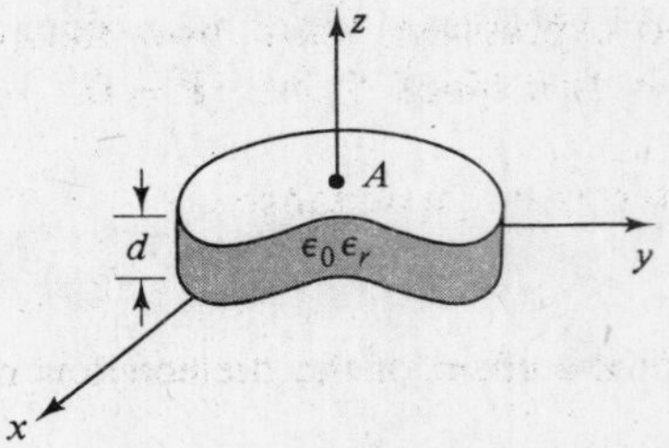

Fig. 7-6

Notice that the result does not depend on the shape of the plates.

7.6 MULTIPLE-DIELECTRIC CAPACITORS

When two dielectrics are present with the interface parallel to **E** and **D**, as in Fig. 7-7(*a*), the capacitance can be found by treating the arrangement as two capacitors in parallel:

$$C_{eq} = C_1 + C_2$$

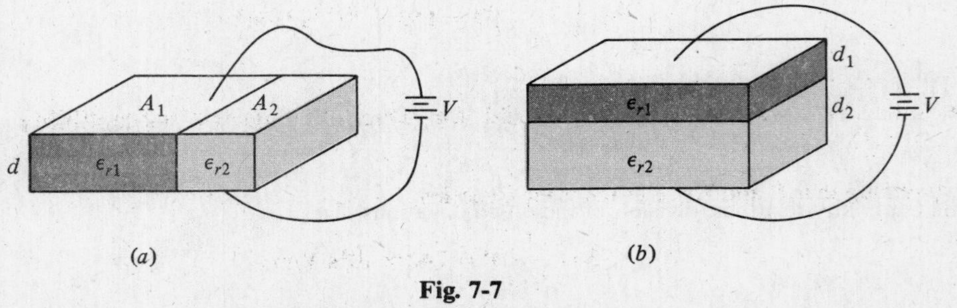

Fig. 7-7

[see Problem 7.8(*a*)]. Of course, the result may be extended to any number of side-by-side dielectrics: *the equivalent capacitance is the sum of the individual capacitances.*

When the dielectric interface is normal to **D** and **E**, as in Fig. 7-7(*b*), the capacitance can be found by treating the arrangement as two capacitors in series:

$$\frac{1}{C_{eq}} = \frac{1}{C_1} + \frac{1}{C_2}$$

[see Problem 7.8(*b*)]. The result can be extended to any number of stacked dielectrics: *the reciprocal of the equivalent capacitance is the sum of reciprocals of the individual capacitances.*

7.7 ENERGY STORED IN A CAPACITOR

The result of Problem 5.15 gives as the energy stored in a capacitor

$$W_E = \frac{1}{2}\int \mathbf{D}\cdot\mathbf{E}\,dv$$

where the integration may be taken over the space between the conductors with fringing neglected. If this space is occupied by a dielectric of relative permittivity ϵ_r, then

$$\mathbf{D} = \epsilon_0\mathbf{E} + \mathbf{P} = \epsilon_0\epsilon_r\mathbf{E}$$

and so

$$W_E = \frac{1}{2}\int(\epsilon_0 E^2 + \mathbf{P}\cdot\mathbf{E})\,dv = \frac{1}{2}\int \epsilon_0\epsilon_r E^2\,dv$$

The two expressions show how the presence of a dielectric results in an increase in stored energy over the free-space value ($\mathbf{P} = 0$, $\epsilon_r = 1$), either through the term $\mathbf{P}\cdot\mathbf{E}$ or through the factor $\epsilon_r > 1$.

In terms of capacitance,

$$W_E = \tfrac{1}{2}CV^2$$

and here the effect of the dielectric is reflected in C, which is directly proportional to ϵ_r.

Solved Problems

7.1. Find the polarization **P** in a dielectric material with $\epsilon_r = 2.8$ if $\mathbf{D} = 3.0\times10^{-7}\mathbf{a}$ C/m^2.

Assuming the material to be homogeneous and isotropic,

$$\mathbf{P} = \chi_e\epsilon_0\mathbf{E}$$

Since $\mathbf{D} = \epsilon_0\epsilon_r\mathbf{E}$ and $\chi_e = \epsilon_r - 1$,

$$\mathbf{P} = \left(\frac{\epsilon_r - 1}{\epsilon_r}\right)\mathbf{D} = 1.93\times10^{-7}\mathbf{a}\ \text{C/m}^2$$

7.2. Determine the value of **E** in a material for which the electric susceptibility is 3.5 and $\mathbf{P} = 2.3\times10^{-7}\mathbf{a}$ C/m^2.

Assuming that **P** and **E** are in the same direction,

$$\mathbf{E} = \frac{1}{\chi_e\epsilon_0}\mathbf{P} = 7.42\times10^{3}\mathbf{a}\ \text{V/m}$$

7.3. Two point charges in a dielectric medium where $\epsilon_r = 5.2$ interact with a force of 8.6×10^{-3} N. What force could be expected if the charges were in free space?

Coulomb's law, $F = Q_1Q_2/(4\pi\epsilon_0\epsilon_r d^2)$, shows that the force is inversely proportional to ϵ_r. In free space the force will have its maximum value.

$$F = \epsilon_r(8.6 \times 10^{-3}) = 4.47 \times 10^{-2}\ \text{N}$$

7.4. Region 1, defined by $x < 0$, is free space, while region 2, $x > 0$, is a dielectric material for which $\epsilon_{r2} = 2.4$. See Fig. 7-8. Given

$$\mathbf{D}_1 = 3\mathbf{a}_x - 4\mathbf{a}_y + 6\mathbf{a}_z \quad \text{C/m}^2$$

find $\mathbf{E}_2$ and the angles θ_1 and θ_2.

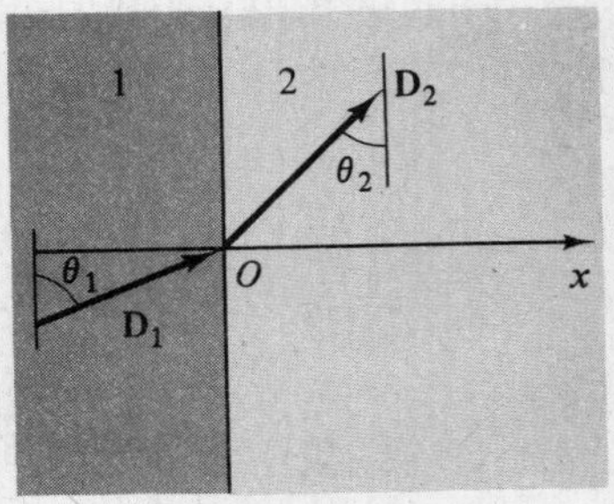

Fig. 7-8

The x components are normal to the interface; D_n and E_t are continuous.

$$\mathbf{D}_1 = 3\mathbf{a}_x - 4\mathbf{a}_y + 6\mathbf{a}_z \qquad \mathbf{E}_1 = \frac{3}{\epsilon_0}\mathbf{a}_x - \frac{4}{\epsilon_0}\mathbf{a}_y + \frac{6}{\epsilon_0}\mathbf{a}_z$$

$$\mathbf{D}_2 = 3\mathbf{a}_x + D_{y2}\mathbf{a}_y + D_{z2}\mathbf{a}_z \qquad \mathbf{E}_2 = E_{x2}\mathbf{a}_x - \frac{4}{\epsilon_0}\mathbf{a}_y + \frac{6}{\epsilon_0}\mathbf{a}_z$$

Then $\mathbf{D}_2 = \epsilon_0\epsilon_{r2}\mathbf{E}_2$ gives

$$3\mathbf{a}_x + D_{y2}\mathbf{a}_y + D_{z2}\mathbf{a}_z = \epsilon_0\epsilon_{r2}E_{x2}\mathbf{a}_x - 4\epsilon_{r2}\mathbf{a}_y + 6\epsilon_{r2}\mathbf{a}_z$$

whence

$$E_{x2} = \frac{3}{\epsilon_0\epsilon_{r2}} = \frac{1.25}{\epsilon_0} \qquad D_{y2} = -4\epsilon_{r2} = -9.6 \qquad D_{z2} = 6\epsilon_{r2} = 14.4$$

To find the angles:

$$\mathbf{D}_1 \cdot \mathbf{a}_x = |\mathbf{D}_1| \cos(90° - \theta_1)$$
$$3 = \sqrt{61}\sin\theta_1$$
$$\theta_1 = 22.6°$$

Similarly, $\theta_2 = 9.83°$.

7.5. In the free-space region $x < 0$ the electric field intensity is $\mathbf{E}_1 = 3\mathbf{a}_x + 5\mathbf{a}_y - 3\mathbf{a}_z$ V/m. The region $x > 0$ is a dielectric for which $\epsilon_{r2} = 3.6$. Find the angle θ_2 that the field in the dielectric makes with the $x = 0$ plane.

The angle made by $\mathbf{E}_1$ is found from

$$\mathbf{E}_1 \cdot \mathbf{a}_x = |\mathbf{E}_1| \cos(90° - \theta_1)$$
$$3 = \sqrt{43}\sin\theta_1$$
$$\theta_1 = 27.2°$$

Then, by the formula developed in Example 1, Section 7.4,

$$\tan\theta_2 = \frac{1}{\epsilon_{r2}}\tan\theta_1 = 0.1428$$

and $\theta_2 = 8.13°$.

7.6. A dielectric–free space interface has the equation $3x + 2y + z = 12$ m. The origin side of the interface has $\epsilon_{r1} = 3.0$ and $\mathbf{E}_1 = 2\mathbf{a}_x + 5\mathbf{a}_z$ V/m. Find $\mathbf{E}_2$.

The interface is indicated in Fig. 7-9 by its intersections with the axes. The unit normal vector on the free-space side is

$$\mathbf{a}_n = \frac{3\mathbf{a}_x + 2\mathbf{a}_y + \mathbf{a}_z}{\sqrt{14}}$$

The projection of $\mathbf{E}_1$ on $\mathbf{a}_n$ is the normal component of $\mathbf{E}$ at the interface.

$$\mathbf{E}_1 \cdot \mathbf{a}_n = \frac{11}{\sqrt{14}}$$

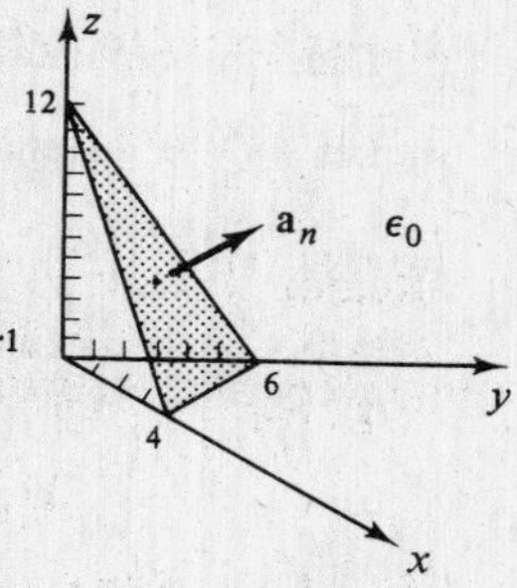

Fig. 7-9

Then

$$\mathbf{E}_{n1} = \frac{11}{\sqrt{14}}\mathbf{a}_n = 2.36\,\mathbf{a}_x + 1.57\,\mathbf{a}_y + 0.79\,\mathbf{a}_z$$
$$\mathbf{E}_{t1} = \mathbf{E}_1 - \mathbf{E}_{n1} = -0.36\,\mathbf{a}_x - 1.57\,\mathbf{a}_y + 4.21\,\mathbf{a}_z = \mathbf{E}_{t2}$$
$$\mathbf{D}_{n1} = \epsilon_0\epsilon_{r1}\mathbf{E}_{n1} = \epsilon_0(7.08\,\mathbf{a}_x + 4.71\,\mathbf{a}_y + 2.37\,\mathbf{a}_z) = \mathbf{D}_{n2}$$
$$\mathbf{E}_{n2} = \frac{1}{\epsilon_0}\mathbf{D}_{n2} = 7.08\,\mathbf{a}_x + 4.71\,\mathbf{a}_y + 2.37\,\mathbf{a}_z$$

and finally

$$\mathbf{E}_2 = \mathbf{E}_{n2} + \mathbf{E}_{t2} = 6.72\,\mathbf{a}_x + 3.14\,\mathbf{a}_y + 6.58\,\mathbf{a}_z \quad \text{V/m}$$

7.7. Figure 7-10 shows a planar dielectric slab with free space on either side. Assuming a constant field $\mathbf{E}_2$ within the slab, show that $\mathbf{E}_3 = \mathbf{E}_1$.

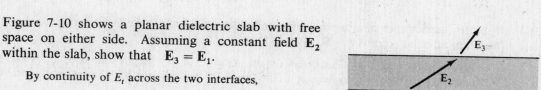

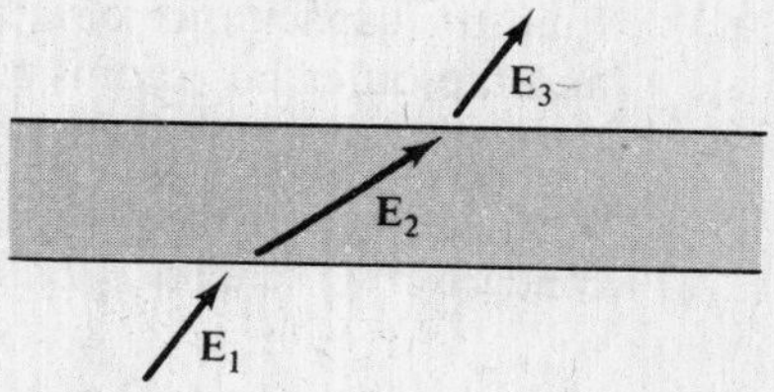

Fig. 7-10

By continuity of E_t across the two interfaces,

$$E_{t3} = E_{t1}$$

By continuity of D_n across the two interfaces (no surface charges),

$$D_{n3} = D_{n1} \qquad \text{and so} \qquad E_{n3} = E_{n1}$$

Consequently, $\mathbf{E}_3 = \mathbf{E}_1$.

7.8. (*a*) Show that the capacitor of Fig. 7-7(*a*) has capacitance

$$C_{eq} = \frac{\epsilon_0\epsilon_{r1}A_1}{d} + \frac{\epsilon_0\epsilon_{r2}A_2}{d} = C_1 + C_2$$

(*b*) Show that the capacitor of Fig. 7-7(*b*) has reciprocal capacitance

$$\frac{1}{C_{eq}} = \frac{1}{\epsilon_0\epsilon_{r1}A/d_1} + \frac{1}{\epsilon_0\epsilon_{r2}A/d_2} = \frac{1}{C_1} + \frac{1}{C_2}$$

(*a*) Because the voltage difference V is common to the two dielectrics,

$$\mathbf{E}_1 = \mathbf{E}_2 = \frac{V}{d}\mathbf{a}_n \qquad \text{and} \qquad \frac{\mathbf{D}_1}{\epsilon_0\epsilon_{r1}} = \frac{\mathbf{D}_2}{\epsilon_0\epsilon_{r2}} = \frac{V}{d}\mathbf{a}_n$$

where $\mathbf{a}_n$ is the downward normal to the upper plate. Since $D_n = \rho_s$, the charge densities on the two sections of the upper plate are

$$\rho_{s1} = \frac{V}{d}\epsilon_0\epsilon_{r1} \qquad \rho_{s2} = \frac{V}{d}\epsilon_0\epsilon_{r2}$$

and the total charge is

$$Q = \rho_{s1} A_1 + \rho_{s2} A_2 = V\left(\frac{\epsilon_0 \epsilon_{r1} A_1}{d} + \frac{\epsilon_0 \epsilon_{r2} A_2}{d}\right)$$

Thus, the capacitance of the system, $C_{eq} = Q/V$, has the asserted form.

(*b*) Let $+Q$ be the charge on the upper plate. Then

$$\mathbf{D} = \frac{Q}{A}\mathbf{a}_n$$

everywhere between the plates, so that

$$\mathbf{E}_1 = \frac{Q}{\epsilon_0 \epsilon_{r1} A}\mathbf{a}_n \qquad \mathbf{E}_2 = \frac{Q}{\epsilon_0 \epsilon_{r2} A}\mathbf{a}_n$$

The voltage differences across the two dielectrics are then

$$V_1 = E_1 d_1 = \frac{Qd_1}{\epsilon_0 \epsilon_{r1} A} \qquad V_2 = E_2 d_2 = \frac{Qd_2}{\epsilon_0 \epsilon_{r2} A}$$

and

$$V = V_1 + V_2 = Q\left(\frac{1}{\epsilon_0 \epsilon_{r1} A/d_1} + \frac{1}{\epsilon_0 \epsilon_{r2} A/d_2}\right)$$

From this it is seen that $1/C_{eq} = V/Q$ has the asserted form.

7.9. Find the capacitance of a coaxial capacitor of length L, where the inner conductor has radius a and the outer has radius b. See Fig. 7-11.

With fringing neglected, Gauss's law requires that $D \propto 1/r$ between the conductors (see Problem 6.24). At $r = a$, $D = \rho_s$, where ρ_s is the (assumed positive) surface charge density on the inner conductor. Therefore,

$$\mathbf{D} = \rho_s \frac{a}{r}\mathbf{a}_r \qquad \mathbf{E} = \frac{\rho_s a}{\epsilon_0 \epsilon_r r}\mathbf{a}_r$$

and the voltage difference between the conductors is

$$V_{ab} = -\int_b^a \left(\frac{\rho_s a}{\epsilon_0 \epsilon_r r}\mathbf{a}_r\right) \cdot dr\,\mathbf{a}_r = \frac{\rho_s a}{\epsilon_0 \epsilon_r} \ln\frac{b}{a}$$

The total charge on the inner conductor is $Q = \rho_s(2\pi a L)$, and so

$$C = \frac{Q}{V} = \frac{2\pi\epsilon_0 \epsilon_r L}{\ln(b/a)}$$

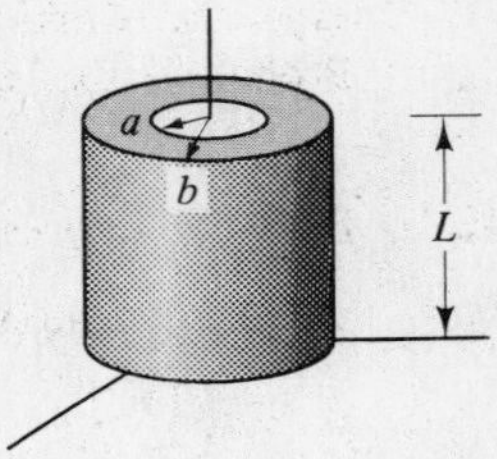

Fig. 7-11

7.10. In the capacitor shown in Fig. 7-12, the region between the plates is filled with a dielectric having $\epsilon_r = 4.5$. Find the capacitance.

With fringing neglected, the $\mathbf{D}$ field between the plates should, in cylindrical coordinates, be of the form $\mathbf{D} = D_\phi \mathbf{a}_\phi$, where D_ϕ depends only on r. Then, if the voltage of the plate $\phi = \alpha$ with respect to the plate $\phi = 0$ is V_0,

$$V_0 = -\int \mathbf{E} \cdot d\mathbf{l} = -\int_0^\alpha \left(\frac{D_\phi}{\epsilon_0 \epsilon_r}\mathbf{a}_\phi\right) \cdot (r\,d\phi\,\mathbf{a}_\phi) = -\frac{D_\phi r}{\epsilon_0 \epsilon_r}\int_0^\alpha d\phi = -\frac{D_\phi r\alpha}{\epsilon_0 \epsilon_r}$$

Thus, $D_\phi = -\epsilon_0 \epsilon_r V_0 / r\alpha$, and the charge density on the plate $\phi = \alpha$ is

$$\rho_s = D_n = -D_\phi = \frac{\epsilon_0 \epsilon_r V_0}{r\alpha}$$

The total charge on the plate is then given by

$$Q = \int \rho_s \, dS = \int_0^h \int_{r_1}^{r_2} \frac{\epsilon_0 \epsilon_r V_0}{r\alpha} \, dr \, dz$$

$$= \frac{\epsilon_0 \epsilon_r V_0 h}{\alpha} \ln \frac{r_2}{r_1}$$

Hence $\quad C = \frac{Q}{V_0} = \frac{\epsilon_0 \epsilon_r h}{\alpha} \ln \frac{r_2}{r_1}$

When the numerical values are substituted (with α converted to radians), one obtains $C = 7.76$ pF.

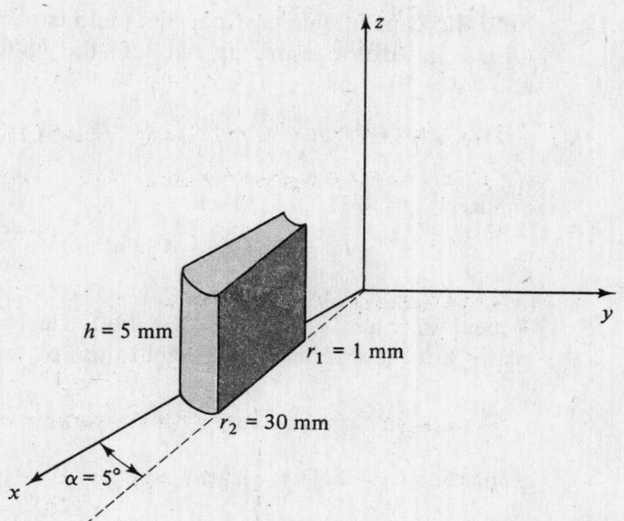

Fig. 7-12

7.11. Referring to Problem 7.10, find the separation d which results in the same capacitance when the plates are brought into parallel arrangement, with the same dielectric in between.

With the plates parallel

$$C = \frac{\epsilon_0 \epsilon_r A}{d}$$

so that

$$d = \frac{\epsilon_0 \epsilon_r A}{C} = \frac{\epsilon_0 \epsilon_r h(r_2 - r_1)}{(\epsilon_0 \epsilon_r h/\alpha)[\ln (r_2/r_1)]} = \frac{\alpha(r_2 - r_1)}{\ln (r_2/r_1)}$$

Notice that the numerator on the right is the difference of the arc lengths at the two ends of the capacitor, while the denominator is the logarithm of the ratio of these arc lengths. For the data of Problem 7.10, $\alpha r_1 = 0.087$ mm, $\alpha r_2 = 2.62$ mm, and $d = 0.74$ mm.

7.12. Find the capacitance of an isolated spherical shell of radius a.

The potential of such a conductor with a zero reference at infinity is (see Problem 2.35)

$$V = \frac{Q}{4\pi\epsilon_0 a}$$

Then

$$C = \frac{Q}{V} = 4\pi\epsilon_0 a$$

7.13. Find the capacitance between two spherical shells of radius a separated by a distance $d \gg a$.

As an approximation, the result of Problem 7.12 for the capacitance of a single spherical shell, $4\pi\epsilon_0 a$, may be used. From Fig. 7-13 two such identical capacitors appear to be in series.

$$\frac{1}{C} = \frac{1}{C_1} + \frac{1}{C_2}$$

$$C = \frac{C_1 C_2}{C_1 + C_2} = 2\pi\epsilon_0 a$$

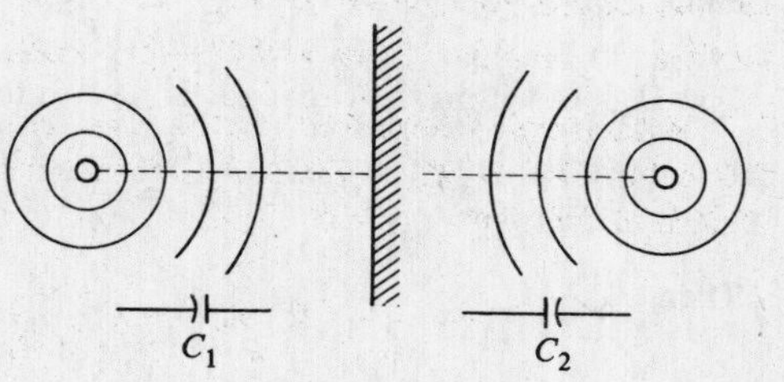

Fig. 7-13

7.14. Find the capacitance of a parallel-plate capacitor containing two dielectrics, $\epsilon_{r1} = 1.5$ and $\epsilon_{r2} = 3.5$, each comprising one-half the volume, as shown in Fig. 7-14. Here, $A = 2\ \text{m}^2$ and $d = 10^{-3}$ m.

$$C_1 = \frac{\epsilon_0 \epsilon_{r1} A_1}{d} = \frac{(8.854 \times 10^{-12})(1.5)1}{10^{-3}} = 13.3\ \text{nF}$$

Similarly, $C_2 = 31.0$ nF. Then

$$C = C_1 + C_2 = 44.3\ \text{nF}$$

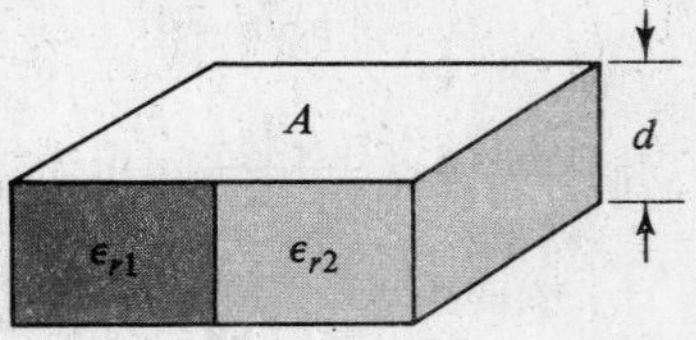

Fig. 7-14

7.15. Repeat Problem 7.14 if the two dielectrics each occupy one-half of the space between the plates but the interface is parallel to the plates.

$$C_1 = \frac{\epsilon_0 \epsilon_r A}{d_1} = \frac{\epsilon_0 \epsilon_r A}{d/2} = \frac{(8.854 \times 10^{-12})(1.5)2}{10^{-3}/2} = 53.1\ \text{nF}$$

Similarly, $C_2 = 124$ nF. Then

$$C = \frac{C_1 C_2}{C_1 + C_2} = 37.2\ \text{nF}$$

7.16. In the cylindrical capacitor shown in Fig. 7-15 each dielectric occupies one-half the volume. Find the capacitance.

The dielectric interface is parallel to **D** and **E**, so the configuration may be treated as two capacitors in parallel. Since each capacitor carries half as much charge as a full cylinder would carry, the result of Problem 7.9 gives

$$C = C_1 + C_2 = \frac{\pi \epsilon_0 \epsilon_{r1} L}{\ln(b/a)} + \frac{\pi \epsilon_0 \epsilon_{r2} L}{\ln(b/a)} = \frac{2\pi \epsilon_0 \epsilon_{r\,\text{avg}} L}{\ln(b/a)}$$

where $\epsilon_{r\,\text{avg}} = \frac{1}{2}(\epsilon_{r1} + \epsilon_{r2})$. The two dielectrics act like a single dielectric having the average relative permittivity.

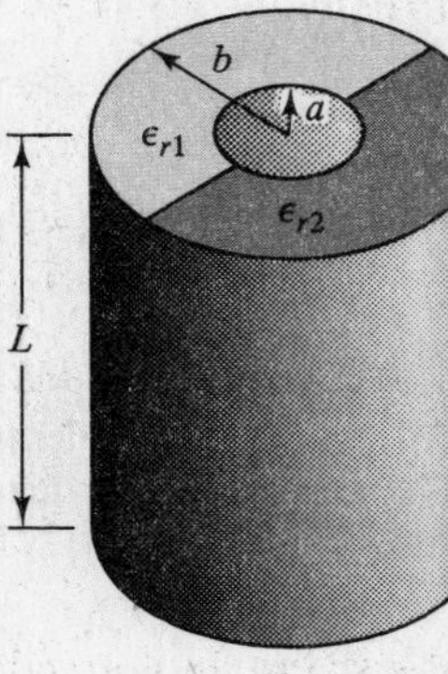

Fig. 7-15

7.17. Find the voltage across each dielectric in the capacitor shown in Fig. 7-16 when the applied voltage is 200 V.

$$C_1 = \frac{\epsilon_0 5(1)}{10^{-3}} = 5000\epsilon_0$$

$$C_2 = 1000\epsilon_0/3$$

and $$C = \frac{C_1 C_2}{C_1 + C_2} = 312.5\,\epsilon_0 = 2.77 \times 10^{-9}\ \text{F}$$

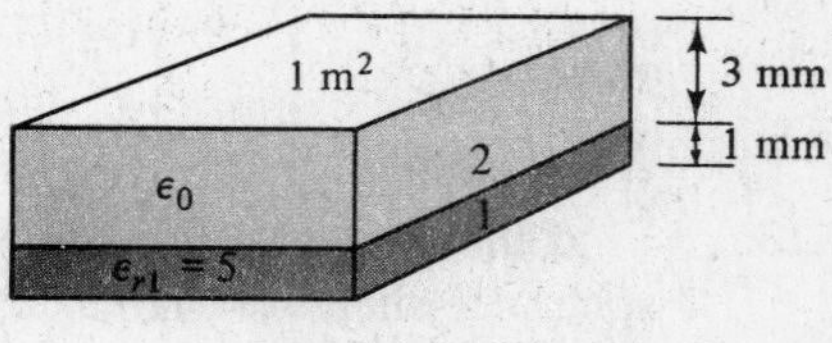

Fig. 7-16

The **D** field within the capacitor is now found from

$$D_n = \rho_s = \frac{Q}{A} = \frac{CV}{A} = \frac{(2.77 \times 10^{-9})(200)}{1} = 5.54 \times 10^{-7}\ \text{C/m}^2$$

Then,

$$E_1 = \frac{D}{\epsilon_0 \epsilon_{r1}} = 1.25 \times 10^4\ \text{V/m} \qquad E_2 = \frac{D}{\epsilon_0} = 6.25 \times 10^4\ \text{V/m}$$

from which

$$V_1 = E_1 d_1 = 12.5\ \text{V} \qquad V_2 = E_2 d_2 = 187.5\ \text{V}$$

7.18. Find the voltage drop across each dielectric in Fig. 7-17, where $\epsilon_{r1} = 2.0$ and $\epsilon_{r2} = 5.0$. The inner conductor is at $r_1 = 2$ cm and the outer at $r_2 = 2.5$ cm, with the dielectric interface halfway between.

The voltage division is the same as it would be for full right circular cylinders. The segment shown, with angle α, will have a capacitance $\alpha/2\pi$ times that of the complete coaxial capacitor. From Problem 7.9,

$$C_1 = \left(\frac{\alpha}{2\pi}\right)\frac{2\pi\epsilon_0\epsilon_{r1}L}{\ln\left(\dfrac{2.25}{2.0}\right)} = \alpha L(1.5 \times 10^{-10}) \quad \text{(F)}$$

$$C_2 = \alpha L(4.2 \times 10^{-10}) \quad \text{(F)}$$

Since $Q = C_1V_1 = C_2V_2$ and $V_1 + V_2 = V$, it follows that

$$V_1 = \frac{C_2}{C_1 + C_2}V = \frac{4.2}{1.5 + 4.2}(100) = 74 \text{ V}$$

$$V_2 = \frac{C_1}{C_1 + C_2}V = \frac{1.5}{1.5 + 4.2}(100) = 26 \text{ V}$$

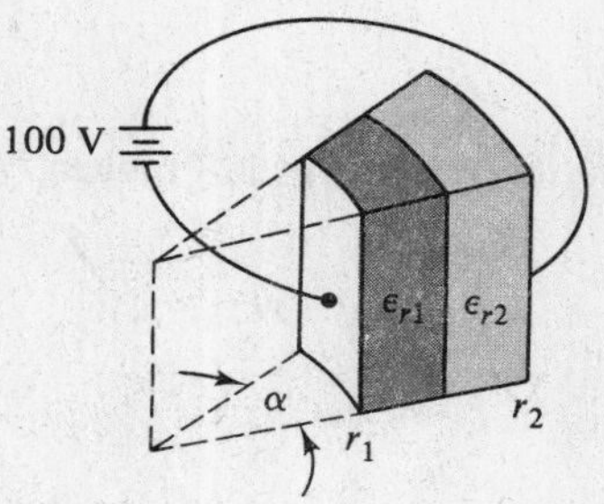

Fig. 7-17

7.19. A parallel-plate capacitor with free space between the plates is connected to a constant source of voltage. Determine how W_E, D, E, C, Q, ρ_s and V change as a dielectric of $\epsilon_r = 2$ is inserted between the plates.

Relationship	**Explanation**
$V_2 = V_1$	the source V remains connected
$E_2 = E_1$	since $E = V/d$
$W_2 = 2W_1$	$W = \frac{1}{2}\int \epsilon_0\epsilon_r E^2\,dv$
$C_2 = 2C_1$	$C = \epsilon_0\epsilon_r A/d$
$D_2 = 2D_1$	$D = \epsilon_0\epsilon_r E$
$\rho_{s2} = 2\rho_{s1}$	$\rho_s = D_n$
$Q_2 = 2Q_1$	$Q = \rho_s A$

In a problem of this type it is advisable first to identify those quantities which remain constant.

7.20. A free-space parallel-plate capacitor is charged by momentary connection to a voltage source V, which is then removed. Determine how W_E, D, E, C, Q, ρ_s and V change as the plates are moved apart to a separation distance $d_2 = 2d_1$ without disturbing the charge.

Relationship	**Explanation**
$Q_2 = Q_1$	the total charge is unchanged
$\rho_{s2} = \rho_{s1}$	$\rho_s = Q/A$
$D_2 = D_1$	$D_n = \rho_s$
$E_2 = E_1$	$E = D/\epsilon_0$
$W_2 = 2W_1$	$W = \frac{1}{2}\int \epsilon_0 E^2\,dv$, and the volume is doubled
$C_2 = \frac{1}{2}C_1$	$C = \epsilon_0 A/d$
$V_2 = 2V_1$	$V = Q/C$

7.21. A parallel-plate capacitor with a separation $d = 1.0$ cm has 29 000 V applied when free space is the only dielectric. Assume that air has a dielectric strength of 30 000 V/cm. Show why the air breaks down when a thin piece of glass ($\epsilon_r = 6.5$) with a dielectric strength of 290 000 V/cm and thickness $d_2 = 0.20$ cm is inserted as shown in Fig. 7-18.

The problem becomes one of two capacitors in series,

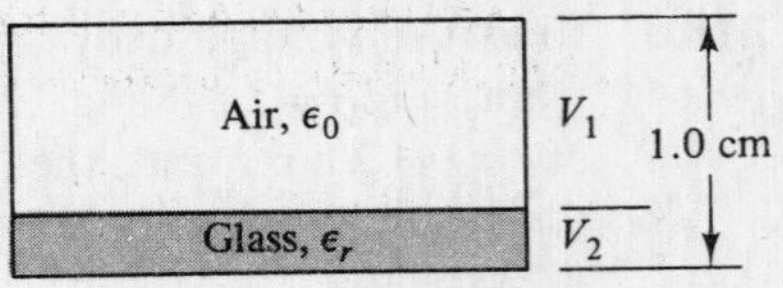

Fig. 7-18

$$C_1 = \frac{\epsilon_0 A}{8 \times 10^{-3}} = 125\epsilon_0 A$$

$$C_2 = \frac{\epsilon_0 \epsilon_r A}{2 \times 10^{-3}} = 3250\epsilon_0 A$$

Then, as in Problem 7.18,

$$V_1 = \frac{3250}{125 + 3250}(29\,000) = 27\,926 \text{ V}$$

so that

$$E_1 = \frac{27\,933 \text{ V}}{0.80 \text{ cm}} = 34\,907 \text{ V/cm}$$

which exceeds the dielectric strength of air.

7.22. Find the capacitance per unit length between a cylindrical conductor of radius $a = 2.5$ cm and a ground plane parallel to the conductor axis and a distance $h = 6.0$ m from it.

A useful technique in problems of this kind is the *method of images*. Take the mirror image of the conductor in the ground plane, and let this image conductor carry the negative of the charge distribution on the actual conductor. Now suppose the ground plane is removed. It is clear that the electric field of the two conductors obeys the right boundary condition at the actual conductor, and, by symmetry, has an equipotential surface (Section 5.2) where the ground plane was. Thus, this field is *the* field in the region between the actual conductor and the ground plane.

Approximating the actual and image charge distributions by line charges $+\rho_\ell$ and $-\rho_\ell$, respectively, at the conductor centers, one has (see Fig. 7-19):

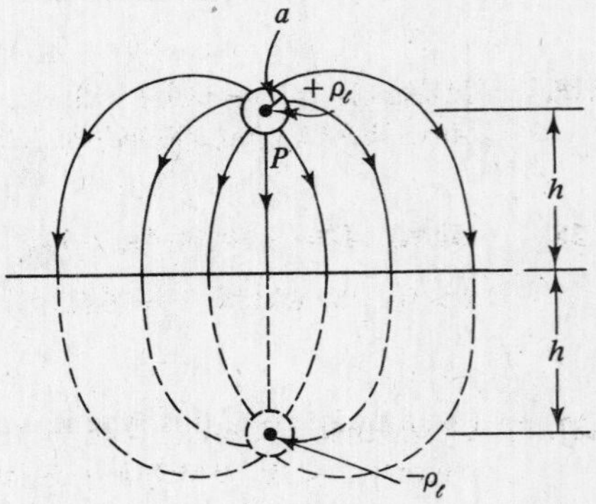

Fig. 7-19

$$\text{potential at radius } a \text{ due to } +\rho_\ell = -\left(\frac{+\rho_\ell}{2\pi\epsilon_0}\right)\ln a$$

$$\text{potential at point } P \text{ due to } -\rho_\ell = -\left(\frac{-\rho_\ell}{2\pi\epsilon_0}\right)\ln(2h - a)$$

The potential due to $-\rho_\ell$ is *not* constant over $r = a$, the surface of the actual conductor. But it is very nearly so if $a \ll h$. To this approximation, then, the total potential of the actual conductor is

$$V_a = -\frac{\rho_\ell}{2\pi\epsilon_0}\ln a + \frac{\rho_\ell}{2\pi\epsilon_0}\ln(2h - a) \approx -\frac{\rho_\ell}{2\pi\epsilon_0}\ln a + \frac{\rho_\ell}{2\pi\epsilon_0}\ln 2h = \frac{\rho_\ell}{2\pi\epsilon_0}\ln\frac{2h}{a}$$

Similarly, the potential of the image conductor is $-V_a$. Thus, the potential difference between the conductors is $2V_a$, so that the potential difference between the actual conductor and the ground plane is $\frac{1}{2}(2V_a) = V_a$. The desired capacitance per unit length is then

$$\frac{C}{L} = \frac{Q/L}{V_a} = \frac{\rho_\ell}{V_a} = \frac{2\pi\epsilon_0}{\ln(2h/a)}$$

For the given values of a and h, $C/L = 9.0$ pF/m.

The above expression for C/L is not exact, but provides a good approximation when $a \ll h$ (the practical case). An exact solution gives

$$\left(\frac{C}{L}\right)_{\text{exact}} = \frac{2\pi\epsilon_0}{\ln\left(\dfrac{h + \sqrt{h^2 - a^2}}{a}\right)}$$

Observe that C/L for the source–image system (more generally, for any pair of parallel cylindrical conductors with center-to-center separation $2h$) is one-half the value found above (same charge, twice the voltage). That is, with $d = 2h$,

$$\frac{C}{L} = \frac{\pi\epsilon_0}{\ln\left(\dfrac{d + \sqrt{d^2 - 4a^2}}{2a}\right)} \approx \frac{\pi\epsilon_0}{\ln(d/a)}$$

Supplementary Problems

7.23. Find the magnitude of **D** in a dielectric material for which $\chi_e = 1.6$ and $P = 3.05 \times 10^{-7}\ \text{C/m}^2$.
Ans. $4.96 \times 10^{-7}\ \text{C/m}^2$

7.24. Find the magnitudes of **D**, **P** and ϵ_r for a dielectric material in which $E = 0.15$ MV/m and $\chi_e = 4.25$.
Ans. 6.97 μC/m^2, 5.64 μC/m^2, 5.25

7.25. In a dielectric material with $\epsilon_r = 3.6$, $D = 285\ \text{nC/m}^2$. Find the magnitudes of **E**, **P** and χ_e.
Ans. 8.94 kV/m, 206 nC/m^2, 2.6

7.26. Given $\mathbf{E} = -3\mathbf{a}_x + 4\mathbf{a}_y - 2\mathbf{a}_z$ V/m in the region $z < 0$, where $\epsilon_r = 2.0$, find **E** in the region $z > 0$, for which $\epsilon_r = 6.5$. *Ans.* $-3\mathbf{a}_x + 4\mathbf{a}_y - \dfrac{4}{6.5}\mathbf{a}_z$ V/m

7.27. Given that $\mathbf{D} = 2\mathbf{a}_x - 4\mathbf{a}_y + 1.5\,\mathbf{a}_z\ \text{C/m}^2$ in the region $x > 0$, which is free space, find **P** in the region $x < 0$, which is a dielectric with $\epsilon_r = 5.0$. *Ans.* $1.6\,\mathbf{a}_x - 16\mathbf{a}_y + 6\mathbf{a}_z\ \text{C/m}^2$

7.28. Region 1, $z < 0$ m, is free space where $\mathbf{D} = 5\mathbf{a}_y + 7\mathbf{a}_z\ \text{C/m}^2$. Region 2, $0 < z \leq 1$ m, has $\epsilon_r = 2.5$. And region 3, $z > 1$ m, has $\epsilon_r = 3.0$. Find $\mathbf{E}_2$, $\mathbf{P}_2$ and θ_3.
Ans. $\dfrac{1}{\epsilon_0}\left(5\mathbf{a}_y + \dfrac{7}{2.5}\mathbf{a}_z\right)$ (V/m), $7.5\,\mathbf{a}_y + 4.2\,\mathbf{a}_z$ C/m^2, 25.02°

7.29. The plane interface between two dielectrics is given by $3x + z = 5$. On the side including the origin, $\mathbf{D}_1 = (4.5\,\mathbf{a}_x + 3.2\,\mathbf{a}_z)10^{-7}$ and $\epsilon_{r1} = 4.3$, while on the other side, $\epsilon_{r2} = 1.80$. Find E_1, E_2, D_2 and θ_2. *Ans.* 1.45×10^4, 3.37×10^4, 5.37×10^{-7}, 83.06°

7.30. A dielectric interface is described by $4y + 3z = 12$ m. The side including the origin is free space where $\mathbf{D}_1 = \mathbf{a}_x + 3\mathbf{a}_y + 2\mathbf{a}_z\ \mu\text{C/m}^2$. On the other side, $\epsilon_{r2} = 3.6$. Find D_2 and θ_2.
Ans. 5.14 μC/m^2, 44.4°

7.31. Find the capacitance of a parallel-plate capacitor with a dielectric of $\epsilon_r = 3.0$, area 0.92 m^2 and separation 4.5 mm. *Ans.* 5.43 nF

7.32. A parallel-plate capacitor of 8.0 nF has an area 1.51 m^2 and separation 10 mm. What separation would be required to obtain the same capacitance with free space between the plates? *Ans.* 1.67 mm

7.33. Find the capacitance between the inner and outer curved conductor surfaces shown in Fig. 7-20. Neglect fringing. *Ans.* 6.86 pF

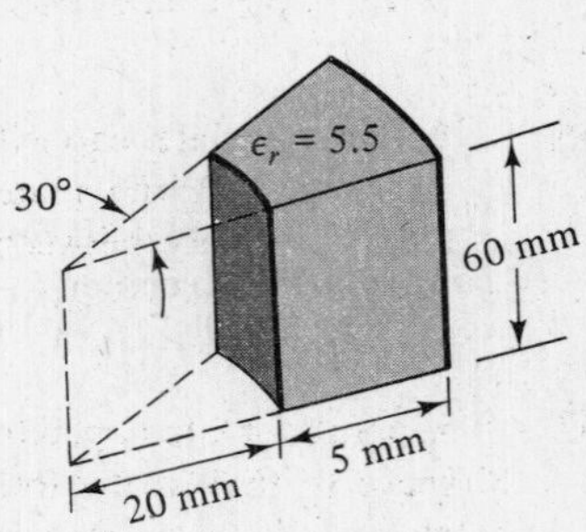

Fig. 7-20

7.34. Find the capacitance per unit length between a cylindrical conductor 2.75 inches in diameter and a parallel plane 28 ft from the conductor axis. *Ans.* 8.99 pF/m (note units)

7.35. Double the conductor diameter in Problem 7.34 and find the capacitance per unit length. *Ans.* 10.1 pF/m

7.36. Find the capacitance per unit length between two parallel cylindrical conductors in air, of radius 1.5 cm and with a center-to-center separation of 85 cm. *Ans.* 6.92 pF/m

7.37. A parallel-plate capacitor with area 0.30 m^2 and separation 5.5 mm contains three dielectrics with interfaces normal to **E** and **D**, as follows: $\epsilon_{r1} = 3.0$, $d_1 = 1.0$ mm; $\epsilon_{r2} = 4.0$, $d_2 = 2.0$ mm; $\epsilon_{r3} = 6.0$, $d_3 = 2.5$ mm. Find the capacitance. *Ans.* 2.12 nF

7.38. With a potential of 1000 V applied to the capacitor of Problem 7.37, find the potential difference and potential gradient (electric field intensity) in each dielectric. *Ans.* 267 V, 267 kV/m; 400 V, 200 kV/m; 333 V, 133 kV/m

7.39. Find the capacitance per unit length of a coaxial conductor with outer radius 4 mm and inner radius 0.5 mm if the dielectric has $\epsilon_r = 5.2$. *Ans.* 139 pF/m

7.40. Find the capacitance per unit length of a cable with an inside conductor of radius 0.75 cm and a cylindrical shield of radius 2.25 cm if the dielectric has $\epsilon_r = 2.70$. *Ans.* 137 pF/m

7.41. The coaxial cable in Fig. 7-21 has an inner conductor radius of 0.5 mm and an outer conductor radius of 5 mm. Find the capacitance per unit length with spacers as shown. *Ans.* 45.9 pF/m

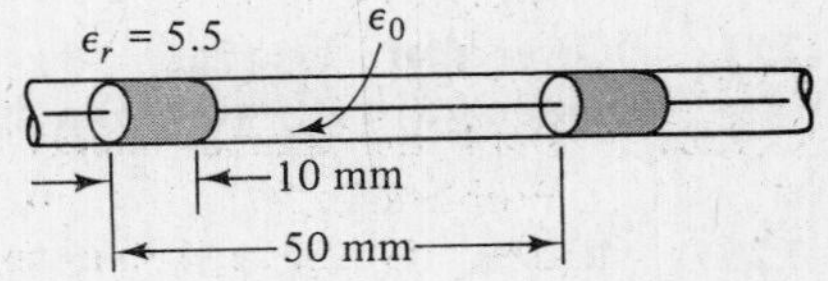

Fig. 7-21

7.42. A parallel-plate capacitor with free space between the plates is charged by momentarily connecting it to a constant 200 V source. After removal from the source a dielectric of $\epsilon_r = 2.0$ is inserted, completely filling the space. Compare the values of W_E, D, E, ρ_s, Q, V and C after insertion of the dielectric to the values before. *Partial Ans.* $V_2 = \frac{1}{2}V_1$

7.43. A parallel-plate capacitor has its dielectric changed from $\epsilon_{r1} = 2.0$ to $\epsilon_{r2} = 6.0$. It is noted that the stored energy remains fixed: $W_2 = W_1$. Examine the changes, if any, in V, C, D, E, Q and ρ_s. *Partial Ans.* $\rho_{s2} = \sqrt{3}\,\rho_{s1}$

7.44. A parallel-plate capacitor with free space between the plates remains connected to a constant voltage source while the plates are moved closer together, from separation d to $\frac{1}{2}d$. Examine the changes in Q, ρ_s, C, D, E and W_E. *Partial Ans.* $D_2 = 2D_1$

7.45. A parallel-plate capacitor with free space between the plates remains connected to a constant voltage source while the plates are moved farther apart, from separation d to $2d$. Express the changes in D, E, Q, ρ_s, C and W_E. *Partial Ans.* $D_2 = \frac{1}{2}D_1$

7.46. A parallel-plate capacitor has free space as the dielectric and a separation d. Without disturbing the charge Q, the plates are moved closer together, to $d/2$, with a dielectric of $\epsilon_r = 3$ completely filling the space between the plates. Express the changes in D, E, V, C and W_E. *Partial Ans.* $V_2 = \frac{1}{6}V_1$

7.47. A parallel-plate capacitor has free space between the plates. Compare the voltage gradient in this free space to that in the free space when a sheet of mica, $\epsilon_r = 5.4$, fills 20% of the distance between the plates. Assume the same applied voltage in each case. *Ans.* 0.84

7.48. A shielded power cable operates at a voltage of 12.5 kV on the inner conductor with respect to the cylindrical shield. There are two insulations; the first has $\epsilon_{r1} = 6.0$ and is from the inner conductor at $r = 0.8$ cm to $r = 1.0$ cm, while the second has $\epsilon_{r2} = 3.0$ and is from $r = 1.0$ cm to $r = 3.0$ cm, the inside surface of the shield. Find the maximum voltage gradient in each insulation. *Ans.* 0.645 MV/m, 1.03 MV/m

7.49. A shielded power cable has a polyethylene insulation for which $\epsilon_r = 2.26$ and the dielectric strength is 18.1 MV/m. What is the upper limit of voltage on the inner conductor with respect to the shield when the inner conductor has a radius 1 cm and the inner side of the concentric shield is at a radius of 8 cm? *Ans.* 0.376 MV

7.50. For the coaxial capacitor of Fig. 7-15, $a = 3$ cm, $b = 12$ cm, $\epsilon_{r1} = 2.50$, $\epsilon_{r2} = 4.0$. Find $\mathbf{E}_1$, $\mathbf{E}_2$, $\mathbf{D}_1$, and $\mathbf{D}_2$ if the voltage difference is 50 V. *Partial Ans.* $\mathbf{E}_2 = \pm(36.1/r)\mathbf{a}_r$ (V/m)

7.51. In Fig. 7-22, the center conductor, $r_1 = 1$ mm, is at 100 V with respect to the outer conductor at $r_3 = 100$ mm. The region $1 < r < 50$ mm is free space, while $50 < r < 100$ mm is a dielectric with $\epsilon_r = 2.0$. Find the voltage across each region. *Ans.* 91.8 V, 8.2 V

7.52. Find the stored energy per unit length in the two regions of Problem 7.51. *Ans.* 59.9 nJ/m, 5.30 nJ/m

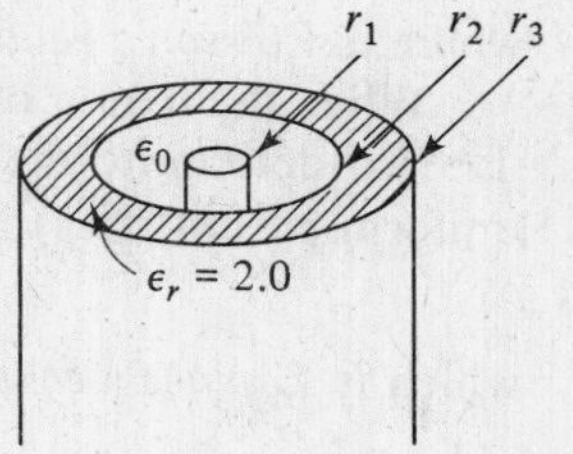

Fig. 7-22

Chapter 8

Laplace's Equation

8.1 INTRODUCTION

Electric field intensity **E** was determined in Chapter 2 by summation or integration of point charges, line charges and other charge configurations. In Chapter 3, Gauss's law was used to obtain **D**, which then gave **E**. While these two approaches are of value to an understanding of electromagnetic field theory, they both tend to be impractical because charge distributions are not usually known. The method of Chapter 5, where **E** was found to be the negative of the gradient of V, requires that the potential function throughout the region be known. But it is generally not known. Instead, conducting materials in the form of planes, curved surfaces or lines are usually specified and the voltage on one is known with respect to some reference, often one of the other conductors. Laplace's equation then provides a method whereby the potential function V can be obtained subject to the conditions on the bounding conductors.

8.2 POISSON'S EQUATION AND LAPLACE'S EQUATION

In Section 4.3 one of Maxwell's equations, $\nabla \cdot \mathbf{D} = \rho$, was developed. Substituting $\epsilon\mathbf{E} = \mathbf{D}$ and $-\nabla V = \mathbf{E}$,

$$\nabla \cdot \epsilon(-\nabla V) = \rho$$

If throughout the region of interest the medium is homogeneous, then ϵ may be removed from the partial derivatives involved in the divergence, giving

$$\nabla \cdot \nabla V = -\frac{\rho}{\epsilon} \qquad \text{or} \qquad \nabla^2 V = -\frac{\rho}{\epsilon}$$

which is *Poisson's equation.*

When the region of interest contains charges in a known distribution ρ, Poisson's equation can be used to determine the potential function. Very often the region is charge-free (as well as being of uniform permittivity). Poisson's equation then becomes

$$\nabla^2 V = 0$$

which is *Laplace's equation.*

8.3 EXPLICIT FORMS OF LAPLACE'S EQUATION

Since the left side of Laplace's equation is the *divergence of the gradient* of V, these two operations can be used to arrive at the form of the equation in a particular coordinate system.

Cartesian Coordinates.

$$\nabla V = \frac{\partial V}{\partial x}\mathbf{a}_x + \frac{\partial V}{\partial y}\mathbf{a}_y + \frac{\partial V}{\partial z}\mathbf{a}_z$$

and, for a general vector field **A**,

$$\nabla \cdot \mathbf{A} = \frac{\partial A_x}{\partial x} + \frac{\partial A_y}{\partial y} + \frac{\partial A_z}{\partial z}$$

Hence, Laplace's equation is

$$\nabla^2 V = \frac{\partial^2 V}{\partial x^2} + \frac{\partial^2 V}{\partial y^2} + \frac{\partial^2 V}{\partial z^2} = 0$$

Cylindrical Coordinates.

$$\nabla V = \frac{\partial V}{\partial r}\mathbf{a}_r + \frac{1}{r}\frac{\partial V}{\partial \phi}\mathbf{a}_\phi + \frac{\partial V}{\partial z}\mathbf{a}_z$$

and

$$\nabla \cdot \mathbf{A} = \frac{1}{r}\frac{\partial}{\partial r}(rA_r) + \frac{1}{r}\frac{\partial A_\phi}{\partial \phi} + \frac{\partial A_z}{\partial z}$$

so that Laplace's equation is

$$\nabla^2 V = \frac{1}{r}\frac{\partial}{\partial r}\left(r\frac{\partial V}{\partial r}\right) + \frac{1}{r^2}\frac{\partial^2 V}{\partial \phi^2} + \frac{\partial^2 V}{\partial z^2} = 0$$

Spherical Coordinates

$$\nabla V = \frac{\partial V}{\partial r}\mathbf{a}_r + \frac{1}{r}\frac{\partial V}{\partial \theta}\mathbf{a}_\theta + \frac{1}{r\sin\theta}\frac{\partial V}{\partial \phi}\mathbf{a}_\phi$$

and

$$\nabla \cdot \mathbf{A} = \frac{1}{r^2}\frac{\partial}{\partial r}(r^2 A_r) + \frac{1}{r\sin\theta}\frac{\partial}{\partial \theta}(A_\theta \sin\theta) + \frac{1}{r\sin\theta}\frac{\partial A_\phi}{\partial \phi}$$

so that Laplace's equation is

$$\nabla^2 V = \frac{1}{r^2}\frac{\partial}{\partial r}\left(r^2\frac{\partial V}{\partial r}\right) + \frac{1}{r^2\sin\theta}\frac{\partial}{\partial \theta}\left(\sin\theta\frac{\partial V}{\partial \theta}\right) + \frac{1}{r^2\sin^2\theta}\frac{\partial^2 V}{\partial \phi^2} = 0$$

8.4 UNIQUENESS THEOREM

Any solution to Laplace's equation or Poisson's equation which also satisfies the boundary conditions must be the only solution that exists. It is *unique*. At times there is some confusion on this point due to incomplete boundaries. As an example, consider the conducting plane at $z = 0$, as shown in Fig. 8-1, with a voltage of 100 V. It is clear that both

$$V_1 = 5z + 100$$

and

$$V_2 = 100$$

satisfy Laplace's equation and the requirement that $V = 100$ when $z = 0$. The answer is that a single conducting surface with a voltage specified and no reference given does not form the complete boundary of a properly defined region. Even two finite parallel conducting planes do not form a complete boundary, since the fringing of the field around the edges cannot be determined. However, when parallel planes are specified and it is also stated to *neglect fringing*, then the region between the planes has proper boundaries.

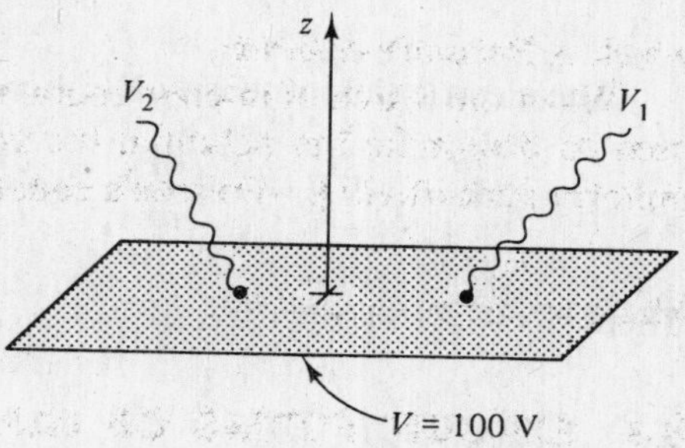

Fig. 8-1

8.5 MEAN VALUE AND MAXIMUM VALUE THEOREMS

Two important properties of the potential in a charge-free region can be obtained from Laplace's equation:

(1) At the center of an included circle or sphere, the potential V is equal to the average of the values it assumes on the circle or sphere. (See Problems 8.1 and 8.2.)

(2) The potential V cannot have a maximum (or a minimum) within the region. (See Problem 8.3.)

It follows from (2) that any maximum of V must occur on the boundary of the region. Now, since V obeys Laplace's equation,

$$\frac{\partial^2 V}{\partial x^2}+\frac{\partial^2 V}{\partial y^2}+\frac{\partial^2 V}{\partial z^2}=0$$

so do $\partial V/\partial x$, $\partial V/\partial y$, and $\partial V/\partial z$. Thus, *the components of the electric field intensity take their maximum values on the boundary.*

8.6 CARTESIAN SOLUTION IN ONE VARIABLE

Consider the parallel conductors of Fig. 8-2, where $V=0$ at $z=0$ and $V=100$ V at $z=d$. Assuming the region between the plates is charge-free,

$$\nabla^2 V=\frac{\partial^2 V}{\partial x^2}+\frac{\partial^2 V}{\partial y^2}+\frac{\partial^2 V}{\partial z^2}=0$$

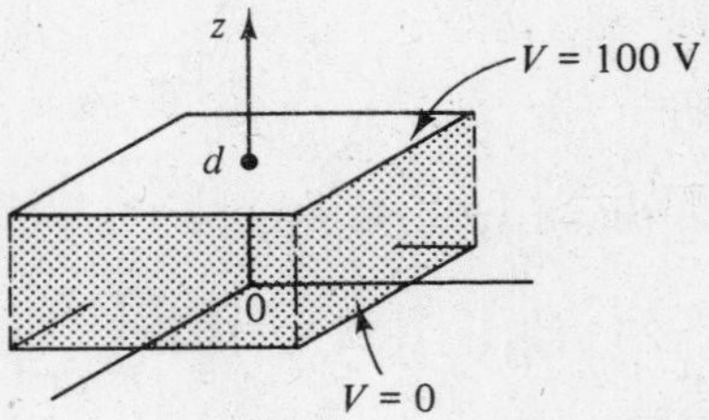

Fig. 8-2

With fringing neglected, the potential can vary only with z. Then

$$\frac{d^2V}{dz^2}=0$$

Integrating,

$$V=Az+B$$

The boundary condition $V=0$ at $z=0$ requires that $B=0$. And $V=100$ at $z=d$ gives $A=100/d$. Thus

$$V=100\left(\frac{z}{d}\right)\quad\text{(V)}$$

The electric field intensity **E** can now be obtained from

$$\mathbf{E}=-\nabla V=-\left(\frac{\partial V}{\partial x}\mathbf{a}_x+\frac{\partial V}{\partial y}\mathbf{a}_y+\frac{\partial V}{\partial z}\mathbf{a}_z\right)=-\frac{\partial}{\partial z}\left(100\frac{z}{d}\right)\mathbf{a}_z=-\frac{100}{d}\mathbf{a}_z\quad\text{(V/m)}$$

Then

$$\mathbf{D}=-\frac{\epsilon 100}{d}\mathbf{a}_z\quad(\text{C/m}^2)$$

At the conductors,

$$\rho_s=D_n=\pm\frac{\epsilon 100}{d}\quad(\text{C/m}^2)$$

where the plus sign applies at $z=d$ and the minus at $z=0$.

8.7 CARTESIAN PRODUCT SOLUTION

When the potential in cartesian coordinates varies in more than one direction, Laplace's equation will contain more than one term. Suppose that V is a function of both x and y, and has the special form $V=X(x)Y(y)$. This will make possible the separation of the variables.

$$\frac{\partial^2(XY)}{\partial x^2}+\frac{\partial^2(XY)}{\partial y^2}=0$$

becomes

$$Y\frac{d^2X}{dx^2}+X\frac{d^2Y}{dy^2}=0\qquad\text{or}\qquad\frac{1}{X}\frac{d^2X}{dx^2}+\frac{1}{Y}\frac{d^2Y}{dy^2}=0$$

Since the first term is independent of y, and the second of x, each may be set equal to a constant. However the constant for one must be the negative of that for the other. Let the constant be a^2.

$$\frac{1}{X}\frac{d^2X}{dx^2} = a^2 \qquad \frac{1}{Y}\frac{d^2Y}{dy^2} = -a^2$$

The general solution for X (for a given a) is

$$X = A_1 e^{ax} + A_2 e^{-ax}$$

or, equivalently,

$$X = A_3 \cosh ax + A_4 \sinh ax$$

and the general solution for Y (for a given a) is

$$Y = B_1 e^{jay} + B_2 e^{-jay}$$

or, equivalently,

$$Y = B_3 \cos ay + B_4 \sin ay$$

Therefore, the potential function in the variables x and y can be written

$$V = (A_1 e^{ax} + A_2 e^{-ax})(B_1 e^{jay} + B_2 e^{-jay})$$

or

$$V = (A_3 \cosh ax + A_4 \sinh ax)(B_3 \cos ay + B_4 \sin ay)$$

Because Laplace's equation is a linear, homogeneous equation, a sum of products of the above form—each product corresponding to a different value of a—is also a solution. The most general solution can be generated in this fashion.

Three-dimensional solutions, $V = X(x)\,Y(y)\,Z(z)$, of similar form can be obtained, but now there are two separation constants.

8.8 CYLINDRICAL PRODUCT SOLUTION

If a solution of the form $V = R(r)\Phi(\phi)Z(z)$ is assumed, Laplace's equation becomes

$$\frac{\Phi Z}{r}\frac{d}{dr}\left(r\frac{dR}{dr}\right) + \frac{RZ}{r^2}\frac{d^2\Phi}{d\phi^2} + R\Phi\frac{d^2Z}{dz^2} = 0$$

Dividing by $R\Phi Z$ and expanding the r-derivative,

$$\frac{1}{R}\frac{d^2R}{dr^2} + \frac{1}{Rr}\frac{dR}{dr} + \frac{1}{r^2\Phi}\frac{d^2\Phi}{d\phi^2} = -\frac{1}{Z}\frac{d^2Z}{dz^2} = -b^2$$

The r and ϕ terms contain no z and the z term contains neither r nor ϕ. They could be set equal to a constant, $-b^2$, as above. Then

$$\frac{1}{Z}\frac{d^2Z}{dz^2} = b^2$$

This equation was encountered in the cartesian product solution. The solution is

$$Z = C_1 \cosh bz + C_2 \sinh bz$$

Now the equation in r and ϕ may be further separated as follows:

$$\frac{r^2}{R}\frac{d^2R}{dr^2} + \frac{r}{R}\frac{dR}{dr} + b^2r^2 = -\frac{1}{\Phi}\frac{d^2\Phi}{d\phi^2} = a^2$$

The resulting equation in ϕ,

$$\frac{1}{\Phi}\frac{d^2\Phi}{d\phi^2} = -a^2$$

has solution

$$\Phi = C_3 \cos a\phi + C_4 \sin a\phi$$

The equation in r,

$$\frac{d^2R}{dr^2} + \frac{1}{r}\frac{dR}{dr} + \left(b^2 - \frac{a^2}{r^2}\right)R = 0$$

is a form of *Bessel's differential equation.* Its solutions are in the form of power series called *Bessel functions.*

$$R = C_5 J_a(br) + C_6 N_a(br)$$

where

$$J_a(br) = \sum_{m=0}^{\infty} \frac{(-1)^m (br/2)^{a+2m}}{m!\,\Gamma(a+m+1)}$$

and

$$N_a(br) = \frac{(\cos a\pi) J_a(br) - J_{-a}(br)}{\sin a\pi}$$

The series $J_a(br)$ is known as a Bessel function of the *first kind, order a*; if $a = n$, an integer, the gamma function in the power series may be replaced by $(n+m)!$. $N_a(br)$ is a Bessel function of the *second kind, order a*; if $a = n$, an integer, $N_n(br)$ is defined as the limit of the above quotient as $a \to n$.

The function $N_a(br)$ behaves like $\ln r$ near $r = 0$ (see Fig. 8-3). Therefore, it is not involved in the solution ($C_6 = 0$) whenever the potential is known to be finite at $r = 0$.

For integral order n and large argument x, the Bessel functions behave like damped sine waves:

$$J_n(x) \approx \sqrt{\frac{2}{\pi x}} \cos\left(x - \frac{\pi}{4} - \frac{n\pi}{2}\right) \qquad N_n(x) \approx \sqrt{\frac{2}{\pi x}} \sin\left(x - \frac{\pi}{4} - \frac{n\pi}{2}\right)$$

See Fig. 8-3.

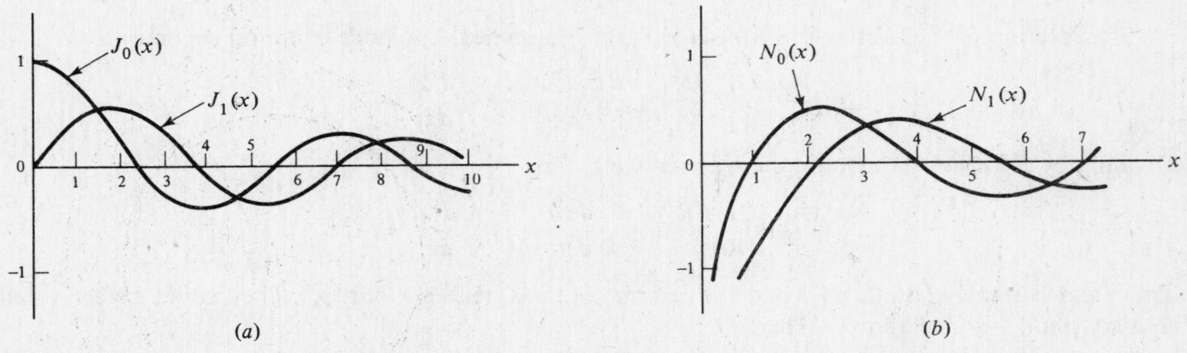

Fig. 8-3

8.9 SPHERICAL PRODUCT SOLUTION

Of particular interest in spherical coordinates are those problems in which V may vary with r and θ but not with ϕ. For a product solution $V = R(r)\Theta(\theta)$, Laplace's equation becomes

$$\left(\frac{r^2}{R}\frac{d^2R}{dr^2} + \frac{2r}{R}\frac{dR}{dr}\right) + \left(\frac{1}{\Theta}\frac{d^2\Theta}{d\theta^2} + \frac{1}{\Theta \tan\theta}\frac{d\Theta}{d\theta}\right) = 0$$

The separation constant is chosen as $n(n+1)$, where n is an integer, for reasons which will become apparent. The two separated equations are

$$r^2\frac{d^2R}{dr^2}+2r\frac{dR}{dr}-n(n+1)R=0$$

and

$$\frac{d^2\Theta}{d\theta^2}+\frac{1}{\tan\theta}\frac{d\Theta}{d\theta}+n(n+1)\Theta=0$$

The equation in r has the solution

$$R=C_1r^n+C_2r^{-(n+1)}$$

The equation in θ possesses (unlike Bessel's equation) a polynomial solution of degree n in the variable $\xi=\cos\theta$, given by

$$P_n(\xi)=\frac{1}{2^n n!}\frac{d^n}{d\xi^n}(\xi^2-1)^n \qquad n=0,1,2,\ldots$$

The polynomial $P_n(\xi)$ is the *Legendre polynomial of order n*. There is a second, independent solution, $Q_n(\xi)$, which is logarithmically infinite at $\xi=\pm1$ (i.e. $\theta=0,\pi$).

Solved Problems

8.1. As shown in Fig. 8-4(a), the potential has the value V_1 on $1/n$ of the circle, and the value 0 on the rest of the circle. Find the potential at the center of the circle. The entire region is charge-free.

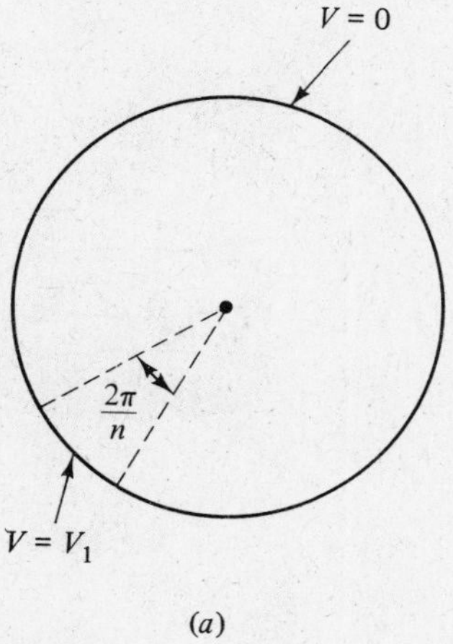

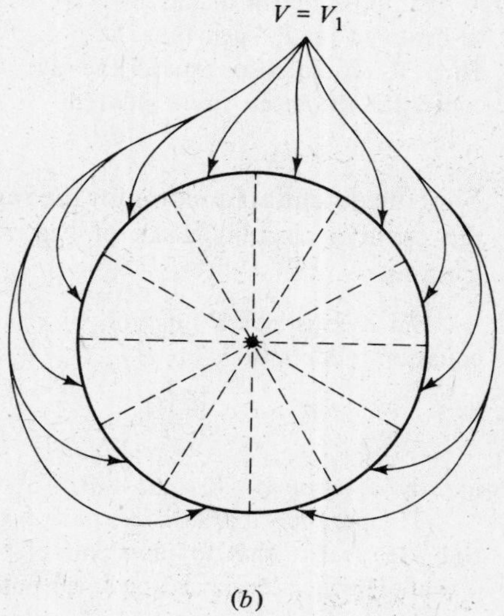

Fig. 8-4

Call the potential at the center V_c. Laplace's equation allows superposition of solutions. If n problems of the type of Fig. 8-4(a) are superposed, the result is the problem shown in Fig. 8-4(b). Because of the rotational symmetry, each subproblem in Fig. 8-4(b) gives the same potential, V_c, at the center of the circle. The total potential at the center is therefore nV_c. But, clearly, the unique solution for Fig. 8-4(*b*) is $V = V_1$ everywhere inside the circle, in particular at the center. Hence,

$$nV_c = V_1 \qquad \text{or} \qquad V_c = \frac{V_1}{n}$$

8.2. Show how the mean value theorem follows from the result of Problem 8.1.

Consider first the special case shown in Fig. 8-5, where the potential assumes n different values on n equal segments of a circle. A superposition of the solutions found in Problem 8.1 gives for the potential at the center

$$V_c = \frac{V_1}{n} + \frac{V_2}{n} + \cdots + \frac{V_n}{n} = \frac{V_1 + V_2 + \cdots + V_n}{n}$$

which is the mean value theorem in this special case. With $\Delta\phi = 2\pi/n$,

$$V_c = \frac{1}{2\pi}(V_1\,\Delta\phi + V_2\,\Delta\phi + \cdots + V_n\,\Delta\phi)$$

Now, letting $n \to \infty$,

$$V_c = \frac{1}{2\pi}\int_0^{2\pi} V(\phi)\,d\phi$$

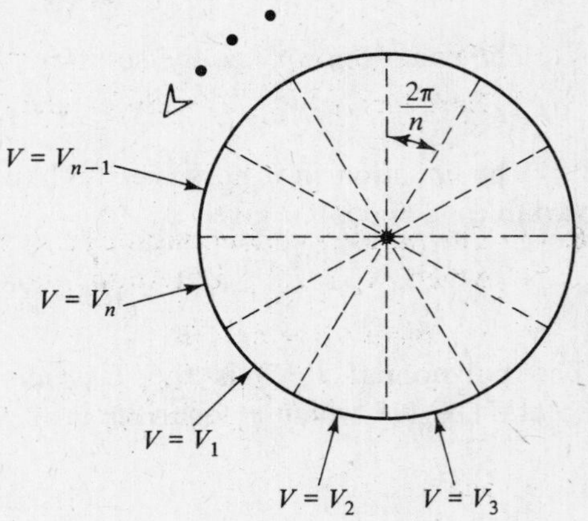

Fig. 8-5

which is the general mean value theorem for a circle.

Exactly the same reasoning, but with solid angles in place of plane angles, establishes the mean value theorem for a sphere.

8.3. Prove that within a charge-free region the potential cannot attain a maximum value.

Suppose that a maximum were attained at an interior point P. Then a very small sphere could be centered on P, such that the potential V_c at P exceeded the potential at each point on the sphere. Then V_c would also exceed the average value of the potential over the sphere. But that would contradict the mean value theorem.

8.4. Find the potential function for the region between the parallel circular disks of Fig. 8-6. Neglect fringing.

Since V is not a function of r or ϕ, Laplace's equation reduces to

$$\frac{d^2V}{dz^2} = 0$$

and the solution is $V = Az + B$.

The parallel circular disks have a potential function identical to that for any pair of parallel planes. For another choice of axes, the linear potential function might be $Ay + B$ or $Ax + B$.

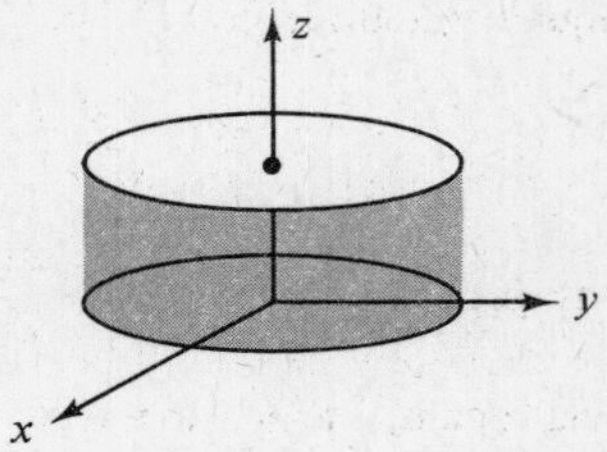

Fig. 8-6

8.5. Two parallel conducting planes in free space are at $y = 0$ and $y = 0.02$ m, and the zero voltage reference is at $y = 0.01$ m. If $\mathbf{D} = 253\mathbf{a}_y$ nC/m^2 between the conductors, determine the conductor voltages.

From Problem 8.4, $V = Ay + B$. Then

$$\mathbf{E} = \frac{\mathbf{D}}{\epsilon_0} = -\nabla V = -A\mathbf{a}_y$$

$$\frac{253 \times 10^{-9}}{8.854 \times 10^{-12}}\mathbf{a}_y = -A\mathbf{a}_y$$

whence $A = -2.86 \times 10^4$ V/m. Then,

$$0 = (-2.86 \times 10^4)(0.01) + B \qquad \text{or} \qquad B = 2.86 \times 10^2 \text{ V}$$

and

$$V = -2.86 \times 10^4 y + 2.86 \times 10^2 \quad \text{(V)}$$

Then, for $y = 0$, $V = 286$ V and for $y = 0.02$, $V = -286$ V.

8.6. The parallel conducting disks in Fig. 8-7 are separated by 5 mm and contain a dielectric for which $\epsilon_r = 2.2$. Determine the charge densities on the disks.

Since $V = Az + B$,

$$A = \frac{\Delta V}{\Delta z} = \frac{250 - 100}{5 \times 10^{-3}} = 3 \times 10^4 \text{ V/m}$$

and

$$\mathbf{E} = -\nabla V = -3 \times 10^4 \mathbf{a}_z \text{ V/m}$$

$$\mathbf{D} = \epsilon_0 \epsilon_r \mathbf{E} = -5.84 \times 10^{-7} \mathbf{a}_z \text{ C/m}^2$$

Since $\mathbf{D}$ is constant between the disks, and $D_n = \rho_s$ at a conductor surface,

$$\rho_s = \pm 5.84 \times 10^{-7} \text{ C/m}^2$$

+ on the upper plate, and − on the lower plate.

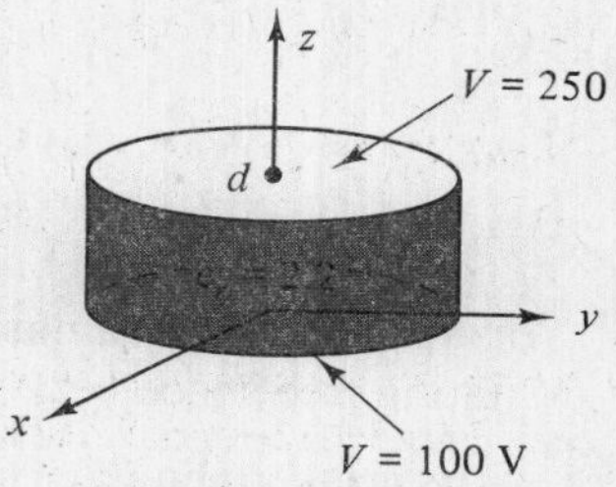

Fig. 8-7

8.7. Find the potential function and the electric field intensity for the region between two concentric right circular cylinders, where $V = 0$ at $r = 1$ mm and $V = 150$ V at $r = 20$ mm. Neglect fringing.

The potential is constant with ϕ and z. Then Laplace's equation reduces to

$$\frac{1}{r}\frac{d}{dr}\left(r\frac{dV}{dr}\right) = 0$$

Integrating once, $$r\frac{dV}{dr} = A$$

and again, $V = A \ln r + B$. Applying the boundary conditions,

$$0 = A \ln 0.001 + B \qquad 150 = A \ln 0.020 + B$$

which give $A = 50.1$, $B = 345.9$. Thus

$$V = 50.1 \ln r + 345.9 \quad \text{(V)}$$

and

$$\mathbf{E} = \frac{50.1}{r}(-\mathbf{a}_r) \quad \text{(V/m)}$$

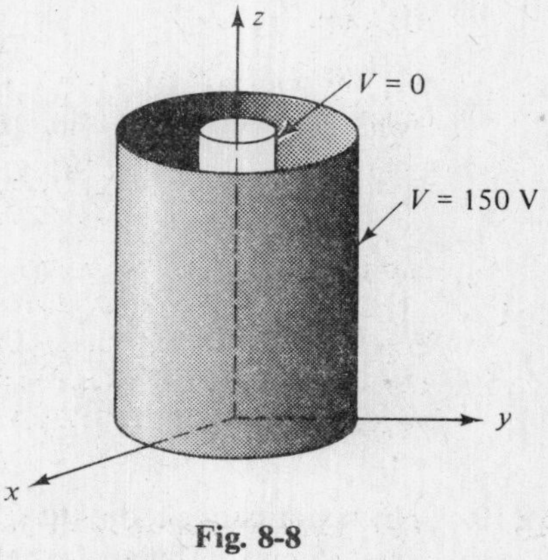

Fig. 8-8

8.8. In cylindrical coordinates two $\phi = \text{const.}$ planes are insulated along the z axis, as shown in Fig. 8-9. Neglect fringing and find the expression for $\mathbf{E}$ between the planes, assuming a potential of 100 V for $\phi = \alpha$ and a zero reference at $\phi = 0$.

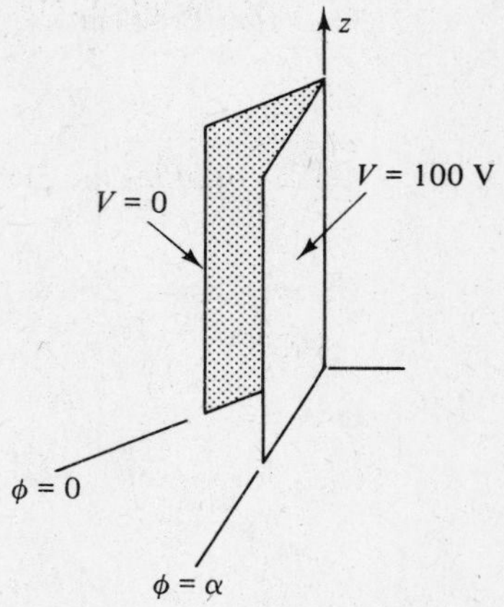

Fig. 8-9

This problem has already been solved in Problem 7.10; here Laplace's equation will be used to obtain the same result.

Since the potential is constant with r and z, Laplace's equation is

$$\frac{1}{r}\frac{d^2V}{d\phi^2} = 0$$

Integrating, $V = A\phi + B$. Applying the boundary conditions,

$$0 = A(0) + B \qquad 100 = A(\alpha) + B$$

whence

$$A = \frac{100}{\alpha} \qquad B = 0$$

Thus

$$V = 100\frac{\phi}{\alpha}\ \text{V}$$

and

$$\mathbf{E} = -\nabla V = -\frac{1}{r}\frac{d}{d\phi}\left(100\frac{\phi}{\alpha}\right)\mathbf{a}_\phi = -\frac{100}{r\alpha}\mathbf{a}_\phi \quad \text{(V/m)}$$

8.9. In spherical coordinates, $V = 0$ for $r = 0.10$ m and $V = 100$ V for $r = 2.0$ m. Assuming free space between these concentric spherical shells, find $\mathbf{E}$ and $\mathbf{D}$.

Since V is not a function of θ or ϕ, Laplace's equation reduces to

$$\frac{1}{r^2}\frac{d}{dr}\left(r^2\frac{dV}{dr}\right) = 0$$

Integrating gives

$$r^2\frac{dV}{dr} = A$$

and a second integration gives

$$V = \frac{-A}{r} + B$$

The boundary conditions give

$$0 = \frac{-A}{0.10} + B \qquad \text{and} \qquad 100 = \frac{-A}{2.00} + B$$

whence $A = 10.53\ \text{V}\cdot\text{m}$, $B = 105.3$ V. Then

$$V = \frac{-10.53}{r} + 105.3 \quad \text{(V)}$$

$$\mathbf{E} = -\nabla V = -\frac{dV}{dr}\mathbf{a}_r = -\frac{10.53}{r^2}\mathbf{a}_r \quad \text{(V/m)}$$

$$\mathbf{D} = \epsilon_0\mathbf{E} = \frac{-9.32\times10^{-11}}{r^2}\mathbf{a}_r \quad (\text{C/m}^2)$$

8.10. In spherical coordinates, $V = -25$ V on a conductor at $r = 2$ cm and $V = 150$ V at $r = 35$ cm. The space between the conductors is a dielectric for which $\epsilon_r = 3.12$. Find the surface charge densities on the conductors.

From Problem 8.9,

$$V = \frac{-A}{r} + B$$

The constants are determined from the boundary conditions

$$-25 = \frac{-A}{0.02} + B \qquad 150 = \frac{-A}{0.35} + B$$

giving

$$V = \frac{-3.71}{r} + 160.61 \quad \text{(V)}$$

$$\mathbf{E} = -\nabla V = -\frac{d}{dr}\left(\frac{-3.71}{r} + 160.61\right)\mathbf{a}_r = \frac{-3.71}{r^2}\,\mathbf{a}_r \quad \text{(V/m)}$$

$$\mathbf{D} = \epsilon_0\,\epsilon_r\,\mathbf{E} = \frac{-0.103}{r^2}\,\mathbf{a}_r \quad (\text{nC/m}^2)$$

On a conductor surface, $D_n = \rho_s$.

$$\text{at} \quad r = 0.02 \text{ m:} \qquad \rho_s = \frac{-0.103}{(0.02)^2} = -256 \text{ nC/m}^2$$

$$\text{at} \quad r = 0.35 \text{ m:} \qquad \rho_s = \frac{+0.103}{(0.35)^2} = +0.837 \text{ nC/m}^2$$

8.11. Solve Laplace's equation for the region between coaxial cones, as shown in Fig. 8-10. A potential V_1 is assumed at θ_1, and $V = 0$ at θ_2. The cone vertices are insulated at $r = 0$.

The potential is constant with r and ϕ. Laplace's equation reduces to

$$\frac{1}{r^2 \sin\theta}\frac{d}{d\theta}\left(\sin\theta\,\frac{dV}{d\theta}\right) = 0$$

Integrating,

$$\sin\theta\left(\frac{dV}{d\theta}\right) = A$$

and

$$V = A \ln\left(\tan\frac{\theta}{2}\right) + B$$

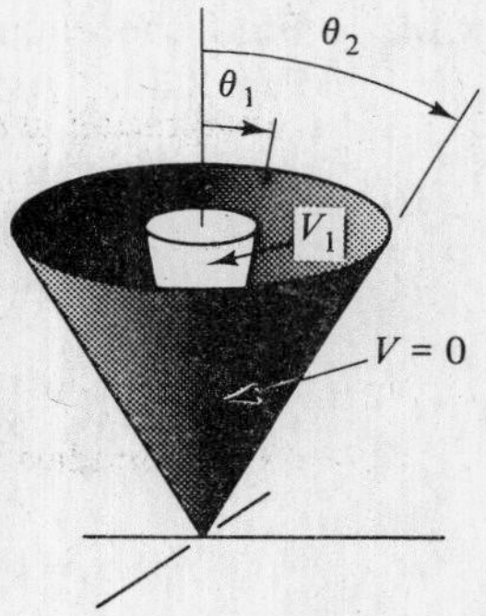

Fig. 8-10

The constants are found from

$$V_1 = A \ln\left(\tan\frac{\theta_1}{2}\right) + B \qquad 0 = A \ln\left(\tan\frac{\theta_2}{2}\right) + B$$

Hence

$$V = V_1 \frac{\ln\left(\tan\frac{\theta}{2}\right) - \ln\left(\tan\frac{\theta_2}{2}\right)}{\ln\left(\tan\frac{\theta_1}{2}\right) - \ln\left(\tan\frac{\theta_2}{2}\right)}$$

8.12. In Problem 8.11, let $\theta_1 = 10°$, $\theta_2 = 30°$ and $V_1 = 100$ V. Find the voltage at $\theta = 20°$. At what angle θ is the voltage 50 V?

Substituting the values in the general potential expression gives

$$V = -89.34\left[\ln\left(\tan\frac{\theta}{2}\right) - \ln 0.268\right] = -89.34 \ln\left(\frac{\tan\frac{\theta}{2}}{0.268}\right)$$

Then, at $\theta = 20°$, $$V = -89.34 \ln\left(\frac{\tan 10°}{0.268}\right) = 37.40 \text{ V}$$

For $V = 50$ V, $$50 = -89.34 \ln\left(\frac{\tan \theta/2}{0.268}\right)$$

Solving gives $\theta = 17.41°$.

8.13. Find the charge distribution on the conducting plane at $\theta_2 = 90°$. See Fig. 8-11.

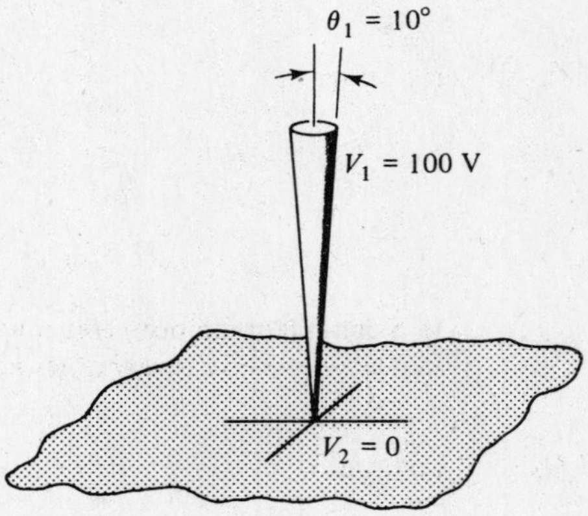

Fig. 8-11

The potential is obtained by substituting $\theta_2 = 90°$, $\theta_1 = 10°$, and $V_1 = 100$ V in the expression of Problem 8.11. Thus

$$V = 100 \frac{\ln\left(\tan\frac{\theta}{2}\right)}{\ln(\tan 5°)}$$

Then

$$\mathbf{E} = -\frac{1}{r}\frac{dV}{d\theta}\,\mathbf{a}_\theta = \frac{-100}{(r\sin\theta)\ln(\tan 5°)}\,\mathbf{a}_\theta = \frac{41.05}{r\sin\theta}\,\mathbf{a}_\theta$$

$$\mathbf{D} = \epsilon_0 \mathbf{E} = \frac{3.63 \times 10^{-10}}{r\sin\theta}\,\mathbf{a}_\theta \quad (\text{C/m}^2)$$

On the plane $\theta = 90°$, $\sin\theta = 1$ the direction of **D** requires that the surface charge on the plane be negative in sign. Hence,

$$\rho_s = -\frac{3.63 \times 10^{-10}}{r} \quad (\text{C/m}^2)$$

8.14. Find the capacitance between the two cones of Fig. 8-12. Assume free space.

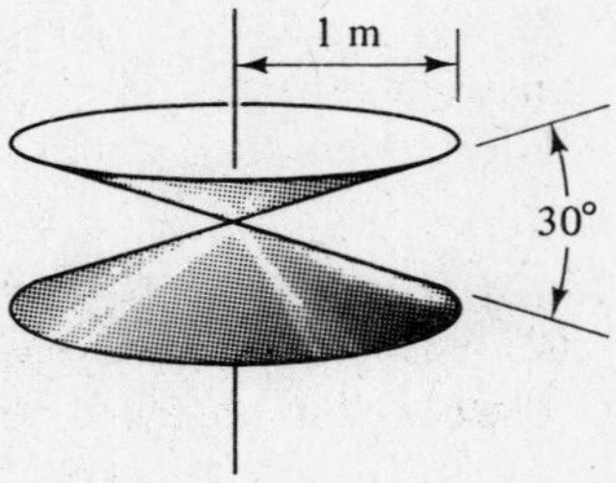

Fig. 8-12

If fringing is neglected, the potential function is given by the expression of Problem 8.11 with $\theta_1 = 75°$, $\theta_2 = 105°$. Thus

$$V = V_1 \frac{\ln\left(\tan\frac{\theta}{2}\right) - \ln(\tan 52.5°)}{\ln(\tan 37.5°) - \ln(\tan 52.5°)}$$
$$= (-1.89\,V_1)\ln\left(\tan\frac{\theta}{2}\right) + \text{const.}$$

from which

$$\mathbf{D} = \epsilon_0 \mathbf{E} = \epsilon_0\left(-\frac{1}{r}\frac{dV}{d\theta}\,\mathbf{a}_\theta\right) = \frac{1.89\,\epsilon_0 V_1}{r\sin\theta}\,\mathbf{a}_\theta$$

The charge density on the upper plate is then

$$\rho_s = D_n = \frac{1.89\,\epsilon_0 V_1}{r\sin 75°}$$

so that the total charge on the upper plate is

$$Q = \int \rho_s\,dS = \int_0^{2\pi}\int_0^{\csc 75°} \frac{1.89\,\epsilon_0 V_1}{r\sin 75°}\, r\sin 75°\,dr\,d\phi = 12.28\,\epsilon_0 V_1$$

and the capacitance is $C = Q/V_1 = 12.28\,\epsilon_0$.

8.15. The region between two concentric right circular cylinders contains a uniform charge density ρ. Use Poisson's equation to find V.

Neglecting fringing, Poisson's equation reduces to

$$\frac{1}{r}\frac{d}{dr}\left(r\frac{dV}{dr}\right) = -\frac{\rho}{\epsilon}$$

$$\frac{d}{dr}\left(r\frac{dV}{dr}\right) = -\frac{\rho r}{\epsilon}$$

Integrating,

$$r\frac{dV}{dr} = -\frac{\rho r^2}{2\epsilon} + A$$

$$\frac{dV}{dr} = -\frac{\rho r}{2\epsilon} + \frac{A}{r}$$

$$V = -\frac{\rho r^2}{4\epsilon} + A\ln r + B$$

Note that static problems involving charge distributions in space are theoretical exercises, since no means exists to hold the charges in position against the coulomb forces.

8.16. The region

$$-\frac{\pi}{2} < \frac{z}{z_0} < \frac{\pi}{2}$$

has a charge density $\rho = 10^{-8}\cos(z/z_0)$ (C/m³). Elsewhere the charge density is zero. Find V and $\mathbf{E}$ from Poisson's equation, and compare with the results given by Gauss's law.

Since V is not a function of x or y, Poisson's equation is

$$\frac{d^2V}{dz^2} = -\frac{\rho}{\epsilon} = -\frac{10^{-8}\cos(z/z_0)}{\epsilon}$$

Integrating twice,

$$V = \frac{10^{-8}z_0^2\cos(z/z_0)}{\epsilon} + Az + B \quad \text{(V)}$$

and

$$\mathbf{E} = -\nabla V = \left(\frac{10^{-8}z_0\sin(z/z_0)}{\epsilon} - A\right)\mathbf{a}_z \quad \text{(V/m)}$$

But, by the symmetry of the charge distribution, the field must vanish on the plane $z = 0$. Therefore $A = 0$ and

$$\mathbf{E} = \frac{10^{-8}z_0\sin(z/z_0)}{\epsilon}\mathbf{a}_z \quad \text{(V/m)}$$

A special Gaussian surface centered about $z = 0$ is shown in Fig. 8-13. $\mathbf{D}$ cuts only the top and bottom surfaces, each of area A. Furthermore, since the charge distribution is symmetrical about $z = 0$, $\mathbf{D}$ must be antisymmetrical about $z = 0$, so that $\mathbf{D}_{\text{top}} = D\mathbf{a}_z$, $\mathbf{D}_{\text{bottom}} = D(-\mathbf{a}_z)$.

$$D\int_{\text{top}} dS + D\int_{\text{bottom}} dS = \int_{-z}^{z}\iint 10^{-8}\cos(z/z_0)\,dx\,dy\,dz$$

$$2DA = 2z_0 A 10^{-8}\sin(z/z_0)$$

or $\quad D = z_0 10^{-8}\sin(z/z_0) \quad$ for $\quad 0 < z < \pi z_0/2$

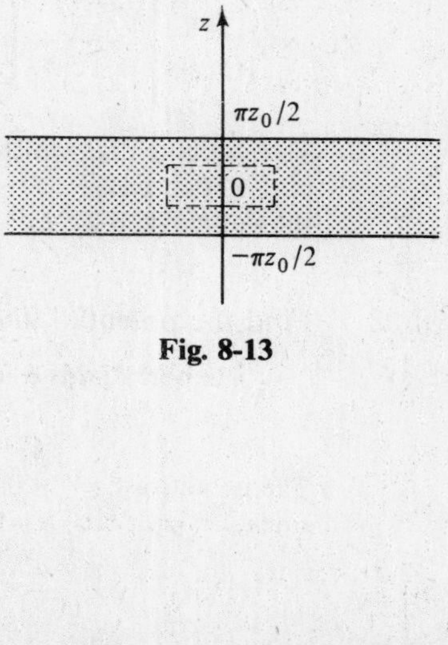

Fig. 8-13

Then, for $-\pi z_0/2 < z < \pi z_0/2$,

$$\mathbf{D} = z_0 10^{-8}\sin(z/z_0)\mathbf{a}_z \quad \text{(C/m}^2\text{)}$$

and $\mathbf{E} = \mathbf{D}/\epsilon$ agrees with the result from Poisson's equation.

8.17. A potential in cylindrical coordinates is a function of r and ϕ but not z. Obtain the separated differential equations for R and Φ, where $V = R(r)\Phi(\phi)$, and solve them. The region is charge-free.

Laplace's equation becomes

$$\Phi\frac{d^2R}{dr^2} + \frac{\Phi}{r}\frac{dR}{dr} + \frac{R}{r^2}\frac{d^2\Phi}{d\phi^2} = 0$$

or

$$\frac{r^2}{R}\frac{d^2R}{dr^2} + \frac{r}{R}\frac{dR}{dr} = -\frac{1}{\Phi}\frac{d^2\Phi}{d\phi^2}$$

The left side is a function of r only, while the right side is a function of ϕ only; therefore, both sides are equal to a constant, a^2.

$$\frac{r^2}{R}\frac{d^2R}{dr^2} + \frac{r}{R}\frac{dR}{dr} = a^2$$

$$\frac{d^2R}{dr^2} + \frac{1}{r}\frac{dR}{dr} - \frac{a^2R}{r^2} = 0$$

with solution $R = C_1r^a + C_2r^{-a}$. Also,

$$-\frac{1}{\Phi}\frac{d^2\Phi}{d\phi^2} = a^2$$

with solution $\Phi = C_3\cos a\phi + C_4\sin a\phi$.

8.18. Given the potential function $V = V_0(\sinh ax)(\sin az)$ (see Section 8.7), determine the shape and location of the surfaces on which $V = 0$ and $V = V_0$. Assume that $a > 0$.

Since the potential is not a function of y, the equipotential surfaces extend to $\pm\infty$ in the y direction. Because $\sin az = 0$ for $z = n\pi/a$, where $n = 0, 1, 2, \ldots$, the planes $z = 0$ and $z = \pi/a$ are at zero potential. Because $\sinh ax = 0$ for $x = 0$, the plane $x = 0$ is also at zero potential. The $V = 0$ equipotential is shown by the heavy line in Fig. 8-14.

The surface on which $V = V_0$ has x and z coordinates which satisfy the equation

$$V_0 = V_0(\sinh ax)(\sin az) \qquad \text{or} \qquad \sinh ax = \frac{1}{\sin az}$$

When values of z between zero and π/a are substituted, the corresponding x coordinates are readily obtained. For example:

az	1.57 1.57	1.02 2.12	0.67 2.47	0.49 2.65	0.28 2.86	0.10 3.04
ax	0.88	1.0	1.25	1.50	2.00	3.00

The equipotential, which is symmetrical about $z = \pi/2a$, is shown in Fig. 8-14. Because V is periodic in z, and because $V(-x, -z) = V(x, z)$, the whole xz plane can be filled with replicas of the strip shown in Fig. 8-14.

8.19. Find the potential function for the region inside the rectangular trough shown in Fig. 8-15.

The potential is a function of x and z, of the form (see Section 8.7)

$$V = (C_1\cosh az + C_2\sinh az)(C_3\cos ax + C_4\sin ax)$$

The conditions $V = 0$ at $x = 0$ and $z = 0$ require the constants C_1 and C_3 to be zero. Then, since $V = 0$ at $x = c$, $a = n\pi/c$, where n is an integer. Replacing C_2C_4 by C, the expression

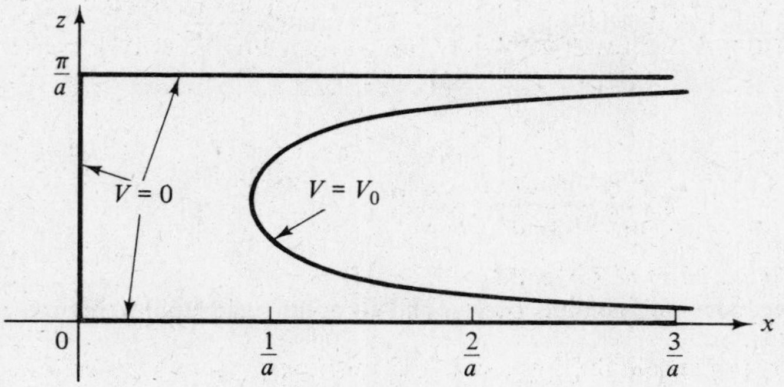

Fig. 8-14

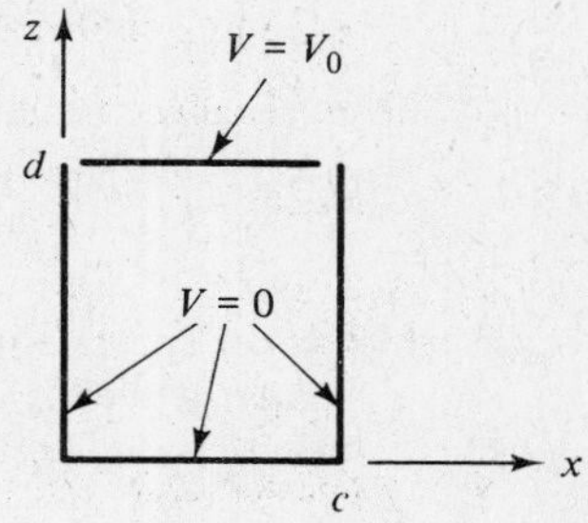

Fig. 8-15

becomes

$$V = C \sinh \frac{n\pi z}{c} \sin \frac{n\pi x}{c}$$

or more generally, by superposition,

$$V = \sum_{n=1}^{\infty} C_n \sinh \frac{n\pi z}{c} \sin \frac{n\pi x}{c}$$

The final boundary condition requires that

$$V_0 = \sum_{n=1}^{\infty} \left(C_n \sinh \frac{n\pi d}{c} \right) \sin \frac{n\pi x}{c} \qquad (0 < x < c)$$

Thus the constants $b_n \equiv C_n \sinh (n\pi d/c)$ are determined as the coefficients in the *Fourier sine series* for $f(x) \equiv V_0$ in the range $0 < x < c$. The well-known formula for the Fourier coefficients,

$$b_n = \frac{2}{c} \int_0^c f(x) \sin \frac{n\pi x}{c} dx \qquad n = 1, 2, 3, \ldots$$

gives

$$b_n = \frac{2V_0}{c} \int_0^c \sin \frac{n\pi x}{c} dx = \begin{cases} 0 & n \text{ even} \\ 4V_0/n\pi & n \text{ odd} \end{cases}$$

The potential function is then

$$V = \sum_{n \text{ odd}} \frac{4V_0}{n\pi} \frac{\sinh (n\pi z/c)}{\sinh (n\pi d/c)} \sin \frac{n\pi x}{c}$$

for $0 < x < c$, $0 < z < d$.

8.20. Identify the spherical product solution

$$V = \frac{C_2}{r^2} P_1(\cos \theta) = \frac{C_2 \cos \theta}{r^2}$$

(Section 8.9, with $C_1 = 0$, $n = 1$) with a point dipole at the origin.

Figure 8-16 shows a finite dipole along the z axis, consisting of a point charge $+Q$ at $z = +d/2$ and a point charge $-Q$ at $z = -d/2$. The quantity $p = Qd$ is the dipole moment (Section 7.1). The potential at point P is

$$V = \frac{Q}{4\pi\epsilon_0 r_1} - \frac{Q}{4\pi\epsilon_0 r_2} = \frac{p}{4\pi\epsilon_0 d} \left(\frac{r_2 - r_1}{r_1 r_2} \right)$$

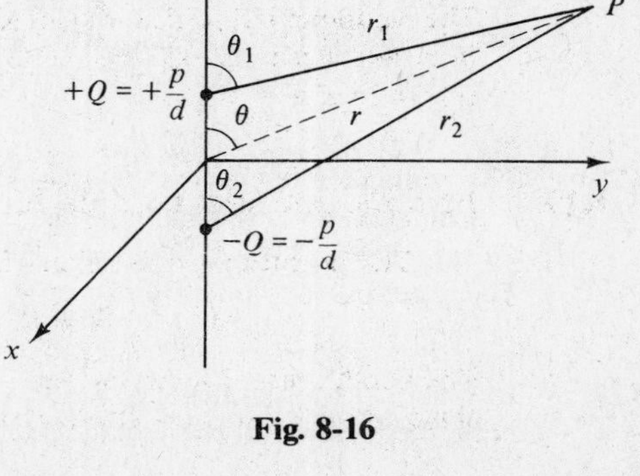

Fig. 8-16

A point dipole at the origin is obtained in the limit as $d \to 0$. For small d,

$$r_2 - r_1 \approx d\cos\theta_2 \approx d\cos\theta \qquad \text{and} \qquad r_1 r_2 \approx r^2$$

Therefore, in the limit,

$$V = \frac{p}{4\pi\epsilon_0}\frac{\cos\theta}{r^2}$$

which is the spherical product solution with $C_2 = p/4\pi\epsilon_0$.

Similarly, the higher-order Legendre polynomials correspond to point quadrupoles, octupoles, etc.

Supplementary Problems

8.21. In cartesian coordinates a potential is a function of x only. At $x = -2.0$ cm, $V = 25.0$ V, and $\mathbf{E} = 1.5 \times 10^3(-\mathbf{a}_x)$ V/m throughout the region. Find V at $x = 3.0$ cm. *Ans.* 100 V

8.22. In cartesian coordinates a plane at $z = 3.0$ cm is the voltage reference. Find the voltage and the charge density on the conductor $z = 0$ if $\mathbf{E} = 6.67 \times 10^3\mathbf{a}_z$ V/m for $z > 0$ and the region contains a dielectric for which $\epsilon_r = 4.5$. *Ans.* 200 V, 266 nC/m^2

8.23. In cylindrical coordinates, $V = 75.0$ V at $r = 5$ mm and $V = 0$ at $r = 60$ mm. Find the voltage at $r = 130$ mm if the potential depends only on r. *Ans.* -23.34 V

8.24. Concentric, right circular, conducting cylinders in free space at $r = 5$ mm and $r = 25$ mm have voltages of zero and V_0, respectively. If $\mathbf{E} = -8.28 \times 10^3\mathbf{a}_r$ V/m at $r = 15$ mm, find V_0 and the charge density on the outer conductor. *Ans.* 200 V, $+44$ nC/m^2

8.25. For concentric conducting cylinders, $V = 75$ V at $r = 1$ mm and $V = 0$ at $r = 20$ mm. Find $\mathbf{D}$ in the region between the cylinders, where $\epsilon_r = 3.6$. *Ans.* $(798/r)\mathbf{a}_r$ (pC/m^2)

8.26. Conducting planes at $\phi = 10°$ and $\phi = 0°$ in cylindrical coordinates have voltages of 75 V and zero, respectively. Obtain $\mathbf{D}$ in the region between the planes, which contains a material for which $\epsilon_r = 1.65$. *Ans.* $(-6.28/r)\mathbf{a}_r$ (nC/m^2)

8.27. Two square conducting planes 50 cm on a side are separated by 2.0 cm along one side and 2.5 cm along the other (Fig. 8-17). Assume a voltage difference and compare the charge density at the center of one plane to that on an identical pair with a uniform separation of 2.0 cm. *Ans.* 0.89

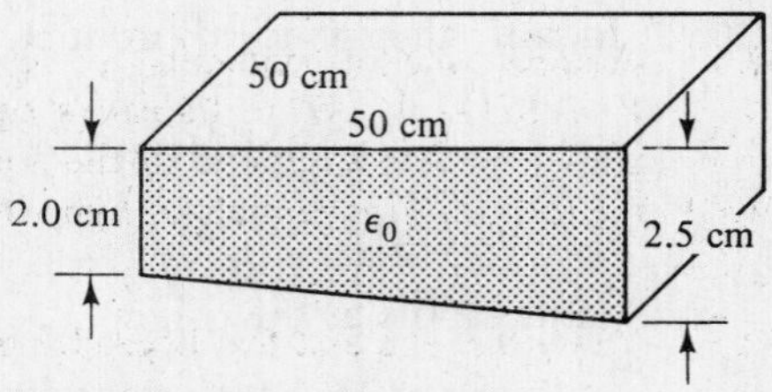

Fig. 8-17

8.28. The voltage reference is at $r = 15$ mm in spherical coordinates and the voltage is V_0 at $r = 200$ mm. Given $\mathbf{E} = -334.7\,\mathbf{a}_r$ V/m at $r = 110$ mm, find V_0. The potential is a function of r only. *Ans.* 250 V

8.29. In spherical coordinates, $V = 865$ V at $r = 50$ cm and $\mathbf{E} = 748.2\,\mathbf{a}_r$ V/m at $r = 85$ cm. Determine the location of the voltage reference if the potential depends only on r. *Ans.* $r = 250$ cm

8.30. With a zero reference at infinity and $V = 45.0$ V at $r = 0.22$ m in spherical coordinates, a dielectric of $\epsilon_r = 1.72$ occupies the region $0.22 < r < 1.00$ m and free space occupies $r > 1.00$ m. Determine D at $r = 1.00 \pm 0$ m.
Ans. 8.55 V/m, 14.7 V/m

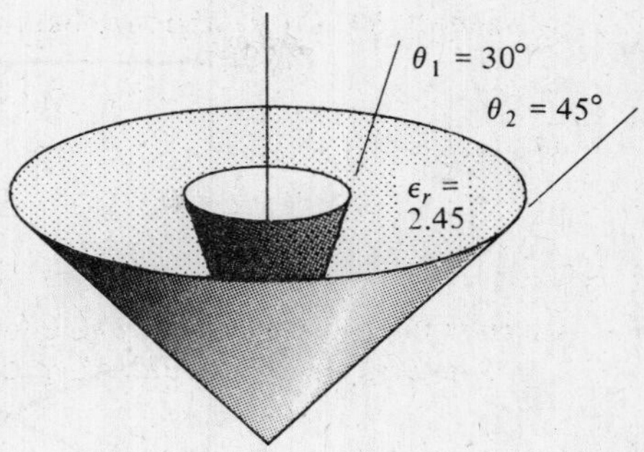

Fig. 8-18

8.31. In Fig. 8-18 the cone at $\theta = 45°$ has a voltage V with respect to the reference at $\theta = 30°$. At $r = 0.25$ m and $\theta = 30°$, $\mathbf{E} = -2.30 \times 10^3 \mathbf{a}_\theta$ V/m. Determine the voltage difference V.
Ans. 125.5 V

8.32. In Problem 8.31 determine the surface charge densities on the conducting cones at 30° and 45°. $\epsilon_r = 2.45$ between the cones.
Ans. $\dfrac{-12.5}{r}$ (nC/m²), $\dfrac{8.84}{r}$ (nC/m²)

8.33. Find E in the region between the two cones shown in Fig. 8-19. *Ans.* $\dfrac{0.288\, V_1}{r \sin\theta}$ (V/m)

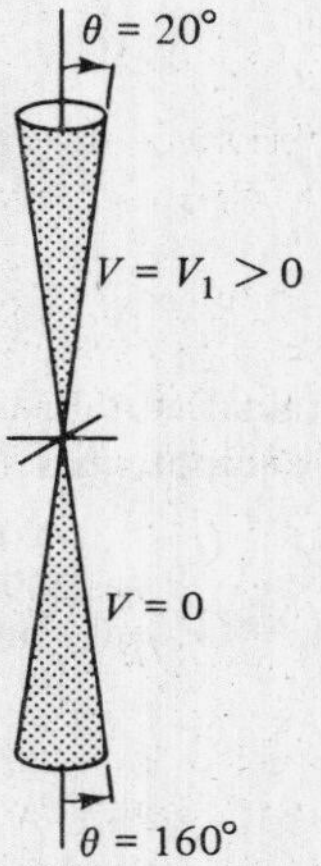

Fig. 8-19

8.34. In cylindrical coordinates, $\rho = 111/r$ (pC/m³). Given that $V = 0$ at $r = 1.0$ m and $V = 50$ V at $r = 3.0$ m due to this charge configuration, find the expression for **E**.
Ans. $\left(12.5 - \dfrac{68.3}{r}\right)\mathbf{a}_r$ (V/m)

8.35. Determine **E** in spherical coordinates from Poisson's equation, assuming a uniform charge density ρ.
Ans. $\left(\dfrac{\rho r}{3\epsilon} - \dfrac{A}{r^2}\right)\mathbf{a}_r$

8.36. Specialize the solution found in Problem 8.35 to the case of a uniformly charged sphere.
Ans. See Problem 2.56.

8.37. Assume that a potential in cylindrical coordinates is a function of r and z but not ϕ, $V = R(r)Z(z)$. Write Laplace's equation and obtain the separated differential equations in r and z. Show that the solutions to the equation in r are Bessel functions and that the solutions in z are in the form of exponentials or hyperbolic functions.

8.38. Verify that the first five Legendre polynomials are:

$$P_0(\cos\theta) = 1$$
$$P_1(\cos\theta) = \cos\theta$$
$$P_2(\cos\theta) = \tfrac{1}{2}(3\cos^2\theta - 1)$$
$$P_3(\cos\theta) = \tfrac{1}{2}(5\cos^3\theta - 3\cos\theta)$$
$$P_4(\cos\theta) = \tfrac{1}{8}(35\cos^4\theta - 30\cos^2\theta + 3)$$

and graph them against $\xi = \cos\theta$. *Ans.* See Fig. 8-20.

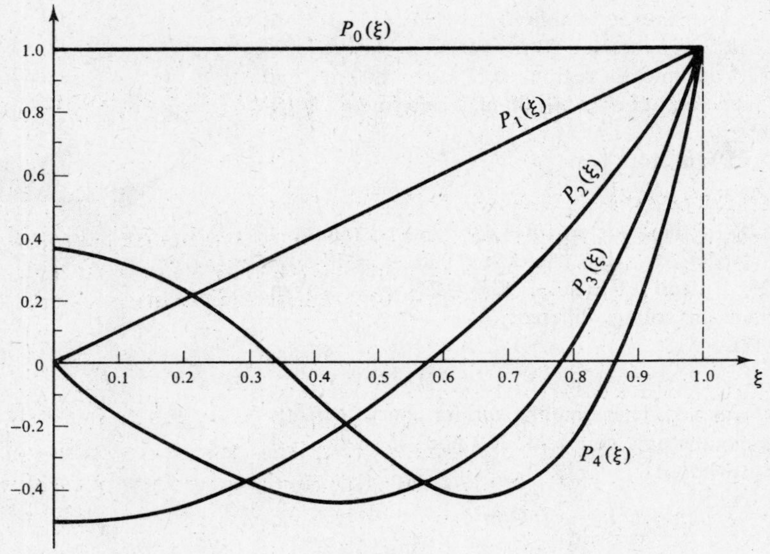

Fig. 8-20

8.39. Obtain **E** for Problem 8.18 and plot several values on Fig. 8-14. Note the orthogonality of **E** and the equipotential surfaces. *Ans.* $\mathbf{E} = -V_0 a[(\cosh ax)(\sin az)\mathbf{a}_x + (\sinh ax)(\cos az)\mathbf{a}_z]$

8.40. Given $V = V_0(\cosh ax)(\sin ay)$, where $a > 0$, determine the shape and location of the surfaces on which $V = 0$ and $V = V_0$. Make a sketch similar to Fig. 8-14. *Ans.* See Fig. 8-21.

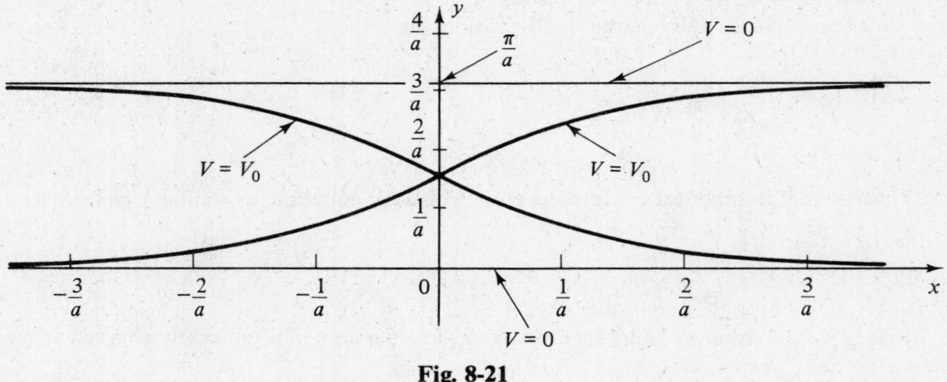

Fig. 8-21

8.41. From the potential function of Problem 8.40, obtain **E** and plot several values on the sketch of the equipotential surfaces, Fig. 8-21.
Ans. $\mathbf{E} = -V_0 a[(\sinh ax)(\sin ay)\mathbf{a}_x + (\cosh ax)(\cos ay)\mathbf{a}_y]$

8.42. Use a superposition of the product solutions found in Problem 8.17 to obtain the potential function for the semicircular strip shown in Fig. 8-22.

Ans. $V = \sum_{n \text{ odd}} \frac{4V_0}{n\pi} \frac{r^n - (a^2/r)^n}{b^n - (a^2/b)^n} \sin n\phi$

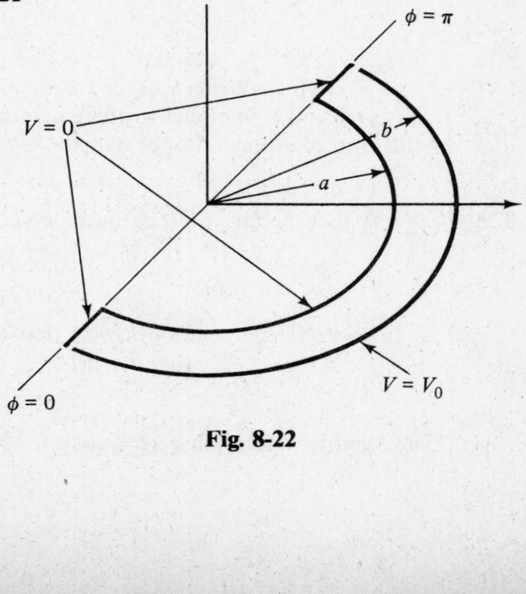

Fig. 8-22

Chapter 9

Ampère's Law and the Magnetic Field

9.1 INTRODUCTION

A static magnetic field can originate from either a constant current or a permanent magnet. This chapter will treat the magnetic fields of constant currents. Time-variable magnetic fields, which coexist with time-variable electric fields, will be examined in Chapters 12 and 13.

9.2 BIOT-SAVART LAW

A differential *magnetic field strength*, $d\mathbf{H}$, results from a differential current element $I\,d\mathbf{l}$. The field varies inversely with the distance squared, is independent of the surrounding medium, and has a direction given by the cross product of $I\,d\mathbf{l}$ and $\mathbf{a}_R$. This relationship is known as the *Biot-Savart law:*

$$d\mathbf{H} = \frac{I\,d\mathbf{l} \times \mathbf{a}_R}{4\pi R^2} \quad \text{(A/m)}$$

The direction of $\mathbf{R}$ must be from the current element to the point at which $d\mathbf{H}$ is to be determined, as shown in Fig. 9-1.

Current elements have no separate existence. All elements making up the complete current filament contribute to $\mathbf{H}$ and must be included. The summation leads to the integral form of the Biot-Savart law:

$$\mathbf{H} = \oint \frac{I\,d\mathbf{l} \times \mathbf{a}_R}{4\pi R^2}$$

The closed line integral simply requires that *all* current elements be included in order to obtain the complete $\mathbf{H}$ (the contour may close at ∞).

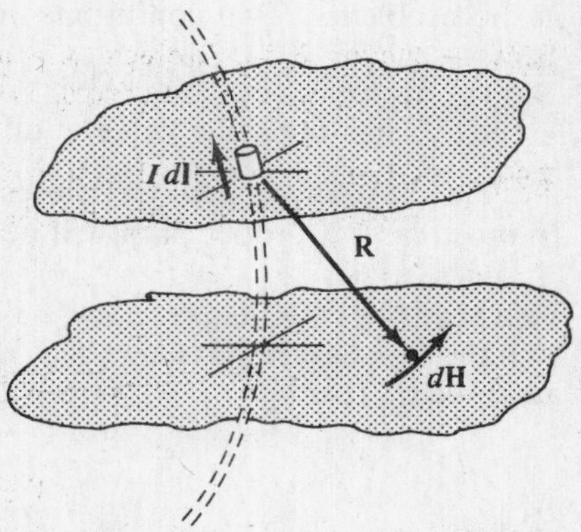

Fig. 9-1

EXAMPLE 1 An infinitely long, straight, filamentary current I along the z axis in cylindrical coordinates is shown in Fig. 9-2. A point in the $z = 0$ plane is selected with no loss in generality. In differential form,

$$d\mathbf{H} = \frac{I\,dz\,\mathbf{a}_z \times (r\mathbf{a}_r - z\mathbf{a}_z)}{4\pi(r^2 + z^2)^{3/2}} = \frac{I\,dz\,r\mathbf{a}_\phi}{4\pi(r^2 + z^2)^{3/2}}$$

The variable of integration is z. Since $\mathbf{a}_\phi$ does not change with z, it may be removed from the integrand before integrating.

$$\mathbf{H} = \left[\int_{-\infty}^{\infty} \frac{Ir\,dz}{4\pi(r^2 + z^2)^{3/2}}\right]\mathbf{a}_\phi = \frac{I}{2\pi r}\mathbf{a}_\phi$$

This important result shows that $\mathbf{H}$ is inversely proportional to the radial distance. The direction is seen to be in agreement with the "right-hand rule" whereby the fingers of the right hand point in the direction of the field when the conductor is grasped such that the right thumb points in the direction of the current.

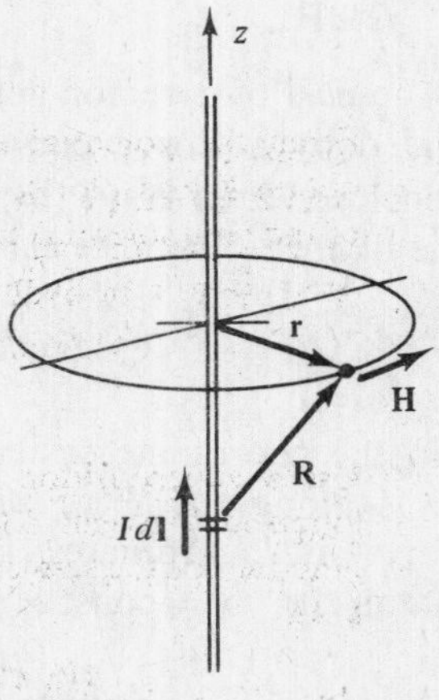

Fig. 9-2

The magnetic fields of sheet currents or volume currents are also given by the integral form of the Biot-Savart law, with $I\,d\mathbf{l}$ replaced by $\mathbf{K}\,dS$ or $\mathbf{J}\,dv$, respectively, and with the integration extended over the entire sheet or volume. A particular case of importance is the infinite plane sheet of constant density $\mathbf{K}$. As shown in Problem 9.3, the field in this case is constant:

$$\mathbf{H} = \tfrac{1}{2}\mathbf{K} \times \mathbf{a}_n$$

9.3 AMPÈRE'S LAW

The line integral of the tangential component of $\mathbf{H}$ around a closed path is equal to the current enclosed by the path.

$$\oint \mathbf{H} \cdot d\mathbf{l} = I_{\text{enc}}$$

This is *Ampère's law*.

At first glance one would think that the law is used to determine the current I by an integration. Instead, the current is usually known and the law provides a method of finding $\mathbf{H}$. This is quite similar to the use of Gauss's law to find $\mathbf{D}$ given the charge distribution.

In order to utilize Ampère's law to determine $\mathbf{H}$ there must be a considerable degree of symmetry in the problem. Two conditions must be met:

1. at each point of the closed path $\mathbf{H}$ is either tangential or normal to the path
2. H has the same value at all points of the path where $\mathbf{H}$ is tangential

The Biot-Savart law can be used to aid in selecting a path which meets the above conditions. In most cases a proper path will be evident.

EXAMPLE 2 Use Ampère's law to obtain $\mathbf{H}$ due to an infinitely long, straight filament of current I.

The Biot-Savart law shows that at each point of the circle in Fig. 9-2 $\mathbf{H}$ is tangential and of the same magnitude. Then

$$\oint \mathbf{H} \cdot d\mathbf{l} = H(2\pi r) = I$$

so that

$$\mathbf{H} = \frac{I}{2\pi r}\,\mathbf{a}_\phi$$

9.4 CURL

The *curl* of a vector field $\mathbf{A}$ is another vector field. Point P in Fig. 9-3 lies in a plane area ΔS bounded by a closed curve C. In the integration that defines the curl, C is traversed such that the enclosed area is on the left. The unit normal $\mathbf{a}_n$, determined by a right-hand rule, is as shown in the figure. Then the *component* of the curl of $\mathbf{A}$ in the direction $\mathbf{a}_n$ is defined as

$$(\text{curl}\,\mathbf{A}) \cdot \mathbf{a}_n \equiv \lim_{\Delta S \to 0} \frac{\oint \mathbf{A} \cdot d\mathbf{l}}{\Delta S}$$

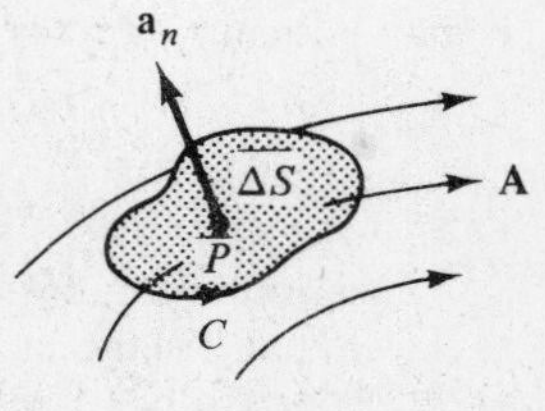

Fig. 9-3

In the coordinate systems, curl $\mathbf{A}$ is completely specified by its components along the three unit vectors. For example, the x component in cartesian coordinates is defined by taking as the contour C a square in the $x = \text{const.}$ plane through P, as shown in Fig. 9-4.

$$(\text{curl}\,\mathbf{A}) \cdot \mathbf{a}_x = \lim_{\Delta y\,\Delta z \to 0} \frac{\oint \mathbf{A} \cdot d\mathbf{l}}{\Delta y\,\Delta z}$$

If $\mathbf{A} = A_x\mathbf{a}_x + A_y\mathbf{a}_y + A_z\mathbf{a}_z$ at the corner of ΔS closest to the origin (point *1*), then

$$\oint = \int_1^2 + \int_2^3 + \int_3^4 + \int_4^1$$
$$= A_y\Delta y + \left(A_z + \frac{\partial A_z}{\partial y}\Delta y\right)\Delta z + \left(A_y + \frac{\partial A_y}{\partial z}\Delta z\right)(-\Delta y) + A_z(-\Delta z)$$
$$= \left(\frac{\partial A_z}{\partial y} - \frac{\partial A_y}{\partial z}\right)\Delta y\,\Delta z$$

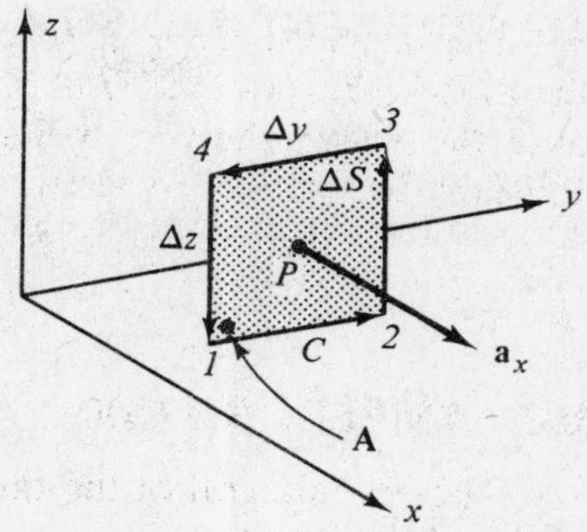

Fig. 9-4

and
$$(\text{curl}\,\mathbf{A})\cdot\mathbf{a}_x = \frac{\partial A_z}{\partial y} - \frac{\partial A_y}{\partial z}$$

The y and z components can be determined in a similar fashion. Combining the three components,

$$\text{curl}\,\mathbf{A} = \left(\frac{\partial A_z}{\partial y} - \frac{\partial A_y}{\partial z}\right)\mathbf{a}_x + \left(\frac{\partial A_x}{\partial z} - \frac{\partial A_z}{\partial x}\right)\mathbf{a}_y + \left(\frac{\partial A_y}{\partial x} - \frac{\partial A_x}{\partial y}\right)\mathbf{a}_z \qquad \text{(cartesian)}$$

A third-order determinant can be written, the expansion of which gives the cartesian curl of **A**.

$$\text{curl}\,\mathbf{A} = \begin{vmatrix} \mathbf{a}_x & \mathbf{a}_y & \mathbf{a}_z \\ \dfrac{\partial}{\partial x} & \dfrac{\partial}{\partial y} & \dfrac{\partial}{\partial z} \\ A_x & A_y & A_z \end{vmatrix}$$

The elements of the second row are the components of the del operator. This suggests (see Section 1.2) that $\nabla \times \mathbf{A}$ can be written for curl **A**. As with other expressions from vector analysis, this convenient notation is used for curl **A** in other coordinate systems, even though ∇ is defined only in cartesian coordinates.

Expressions for curl **A** in cylindrical and spherical coordinates can be derived in the same manner as above, though with more difficulty.

$$\text{curl}\,\mathbf{A} = \left(\frac{1}{r}\frac{\partial A_z}{\partial \phi} - \frac{\partial A_\phi}{\partial z}\right)\mathbf{a}_r + \left(\frac{\partial A_r}{\partial z} - \frac{\partial A_z}{\partial r}\right)\mathbf{a}_\phi + \frac{1}{r}\left[\frac{\partial(rA_\phi)}{\partial r} - \frac{\partial A_r}{\partial \phi}\right]\mathbf{a}_z \qquad \text{(cylindrical)}$$

$$\text{curl}\,\mathbf{A} = \frac{1}{r\sin\theta}\left[\frac{\partial(A_\phi \sin\theta)}{\partial\theta} - \frac{\partial A_\theta}{\partial\phi}\right]\mathbf{a}_r + \frac{1}{r}\left[\frac{1}{\sin\theta}\frac{\partial A_r}{\partial\phi} - \frac{\partial(rA_\phi)}{\partial r}\right]\mathbf{a}_\theta + \frac{1}{r}\left[\frac{\partial(rA_\theta)}{\partial r} - \frac{\partial A_r}{\partial\theta}\right]\mathbf{a}_\phi \qquad \text{(spherical)}$$

Frequently useful are two properties of the curl operator:

(1) *the divergence of a curl is zero*; that is,

$$\nabla\cdot(\nabla\times\mathbf{A}) = 0$$

for any vector field **A**

(2) *the curl of a gradient is zero*; that is,

$$\nabla\times(\nabla f) = 0$$

for any scalar function of position f (see Problem 9.22)

Under static conditions, $\mathbf{E} = -\nabla V$, and so, from (2),

$$\nabla\times\mathbf{E} = 0$$

9.5 CURRENT DENSITY J AND $\nabla \times \mathbf{H}$

The x component of $\nabla \times \mathbf{H}$ is determined by $\oint \mathbf{H} \cdot d\mathbf{l}$, where the path lies in a plane normal to the x axis. Ampère's law states that this integral is equal to the current enclosed. The direction is $\mathbf{a}_x$, so the current can be called I_x. Thus,

$$(\text{curl}\,\mathbf{H}) \cdot \mathbf{a}_x = \lim_{\Delta S \to 0} \frac{I_x}{\Delta S} = J_x$$

the x component of the current density $\mathbf{J}$. Likewise for the y and z directions. Consequently,

$$\nabla \times \mathbf{H} = \mathbf{J}$$

This important result is one of Maxwell's equations for static fields. If $\mathbf{H}$ is known throughout a particular region, then $\nabla \times \mathbf{H}$ will produce $\mathbf{J}$ for that region. See Problem 9.15.

9.6 MAGNETIC FLUX DENSITY B

Like $\mathbf{D}$, the magnetic field strength $\mathbf{H}$ depends only on (moving) charges and is independent of the medium. The force field associated with $\mathbf{H}$ is the *magnetic flux density* $\mathbf{B}$, which is given by

$$\mathbf{B} = \mu \mathbf{H}$$

where $\mu = \mu_0 \mu_r$ is the *permeability* of the medium. The unit of $\mathbf{B}$ is the *tesla*,

$$1\,\text{T} = 1\,\frac{\text{N}}{\text{A} \cdot \text{m}}$$

The free-space permeability μ_0 has a numerical value of $4\pi \times 10^{-7}$ and has the units *henries per meter*, H/m; μ_r, the *relative permeability* of the medium, is a pure number very near to unity, except for a small group of *ferromagnetic* materials which will be treated in Chapter 11.

Magnetic flux, Φ, through a surface is defined as

$$\Phi = \int_S \mathbf{B} \cdot d\mathbf{S}$$

The sign on Φ may be positive or negative depending upon the choice of the surface normal in $d\mathbf{S}$. The unit of magnetic flux is the *weber*, Wb. The various magnetic units are related by:

$$1\,\text{T} = 1\,\text{Wb/m}^2 \qquad 1\,\text{H} = 1\,\text{Wb/A}$$

EXAMPLE 3 Find the flux crossing the portion of the plane $\phi = \pi/4$ defined by $0.01 < r < 0.05$ m and $0 < z < 2$ m (see Fig. 9-5). A current filament of 2.50 A along the z axis is in the $\mathbf{a}_z$ direction.

$$\mathbf{B} = \mu_0 \mathbf{H} = \frac{\mu_0 I}{2\pi r}\,\mathbf{a}_\phi$$

$$d\mathbf{S} = dr\,dz\,\mathbf{a}_\phi$$

$$\Phi = \int_0^2 \int_{0.01}^{0.05} \frac{\mu_0 I}{2\pi r}\,\mathbf{a}_\phi \cdot dr\,dz\,\mathbf{a}_\phi$$

$$= \frac{2\mu_0 I}{2\pi} \ln \frac{0.05}{0.01}$$

$$= 1.61 \times 10^{-6}\,\text{Wb} \quad \text{or} \quad 1.61\,\mu\text{Wb}$$

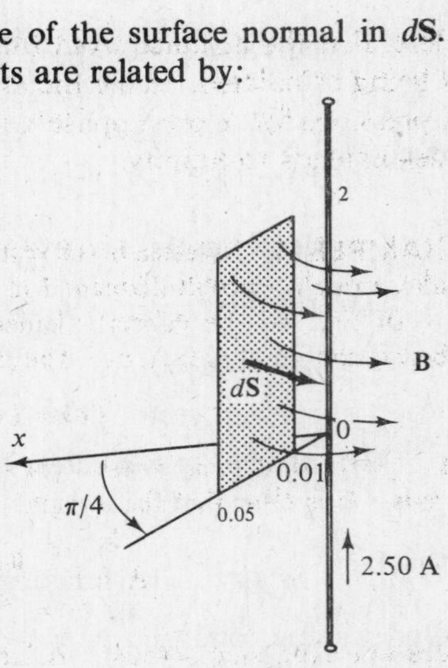

Fig. 9-5

It should be observed that the lines of magnetic flux Φ are closed curves, with no starting point or termination point. This is in contrast with electric flux Ψ, which originates on positive charge and terminates on negative charge. In Fig. 9-6 all of the magnetic flux Φ that enters the closed surface must leave the surface. Thus **B** fields have no sources or sinks, which is mathematically expressed by

$$\nabla \cdot \mathbf{B} = 0$$

(see Section 4.1).

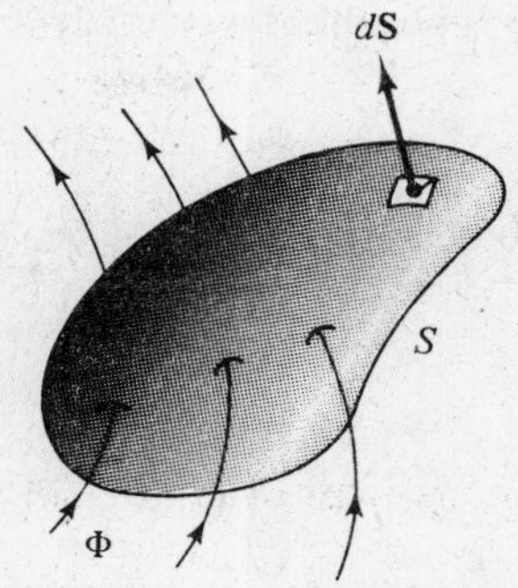

Fig. 9-6

9.7 VECTOR MAGNETIC POTENTIAL A

Electric field intensity **E** was first obtained from known charge configurations. Later, electric potential V was developed and it was found that **E** could be obtained as the negative gradient of V, i.e. $\mathbf{E} = -\nabla V$. Laplace's equation provided a method of obtaining V from known potentials on the boundary conductors. Similarly, a *vector magnetic potential*, **A**, defined such that

$$\nabla \times \mathbf{A} = \mathbf{B}$$

serves as an intermediate quantity, from which **B**, and hence **H**, can be calculated. Note that the definition of **A** is consistent with the requirement that $\nabla \cdot \mathbf{B} = 0$. The units of **A** are Wb/m or T · m.

If the additional condition

$$\nabla \cdot \mathbf{A} = 0$$

is imposed, then vector magnetic potential **A** can be determined from the known currents in the region of interest. For the three standard current configurations the expressions are as follows.

current filament: $$\mathbf{A} = \oint \frac{\mu I \, d\mathbf{l}}{4\pi R}$$

sheet current: $$\mathbf{A} = \int_S \frac{\mu \mathbf{K} \, dS}{4\pi R}$$

volume current: $$\mathbf{A} = \int_v \frac{\mu \mathbf{J} \, dv}{4\pi R}$$

Here, R is the distance from the current element to the point at which the vector magnetic potential is being calculated. Like the analogous integral for the electric potential (see Section 5.4), the above expressions for **A** presuppose a zero level at infinity; they cannot be applied if the current distribution itself extends to infinity.

EXAMPLE 4 Investigate the vector magnetic potential for the infinite, straight, current filament I in free space.

In Fig. 9-7 the current filament is along the z axis and the observation point is (x, y, z). The particular current element

$$I \, d\mathbf{l} = I \, d\ell \, \mathbf{a}_z$$

at $\ell = 0$ is shown, where ℓ is the running variable along the z axis. It is clear that the integral

$$\mathbf{A} = \int_{-\infty}^{\infty} \frac{\mu_0 I \, d\ell}{4\pi R} \mathbf{a}_z$$

does not exist, since, when ℓ is large, $R \approx \ell$. This is a case of a current distribution that extends to infinity.

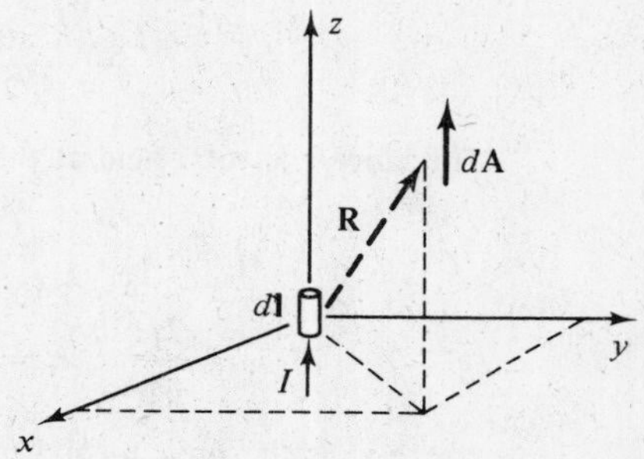

Fig. 9-7

It is possible, however, to consider the *differential* vector potential

$$d\mathbf{A} = \frac{\mu_0 I \, d\ell}{4\pi R} \mathbf{a}_z$$

and to obtain from it the *differential* **B**. Thus, for the particular current element at $\ell = 0$,

$$d\mathbf{A} = \frac{\mu_0 I \, d\ell}{4\pi(x^2 + y^2 + z^2)^{1/2}} \mathbf{a}_z$$

and
$$d\mathbf{B} = \nabla \times d\mathbf{A} = \frac{\mu_0 I \, d\ell}{4\pi} \left[\frac{-y}{(x^2 + y^2 + z^2)^{3/2}} \mathbf{a}_x + \frac{x}{(x^2 + y^2 + z^2)^{3/2}} \mathbf{a}_y \right]$$

This result agrees with that for $d\mathbf{H} = (1/\mu_0)\, d\mathbf{B}$ given by the Biot-Savart law.

For a way of defining **A** for the infinite current filament, see Problem 9.19.

9.8 STOKES' THEOREM

Consider an open surface S whose boundary is a closed curve C. *Stokes' theorem* states that the integral of the tangential component of a vector field **F** around C is equal to the integral of the normal component of curl **F** over S:

$$\oint \mathbf{F} \cdot d\mathbf{l} = \int_S (\nabla \times \mathbf{F}) \cdot d\mathbf{S}$$

If **F** is chosen to be the vector magnetic potential **A**, Stokes' theorem gives

$$\oint \mathbf{A} \cdot d\mathbf{l} = \int_S \mathbf{B} \cdot d\mathbf{S} = \Phi$$

Solved Problems

9.1. Find **H** at the center of a square current loop of side L.

Choose a cartesian coordinate system such that the loop is located as shown in Fig. 9-8. By symmetry, each half-side contributes the same amount to **H** at the center. For the half-side $0 \le x \le L/2$, $y = -L/2$, the Biot-Savart law gives for the field at the origin

$$d\mathbf{H} = \frac{(I\,dx\,\mathbf{a}_x) \times [-x\mathbf{a}_x + (L/2)\mathbf{a}_y]}{4\pi[x^2 + (L/2)^2]^{3/2}} = \frac{I\,dx(L/2)\mathbf{a}_z}{4\pi[x^2 + (L/2)^2]^{3/2}}$$

Therefore, the total field at the origin is

$$\mathbf{H} = 8 \int_0^{L/2} \frac{I\,dx(L/2)\mathbf{a}_z}{4\pi[x^2 + (L/2)^2]^{3/2}} = \frac{2\sqrt{2}\,I}{\pi L} \mathbf{a}_z = \frac{2\sqrt{2}\,I}{\pi L} \mathbf{a}_n$$

where $\mathbf{a}_n$ is the unit normal to the plane of the loop as given by the usual right-hand rule.

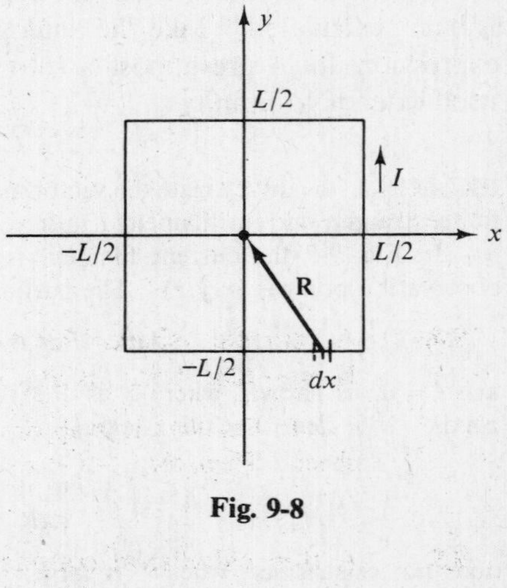

Fig. 9-8

9.2. A current filament of 5.0 A in the $\mathbf{a}_y$ direction is parallel to the y axis at $x = 2$ m, $z = -2$ m. Find **H** at the origin.

The expression for **H** due to a straight current filament applies,

$$\mathbf{H} = \frac{I}{2\pi r}\mathbf{a}_\phi$$

where $r = 2\sqrt{2}$ and (use the right-hand rule)

$$\mathbf{a}_\phi = \frac{\mathbf{a}_x + \mathbf{a}_z}{\sqrt{2}}$$

Thus

$$\mathbf{H} = \frac{5.0}{2\pi(2\sqrt{2})}\left(\frac{\mathbf{a}_x + \mathbf{a}_z}{\sqrt{2}}\right) = (0.281)\left(\frac{\mathbf{a}_x + \mathbf{a}_z}{\sqrt{2}}\right) \text{ A/m}$$

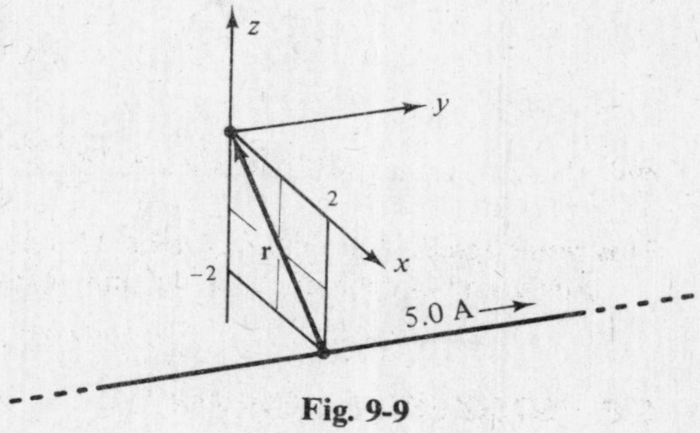

Fig. 9-9

9.3. Determine an expression for **H** due to an infinite plane current sheet of uniform density **K**.

The Biot-Savart law and considerations of symmetry show that **H** has only an x component, which is independent of x and y, if $\mathbf{K} = K\mathbf{a}_y$ (see Fig. 9-10). Applying Ampère's law to the square contour *12341*, and using the fact that **H** must be antisymmetric in z,

$$\oint \mathbf{H}\cdot d\mathbf{l} = (H)(2a) + 0 + (H)(2a) + 0 = (K)(2a) \qquad \text{or} \qquad H = \frac{K}{2}$$

Thus, for all $z > 0$, $\mathbf{H} = (K/2)\mathbf{a}_x$. More generally, for an arbitrary orientation of the current sheet,

$$\mathbf{H} = \tfrac{1}{2}\mathbf{K} \times \mathbf{a}_n$$

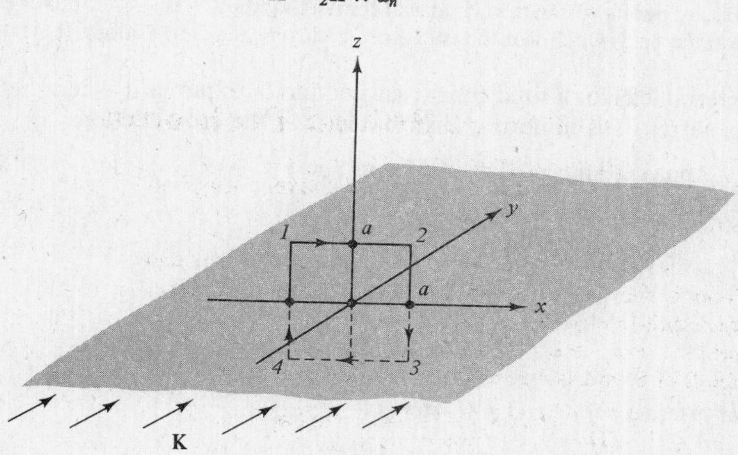

Fig. 9-10

9.4. A current sheet, $\mathbf{K} = 10\mathbf{a}_z$ A/m, lies in the $x = 5$ m plane and a second sheet, $\mathbf{K} = -10\mathbf{a}_z$ A/m, is at $x = -5$ m. Find **H** at all points.

In Fig. 9-11 it is apparent that for points between the sheets, $\mathbf{K} \times \mathbf{a}_n$ results in $-K\mathbf{a}_y$ for each sheet. Then, for $-5 < x < 5$, $\mathbf{H} = 10(-\mathbf{a}_y)$ A/m. Elsewhere $\mathbf{H} = 0$.

9.5. A thin cylindrical conductor of radius a, infinite in length, carries a current I. Find **H** at all points using Ampère's law.

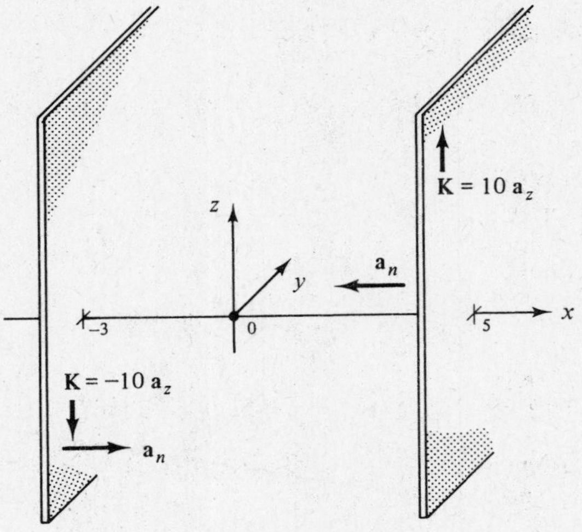

Fig. 9-11

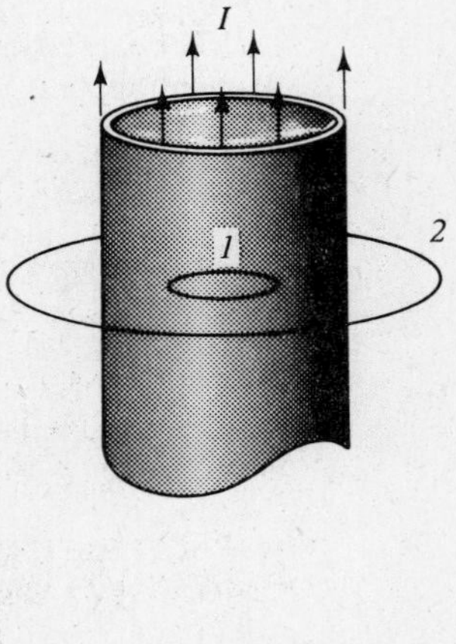

Fig. 9-12

The Biot-Savart law shows that **H** has only a ϕ component. Furthermore, H_ϕ is a function of r only. Proper paths for Ampère's law are concentric circles. For path *1* shown in Fig. 9-12,

$$\oint \mathbf{H} \cdot d\mathbf{l} = 2\pi r H_\phi = I_{\text{enc}} = 0$$

and for path *2*,

$$\oint \mathbf{H} \cdot d\mathbf{l} = 2\pi r H_\phi = I$$

Thus, for points within the cylindrical conducting shell, $\mathbf{H} = 0$, and for external points, $\mathbf{H} = (I/2\pi r)\mathbf{a}_\phi$, the same field which would result from a current filament I along the axis.

9.6. Determine **H** for a solid cylindrical conductor of radius a, where the current I is uniformly distributed over the cross section.

Applying Ampère's law to contour *1* in Fig. 9-13,

$$\oint \mathbf{H} \cdot d\mathbf{l} = I_{\text{enc}}$$

$$H(2\pi r) = I\left(\frac{\pi r^2}{\pi a^2}\right)$$

$$\mathbf{H} = \frac{Ir}{2\pi a^2}\mathbf{a}_\phi$$

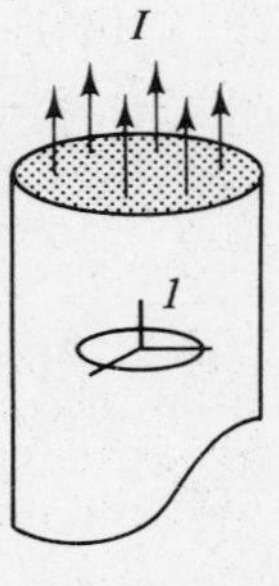

Fig. 9-13

For external points, $\mathbf{H} = (I/2\pi r)\mathbf{a}_\phi$.

9.7. In the region $0 < r < 0.5$ m, in cylindrical coordinates, the current density is

$$\mathbf{J} = 4.5\,e^{-2r}\mathbf{a}_z \quad (\text{A/m}^2)$$

and $\mathbf{J} = 0$ elsewhere. Use Ampère's law to find **H**.

Because the current density is symmetrical about the origin, a circular path may be used in Ampère's law, with the enclosed current given by $\oint \mathbf{J} \cdot d\mathbf{S}$. Thus, for $r < 0.5$ m,

$$H_\phi(2\pi r) = \int_0^{2\pi}\int_0^r 4.5\,e^{-2r} r\,dr\,d\phi$$

$$\mathbf{H} = \frac{1.125}{r}(1 - e^{-2r} - 2re^{-2r})\mathbf{a}_\phi \quad (\text{A/m})$$

For any $r \geq 0.5$ m the enclosed current is the same, $0.594\,\pi$ A. Then

$$H_\phi(2\pi r) = 0.594\,\pi \qquad \text{or} \qquad \mathbf{H} = \frac{0.297}{r}\,\mathbf{a}_\phi \quad \text{(A/m)}$$

9.8. Find **H** on the axis of a circular current loop of radius a. Specialize the result to the center of the loop.

For the point shown in Fig. 9-14,

$$\mathbf{R} = -a\mathbf{a}_r + h\mathbf{a}_z$$

$$d\mathbf{H} = \frac{(Ia\,d\phi\,\mathbf{a}_\phi) \times (-a\mathbf{a}_r + h\mathbf{a}_z)}{4\pi(a^2 + h^2)^{3/2}} = \frac{(Ia\,d\phi)(a\mathbf{a}_z + h\mathbf{a}_r)}{4\pi(a^2 + h^2)^{3/2}}$$

Inspection shows that diametrically opposite current elements produce r components which cancel. Then,

$$\mathbf{H} = \int_0^{2\pi} \frac{Ia^2\,d\phi}{4\pi(a^2 + h^2)^{3/2}}\,\mathbf{a}_z = \frac{Ia^2}{2(a^2 + h^2)^{3/2}}\,\mathbf{a}_z$$

At $h = 0$, $\mathbf{H} = (I/2a)\mathbf{a}_z$.

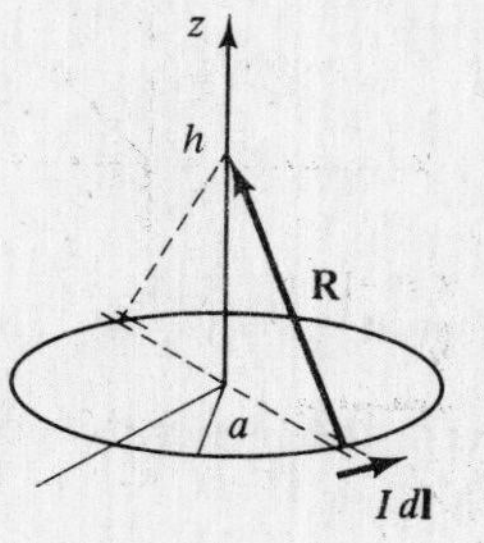

Fig. 9-14

9.9. A current sheet, $\mathbf{K} = 6.0\,\mathbf{a}_x$ A/m, lies in the $z = 0$ plane and a current filament is located at $y = 0$, $z = 4$ m, as shown in Fig. 9-15. Determine I and its direction if $\mathbf{H} = 0$ at $(0, 0, 1.5)$ m.

Due to the current sheet,

$$\mathbf{H} = \tfrac{1}{2}\mathbf{K} \times \mathbf{a}_n = \frac{6.0}{2}(-\mathbf{a}_y)\ \text{A/m}$$

For the field to vanish at $(0, 0, 1.5)$ m, $|\mathbf{H}|$ due to the filament must be 3.0 A/m.

$$|\mathbf{H}| = \frac{I}{2\pi r}$$

$$3.0 = \frac{I}{2\pi(2.5)}$$

$$I = 47.1\ \text{A}$$

To cancel the **H** from the sheet, this current must be in the $\mathbf{a}_x$ direction, as shown in Fig. 9-15.

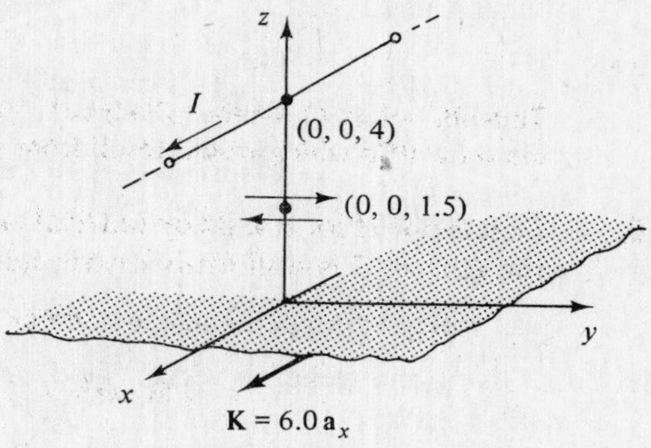

Fig. 9-15

9.10. Given the general vector field $\mathbf{A} = (y\cos ax)\mathbf{a}_x + (y + e^x)\mathbf{a}_z$, find $\nabla \times \mathbf{A}$ at the origin.

$$\nabla \times \mathbf{A} = \begin{vmatrix} \mathbf{a}_x & \mathbf{a}_y & \mathbf{a}_z \\ \dfrac{\partial}{\partial x} & \dfrac{\partial}{\partial y} & \dfrac{\partial}{\partial z} \\ y\cos ax & 0 & y + e^x \end{vmatrix} = \mathbf{a}_x - e^x\mathbf{a}_y - \cos ax\,\mathbf{a}_z$$

At $(0, 0, 0)$, $\nabla \times \mathbf{A} = \mathbf{a}_x - \mathbf{a}_y - \mathbf{a}_z$.

9.11. Calculate the curl of **H** in cartesian coordinates due to a current filament along the z axis with current I in the $\mathbf{a}_z$ direction.

From Example 1,

$$\mathbf{H} = \frac{I}{2\pi r}\,\mathbf{a}_\phi = \frac{I}{2\pi}\left(\frac{-y\mathbf{a}_x + x\mathbf{a}_y}{x^2 + y^2}\right)$$

and so

$$\nabla \times \mathbf{H} = \begin{vmatrix} \mathbf{a}_x & \mathbf{a}_y & \mathbf{a}_z \\ \dfrac{\partial}{\partial x} & \dfrac{\partial}{\partial y} & \dfrac{\partial}{\partial z} \\ \dfrac{-y}{x^2+y^2} & \dfrac{x}{x^2+y^2} & 0 \end{vmatrix}$$

$$= \left[\frac{\partial}{\partial x}\left(\frac{x}{x^2+y^2}\right) - \frac{\partial}{\partial y}\left(\frac{-y}{x^2+y^2}\right)\right]\mathbf{a}_z$$

$$= 0$$

except at $x = y = 0$. This is consistent with $\nabla \times \mathbf{H} = \mathbf{J}$.

9.12. Given the general vector field $\mathbf{A} = 5r\sin\phi\,\mathbf{a}_z$ in cylindrical coordinates, find curl $\mathbf{A}$ at $(2, \pi, 0)$.

Since $\mathbf{A}$ has only a z component, only two partials in the curl expression are nonzero.

$$\nabla \times \mathbf{A} = \frac{1}{r}\frac{\partial}{\partial\phi}(5r\sin\phi)\,\mathbf{a}_r - \frac{\partial}{\partial r}(5r\sin\phi)\,\mathbf{a}_\phi = 5\cos\phi\,\mathbf{a}_r - 5\sin\phi\,\mathbf{a}_\phi$$

Then

$$\nabla \times \mathbf{A}\Big|_{(2,\,\pi,\,0)} = -5\mathbf{a}_r$$

9.13. Given the general vector field $\mathbf{A} = 5e^{-r}\cos\phi\,\mathbf{a}_r - 5\cos\phi\,\mathbf{a}_z$ in cylindrical coordinates, find curl $\mathbf{A}$ at $(2, 3\pi/2, 0)$.

$$\nabla \times \mathbf{A} = \frac{1}{r}\frac{\partial}{\partial\phi}(-5\cos\phi)\,\mathbf{a}_r + \left[\frac{\partial}{\partial z}(5e^{-r}\cos\phi) - \frac{\partial}{\partial r}(-5\cos\phi)\right]\mathbf{a}_\phi - \frac{1}{r}\frac{\partial}{\partial\phi}(5e^{-r}\cos\phi)\,\mathbf{a}_z$$

$$= \left(\frac{5}{r}\sin\phi\right)\mathbf{a}_r + \left(\frac{5}{r}e^{-r}\sin\phi\right)\mathbf{a}_z$$

Then

$$\nabla \times \mathbf{A}\Big|_{(2,\,3\pi/2,\,0)} = -2.50\,\mathbf{a}_r - 0.34\,\mathbf{a}_z$$

9.14. Given the general vector field $\mathbf{A} = 10\sin\theta\,\mathbf{a}_\theta$ in spherical coordinates, find $\nabla \times \mathbf{A}$ at $(2, \pi/2, 0)$.

$$\nabla \times \mathbf{A} = \frac{1}{r\sin\theta}\left[-\frac{\partial}{\partial\phi}(10\sin\theta)\right]\mathbf{a}_r + \frac{1}{r}\frac{\partial}{\partial r}(10r\sin\theta)\,\mathbf{a}_\phi = \frac{10\sin\theta}{r}\,\mathbf{a}_\phi$$

Then

$$\nabla \times \mathbf{A}\Big|_{(2,\,\pi/2,\,0)} = 5\mathbf{a}_\phi$$

9.15. In Problem 9.6 the magnetic field intensity produced by a circular conductor with uniform current density was found. Find, conversely, $\mathbf{J}$ from $\mathbf{H}$.

Given $\mathbf{H} = (Ir/2\pi a^2)\mathbf{a}_\phi$ within the conductor,

$$\mathbf{J} = \nabla \times \mathbf{H} = -\frac{\partial}{\partial z}\left(\frac{Ir}{2\pi a^2}\right)\mathbf{a}_r + \frac{1}{r}\frac{\partial}{\partial r}\left(\frac{Ir^2}{2\pi a^2}\right)\mathbf{a}_z = \frac{I}{\pi a^2}\,\mathbf{a}_z$$

which indeed corresponds to a circular cross section πa^2 and a current I in the $+z$ direction.

Outside the conductor, $\mathbf{H} = (I/2\pi r)\,\mathbf{a}_\phi$, and so

$$\mathbf{J} = \nabla \times \mathbf{H} = -\frac{\partial}{\partial z}\left(\frac{I}{2\pi r}\right)\mathbf{a}_r + \frac{1}{r}\frac{\partial}{\partial r}\left(\frac{I}{2\pi}\right) = 0$$

as expected.

9.16. A circular conductor of radius $r_0 = 1$ cm has an internal field

$$\mathbf{H} = \frac{10^4}{r}\left(\frac{1}{a^2}\sin ar - \frac{r}{a}\cos ar\right)\mathbf{a}_\phi \quad \text{(A/m)}$$

where $a = \pi/2r_0$. Find the total current in the conductor.

There are two methods: (1) to calculate $\mathbf{J} = \nabla \times \mathbf{H}$ and then integrate; (2) to use Ampère's law. The second is simpler here.

$$I_{enc} = \oint_{r=r_0} \mathbf{H}\cdot d\boldsymbol{\ell} = \int_0^{2\pi} \frac{10^4}{r_0}\left(\frac{4r_0^2}{\pi^2}\sin\frac{\pi}{2} - \frac{2r_0^2}{\pi}\cos\frac{\pi}{2}\right) r_0\, d\phi$$
$$= \frac{8\times 10^4 r_0^2}{\pi} = \frac{8}{\pi}\text{ A}$$

9.17. A radial field

$$\mathbf{H} = \frac{2.39\times 10^6}{r}\cos\phi\, \mathbf{a}_r \quad \text{A/m}$$

exits in free space. Find the magnetic flux Φ crossing the surface defined by $-\pi/4 \le \phi \le \pi/4$, $0 \le z \le 1$ m. See Fig. 9-16.

$$\mathbf{B} = \mu_0\mathbf{H} = \frac{3.00}{r}\cos\phi\,\mathbf{a}_r \quad \text{(T)}$$

$$\Phi = \int_0^1\int_{-\pi/4}^{\pi/4}\left(\frac{3.00}{r}\cos\phi\right)\mathbf{a}_r\cdot r\,d\phi\,dz\,\mathbf{a}_r$$
$$= 4.24 \text{ Wb}$$

Since $\mathbf{B}$ is inversely proportional to r (as required by $\nabla\cdot\mathbf{B} = 0$), it makes no difference what radial distance is chosen, the total flux will be the same.

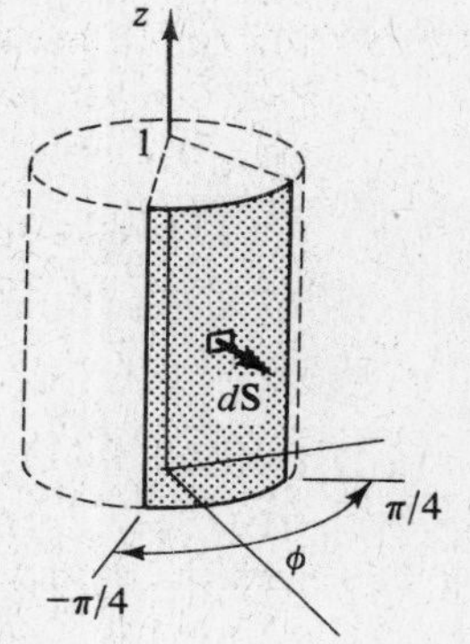

Fig. 9-16

9.18. In cylindrical coordinates, $\mathbf{B} = (2.0/r)\mathbf{a}_\phi$ (T). Determine the magnetic flux Φ crossing the plane surface defined by $0.5 \le r \le 2.5$ m and $0 \le z \le 2.0$ m. See Fig. 9-17.

$$\Phi = \int \mathbf{B}\cdot d\mathbf{S}$$
$$= \int_0^{2.0}\int_{0.5}^{2.5}\frac{2.0}{r}\mathbf{a}_\phi\cdot dr\,dz\,\mathbf{a}_\phi$$
$$= 4.0\left(\ln\frac{2.5}{0.5}\right) = 6.44 \text{ Wb}$$

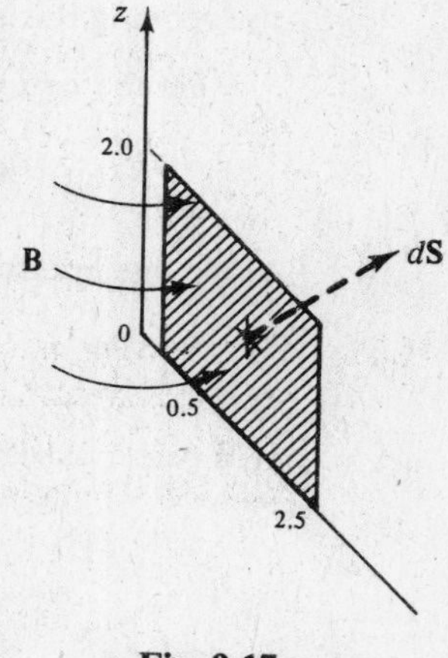

Fig. 9-17

9.19. Obtain the vector magnetic potential $\mathbf{A}$ in the region surrounding an infinitely long, straight, filamentary current I.

As shown in Example 4, the direct expression for $\mathbf{A}$ as an integral cannot be used. However, the relation

$$\nabla\times\mathbf{A} = \mathbf{B} = \frac{\mu_0 I}{2\pi r}\mathbf{a}_\phi$$

may be treated as a vector differential equation for $\mathbf{A}$. Since $\mathbf{B}$ possesses only a ϕ component, only the ϕ component of the cylindrical curl is needed.

$$\frac{\partial A_r}{\partial z} - \frac{\partial A_z}{\partial r} = \frac{\mu_0 I}{2\pi r}$$

It is evident that $\mathbf{A}$ cannot be a function of z, since the filament is uniform with z. Then

$$-\frac{dA_z}{dr} = \frac{\mu_0 I}{2\pi r} \qquad \text{or} \qquad A_z = -\frac{\mu_0 I}{2\pi}\ln r + C$$

The constant of integration permits the location of a zero reference. With $A_z = 0$ at $r = r_0$, the expression becomes

$$\mathbf{A} = \frac{\mu_0 I}{2\pi}\left(\ln\frac{r_0}{r}\right)\mathbf{a}_z$$

9.20. Obtain the vector magnetic potential $\mathbf{A}$ for the current sheet of Problem 9.3.

For $z > 0$,

$$\nabla \times \mathbf{A} = \mathbf{B} = \frac{\mu_0 K}{2}\mathbf{a}_x$$

whence

$$\frac{\partial A_z}{\partial y} - \frac{\partial A_y}{\partial z} = \frac{\mu_0 K}{2}$$

As $\mathbf{A}$ must be independent of x and y,

$$-\frac{dA_y}{dz} = \frac{\mu_0 K}{2} \qquad \text{or} \qquad A_y = -\frac{\mu_0 K}{2}(z - z_0)$$

Thus, for $z > 0$,

$$\mathbf{A} = -\frac{\mu_0 K}{2}(z - z_0)\mathbf{a}_y = -\frac{\mu_0}{2}(z - z_0)\mathbf{K}$$

For $z < 0$, change the sign of the above expression.

9.21. Using the vector magnetic potential found in Problem 9.20, find the magnetic flux crossing the rectangular area shown in Fig. 9-18.

Let the zero reference be at $z_0 = 2$, so that

$$\mathbf{A} = -\frac{\mu_0}{2}(z - 2)\mathbf{K}$$

In the line integral

$$\Phi = \oint \mathbf{A} \cdot d\mathbf{l}$$

$\mathbf{A}$ is perpendicular to the contour on two sides and vanishes on the third $(z = 2)$. Thus,

$$\Phi = \int_{y=0}^{y=2} \mathbf{A} \cdot d\mathbf{l} = -\frac{\mu_0}{2}(1 - 2)\int_0^2 K\,dy = \mu_0 K$$

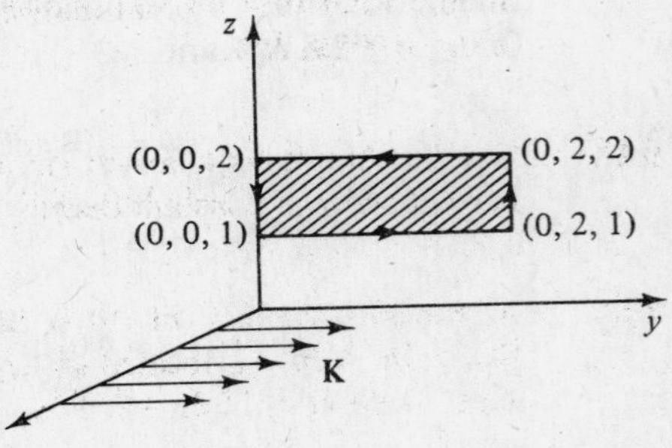

Fig. 9-18

Note how the choice of zero reference simplified the computation. By Stokes' theorem it is $\nabla \times \mathbf{A}$, and not $\mathbf{A}$ itself, that determines Φ; hence the zero reference may be chosen at pleasure.

9.22. Show that the curl of a gradient is zero.

From the definition of curl $\mathbf{A}$ given in Section 9.4, it is seen that curl $\mathbf{A}$ is zero in a region if

$$\oint \mathbf{A} \cdot d\mathbf{l} = 0$$

for every closed path in the region. But if $\mathbf{A} = \nabla f$, where f is a single-valued function,

$$\oint \mathbf{A} \cdot d\mathbf{l} = \oint \nabla f \cdot d\mathbf{l} = \oint df = 0$$

(see Section 5.5).

Supplementary Problems

9.23. Show that the magnetic field due to the finite current element shown in Fig. 9-19 is given by

$$\mathbf{H} = \frac{I}{4\pi r}(\sin\alpha_1 - \sin\alpha_2)\mathbf{a}_\phi$$

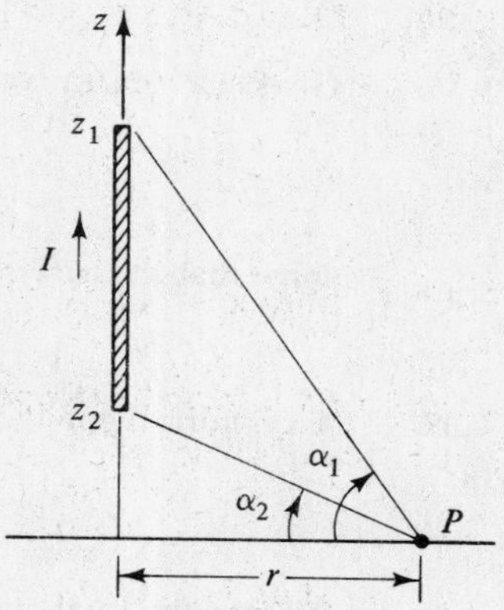

Fig. 9-19

9.24. Obtain $d\mathbf{H}$ at a general point (r, θ, ϕ) in spherical coordinates, due to a differential current element $I\,d\mathbf{l}$ at the origin in the positive z direction.

Ans. $\dfrac{I\,d\ell \sin\theta}{4\pi r^2}\mathbf{a}_\phi$

9.25. Currents in the inner and outer conductors of Fig. 9-20 are uniformly distributed. Use Ampère's law to show that for $b \le r \le c$,

$$\mathbf{H} = \frac{I}{2\pi r}\left(\frac{c^2 - r^2}{c^2 - b^2}\right)\mathbf{a}_\phi$$

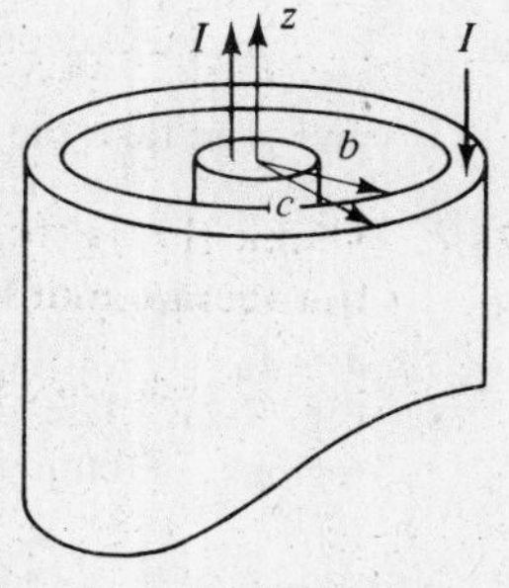

Fig. 9-20

9.26. Two identical circular current loops of radius $r = 3$ m and $I = 20$ A are in parallel planes, separated on their common axis by 10 m. Find $\mathbf{H}$ at a point midway between the two loops.
Ans. $0.908\,\mathbf{a}_n$ A/m

9.27. Use the Biot-Savart law to show that $\mathbf{H} = \frac{1}{2}\mathbf{K} \times \mathbf{a}_n$ for a plane current sheet of constant density $\mathbf{K}$.

9.28. A current filament of 10 A in the $+y$ direction lies along the y axis, and a current sheet, $\mathbf{K} = 2.0\,\mathbf{a}_x$ A/m, is located at $z = 4$ m. Determine $\mathbf{H}$ at the point $(2, 2, 2)$ m.
Ans. $0.398\,\mathbf{a}_x + 1.0\,\mathbf{a}_y - 0.398\,\mathbf{a}_z$ A/m

9.29. Show that the curl of $(x\mathbf{a}_x + y\mathbf{a}_y + z\mathbf{a}_z)/(x^2 + y^2 + z^2)^{3/2}$ is zero. (*Hint*: $\nabla \times \mathbf{E} = 0$.)

9.30. Given the general vector $\mathbf{A} = (-\cos x)(\cos y)\mathbf{a}_z$, find the curl of $\mathbf{A}$ at the origin. *Ans.* 0

9.31. Given the general vector $\mathbf{A} = (\cos x)(\sin y)\mathbf{a}_x + (\sin x)(\cos y)\mathbf{a}_y$, find the curl of $\mathbf{A}$ everywhere.
Ans. 0

9.32. Given the general vector $\mathbf{A} = (\sin 2\phi)\mathbf{a}_\phi$ in cylindrical coordinates, find the curl of $\mathbf{A}$ at $(2, \pi/4, 0)$.
Ans. $0.5\,\mathbf{a}_z$

9.33. Given the general vector $\mathbf{A} = e^{-2z}(\sin \frac{1}{2}\phi)\mathbf{a}_\phi$ in cylindrical coordinates, find the curl of **A** at $(0.800, \pi/3, 0.500)$. *Ans.* $0.368\,\mathbf{a}_r + 0.230\,\mathbf{a}_z$

9.34. Given the general vector $\mathbf{A} = (\sin \phi)\mathbf{a}_r + (\sin \theta)\mathbf{a}_\phi$ in spherical coordinates, find the curl of **A** at the point $(2, \pi/2, 0)$. *Ans.* 0

9.35. Given the general vector $\mathbf{A} = 2.50\,\mathbf{a}_\theta + 5.00\,\mathbf{a}_\phi$ in spherical coordinates, find the curl of **A** at $(2.0, \pi/6, 0)$. *Ans.* $4.33\,\mathbf{a}_r - 2.50\,\mathbf{a}_\theta + 1.25\,\mathbf{a}_\phi$

9.36. Given the general vector

$$\mathbf{A} = \frac{2\cos\theta}{r^3}\,\mathbf{a}_r + \frac{\sin\theta}{r^3}\,\mathbf{a}_\theta$$

show that the curl of **A** is everywhere zero.

9.37. A cylindrical conductor of radius 10^{-2} m has an internal magnetic field

$$\mathbf{H} = (4.77 \times 10^4)\left(\frac{r}{2} - \frac{r^2}{3 \times 10^{-2}}\right)\mathbf{a}_\phi \quad \text{(A/m)}$$

What is the total current in the conductor? *Ans.* 5.0 A

9.38. In cylindrical coordinates, $\mathbf{J} = 10^5(\cos^2 2r)\mathbf{a}_z$ in a certain region. Obtain **H** from this current density and then take the curl of **H** and compare with **J**. *Ans.* $\mathbf{H} = 10^5\left(\frac{r}{4} + \frac{\sin 4r}{8} + \frac{\cos 4r}{32r} - \frac{1}{32r}\right)\mathbf{a}_\phi$

9.39. In cartesian coordinates a constant current density, $\mathbf{J} = J_0\,\mathbf{a}_y$, exists in the region $-a \le z \le a$. See Fig. 9-21. Use Ampère's law to find **H** in all regions. Obtain the curl of **H** and compare with **J**.

$$\textit{Ans.}\quad \mathbf{H} = \begin{cases} J_0\,a\mathbf{a}_x & z > a \\ J_0\,z\mathbf{a}_x & -a \le z \le a \\ -J_0\,a\mathbf{a}_x & z < -a \end{cases}$$

$$\text{curl}\,\mathbf{H} = \mathbf{J}$$

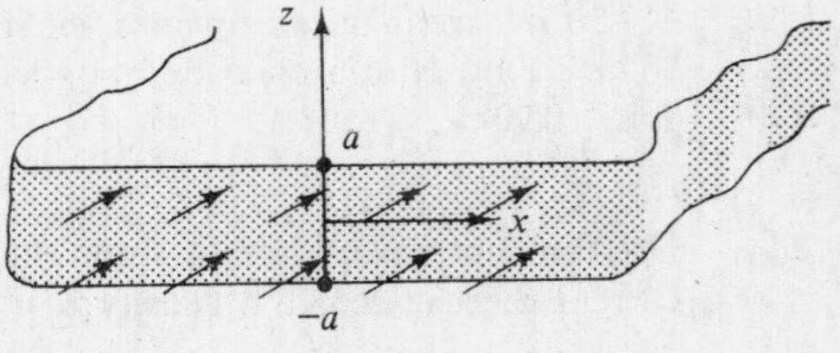

Fig. 9-21

9.40. Compute the total magnetic flux Φ crossing the $z = 0$ plane in cylindrical coordinates for $r \le 5 \times 10^{-2}$ m if

$$\mathbf{B} = \frac{0.2}{r}(\sin^2 \phi)\mathbf{a}_z \quad \text{(T)}$$

Ans. 3.14×10^{-2} Wb

9.41. Given that

$$\mathbf{B} = 2.50\left(\sin\frac{\pi x}{2}\right)e^{-2y}\mathbf{a}_z \quad \text{(T)}$$

find the total magnetic flux crossing the strip $z = 0,\ y \ge 0,\ 0 \le x \le 2$ m. *Ans.* 1.59 Wb

9.42. A coaxial conductor with an inner conductor of radius a and an outer conductor of inner and outer radii b and c, respectively, carries current I in the inner conductor. Find the magnetic flux per unit length crossing a plane $\phi = \text{const.}$ between the conductors. *Ans.* $\dfrac{\mu_0 I}{2\pi} \ln \dfrac{b}{a}$

9.43. One uniform current sheet, $\mathbf{K} = K_0 \mathbf{a}_y$, is at $z = b > 2$ and another, $\mathbf{K} = K_0(-\mathbf{a}_y)$, is at $z = -b$. Find the magnetic flux crossing the area defined by $x = \text{const.}$, $-2 \le z \le 2$, $0 \le y \le L$. Assume free space. *Ans.* $4\mu_0 K_0 L$

9.44. Use the vector magnetic potential from Problem 9.19 to obtain the flux crossing a $\phi = \text{const.}$ plane for $r_1 \le r \le r_0$ and $0 \le z \le L$ due to a current filament I on the z axis.

Ans. $\dfrac{\mu_0 IL}{2\pi} \ln \dfrac{r_0}{r_1}$

9.45. Given that the vector magnetic potential within a cylindrical conductor of radius a is

$$\mathbf{A} = -\frac{\mu_0 I r^2}{4\pi a^2} \mathbf{a}_z$$

find the corresponding **H**. *Ans.* See Problem 9.6.

9.46. One uniform current sheet, $\mathbf{K} = K_0(-\mathbf{a}_y)$, is located at $x = 0$ and another, $\mathbf{K} = K_0 \mathbf{a}_y$, is at $x = a$. Find the vector magnetic potential between the sheets. *Ans.* $(\mu_0 K_0 x + C)\mathbf{a}_y$

9.47. Between the current sheets of Problem 9.46 a portion of a $z = \text{const.}$ plane is defined by $0 \le x \le b$ and $0 \le y \le a$. Find the flux Φ crossing this portion, both from $\int \mathbf{B} \cdot d\mathbf{S}$ and from $\oint \mathbf{A} \cdot d\mathbf{l}$. *Ans.* $ab\mu_0 K_0$

Chapter 10

Forces and Torques in Magnetic Fields

10.1 MAGNETIC FORCE ON PARTICLES

A charged particle *in motion* in a magnetic field experiences a force at right angles to its velocity, with a magnitude proportional to the charge, the velocity and the magnetic flux density. The complete expression is given by the cross product

$$\mathbf{F} = Q\mathbf{U} \times \mathbf{B}$$

Therefore, the direction of a particle in motion can be changed by a magnetic field. The magnitude of the velocity, U, and consequently the kinetic energy, will remain the same. This is in contrast to an electric field, where the force $\mathbf{F} = Q\mathbf{E}$ does work on the particle and therefore changes its kinetic energy.

If the field $\mathbf{B}$ is uniform throughout a region and the particle has an initial velocity normal to the field, the path of the particle is a circle of a certain radius r. The force of the field is of magnitude $F = |Q|UB$ and is directed toward the center of the circle. The centripetal acceleration is of magnitude $\omega^2 r = U^2/r$. Then, by Newton's second law,

$$|Q|UB = m\frac{U^2}{r} \qquad \text{or} \qquad r = \frac{mU}{|Q|B}$$

Observe that r is a measure of the particle's linear momentum, mU.

EXAMPLE 1 Find the force on a particle of mass 1.70×10^{-27} kg and charge 1.60×10^{-19} C if it enters a field $B = 5$ mT with an initial speed of 83.5 km/s.

Unless directions are known for $\mathbf{B}$ and $\mathbf{U}_0$, the particle's initial velocity, the force cannot be calculated. Assuming that $\mathbf{U}_0$ and $\mathbf{B}$ are perpendicular, as shown in Fig. 10-1,

$$\begin{aligned} F &= |Q|UB \\ &= (1.60 \times 10^{-19})(83.5 \times 10^3)(5 \times 10^{-3}) \\ &= 6.68 \times 10^{-17}\ \text{N} \end{aligned}$$

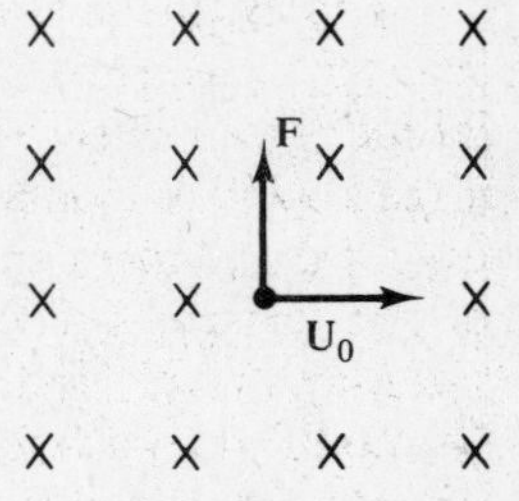

(B into page)

Fig. 10-1

EXAMPLE 2 For the particle of Example 1, find the radius of the circular path and the time required for one revolution.

$$r = \frac{mU}{|Q|B} = \frac{(1.70 \times 10^{-27})(83.5 \times 10^3)}{(1.60 \times 10^{-19})(5 \times 10^{-3})} = 0.177\ \text{m}$$

$$T = \frac{2\pi r}{U} = 13.3\ \mu\text{s}$$

10.2 ELECTRIC AND MAGNETIC FIELDS COMBINED

When both fields are present in a region at the same time, the force on a particle is given by

$$\mathbf{F} = Q(\mathbf{E} + \mathbf{U} \times \mathbf{B})$$

This force, together with the initial conditions, determines the path of the particle.

EXAMPLE 3 A region contains a magnetic flux density $\mathbf{B} = 5.0 \times 10^{-4}\mathbf{a}_z$ T and an electric field $\mathbf{E} = 5.0\,\mathbf{a}_z$ V/m. A proton ($Q_p = 1.602 \times 10^{-19}$ C, $m_p = 1.673 \times 10^{-27}$ kg) enters the fields at the origin with an initial velocity $\mathbf{U}_0 = 2.5 \times 10^5\mathbf{a}_x$ m/s. Describe the proton's motion and give its position after three complete revolutions.

The initial force on the particle is

$$\mathbf{F}_0 = Q(\mathbf{E} + \mathbf{U}_0 \times \mathbf{B}) = Q_p(E\,\mathbf{a}_z - U_0\,B\,\mathbf{a}_y)$$

The z component (electric component) of the force is constant, and produces a constant acceleration in the z direction. Thus the equation of motion in the z direction is

$$z = \tfrac{1}{2}at^2 = \frac{1}{2}\left(\frac{Q_p E}{m_p}\right)t^2$$

The other (magnetic) component, which changes into $-Q_p U\,B\mathbf{a}_r$, produces circular motion perpendicular to the z axis, with period

$$T = \frac{2\pi r}{U} = \frac{2\pi m_p}{Q_p B}$$

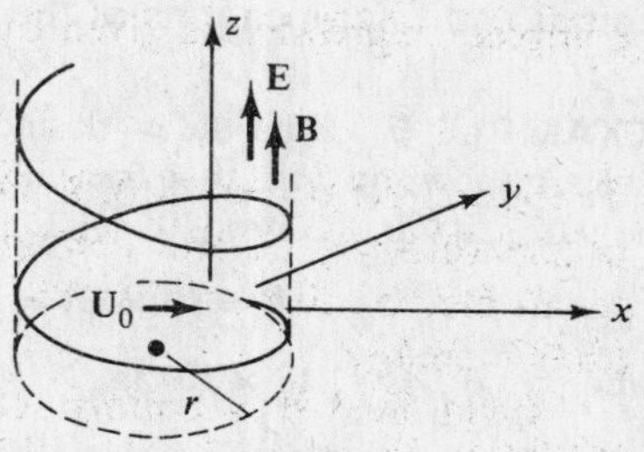

Fig. 10-2

The resultant motion is helical, as shown in Fig. 10-2.

After three revolutions, $x = y = 0$ and

$$z = \frac{1}{2}\left(\frac{Q_p E}{m_p}\right)(3T)^2 = \frac{18\pi^2 E m_p}{Q_p B^2} = 37.0 \text{ m}$$

10.3 MAGNETIC FORCE ON A CURRENT ELEMENT

A frequently encountered situation is that of a current-carrying conductor in an external magnetic field. Since $I = dQ/dt$, the differential force equation may be written

$$d\mathbf{F} = dQ(\mathbf{U} \times \mathbf{B}) = (I\,dt)(\mathbf{U} \times \mathbf{B}) = I(d\mathbf{l} \times \mathbf{B})$$

where $d\mathbf{l} = \mathbf{U}\,dt$ is the elementary length in the direction of the conventional current I. If the conductor is straight and the field is constant along it, the differential force may be integrated to give

$$F = ILB \sin\theta$$

The magnetic force is actually exerted on the electrons that make up the current I. However, since the electrons are confined to the conductor, the force is effectively transferred to the heavy lattice; this transferred force can do work on the conductor as a whole. While this fact provides a reasonable introduction to the behavior of current-carrying conductors in electric machines, certain essential considerations have been omitted. No mention was made, nor will be made in Section 10.4, of the current source and the energy that would be required to maintain a constant current I. Faraday's law of induction (Section 12.3) was not applied. In electric machine theory the result will be modified by these considerations. Conductors in motion in magnetic fields are treated again in Chapter 12; see particularly Problems 12.13 and 12.16.

EXAMPLE 4 Find the force on a straight conductor of length 0.30 m carrying a current of 5.0 A in the $-\mathbf{a}_z$ direction, where the field is $\mathbf{B} = 3.50 \times 10^{-3}(\mathbf{a}_x - \mathbf{a}_y)$ T.

$$\begin{aligned}\mathbf{F} &= I(\mathbf{L} \times \mathbf{B}) \\ &= (5.0)[(0.30)(-\mathbf{a}_z) \times 3.50 \times 10^{-3}(\mathbf{a}_x - \mathbf{a}_y)] \\ &= 7.42 \times 10^{-3}\left(\frac{-\mathbf{a}_x - \mathbf{a}_y}{\sqrt{2}}\right) \text{N}\end{aligned}$$

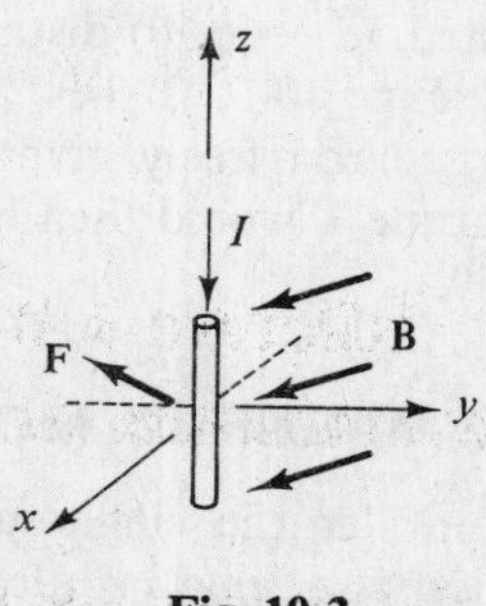

Fig. 10-3

The force, of magnitude 7.42 mN, is at right angles to both the field **B** and the current direction, as shown in Fig. 10-3.

10.4 WORK AND POWER

The magnetic forces on the charged particles and current-carrying conductors examined above result from the field. To counter these forces and establish equilibrium, equal and opposite forces, $\mathbf{F}_a$, would have to be applied. If motion occurs, the work done on the system by the outside agent applying the force is given by the integral

$$W = \int_{\text{initial } \mathbf{l}}^{\text{final } \mathbf{l}} \mathbf{F}_a \cdot d\mathbf{l}$$

A positive result from the integration indicates that work was done by the agent on the system to move the particles or conductor from the initial location to the final, against the field. Because the magnetic force, and hence $\mathbf{F}_a$, is generally nonconservative, the entire path of integration joining the initial and final locations of the conductor must be specified.

EXAMPLE 5 Find the work and power required to move the conductor shown in Fig. 10-4 one full revolution in the direction shown in 0.02 s, if $\mathbf{B} = 2.50 \times 10^{-3}\mathbf{a}_r$ T and the current is 45.0 A.

$$\mathbf{F} = I(\mathbf{l} \times \mathbf{B}) = 1.13 \times 10^{-2}\mathbf{a}_\phi \text{ N}$$

and so $\mathbf{F}_a = -1.13 \times 10^{-2}\mathbf{a}_\phi$ N.

$$W = \int \mathbf{F}_a \cdot d\mathbf{l}$$

$$= \int_0^{2\pi} (-1.13 \times 10^{-2})\mathbf{a}_\phi \cdot r\,d\phi\,\mathbf{a}_\phi$$

$$= -2.13 \times 10^{-3} \text{ J}$$

and $P = W/t = -0.107$ W.

The negative sign means that work is done by the magnetic field in moving the conductor in the direction shown. For motion in the opposite direction, the reversed limits will provide the change of sign, and no attempt to place a sign on $r\,d\phi\,\mathbf{a}_\phi$ should be made.

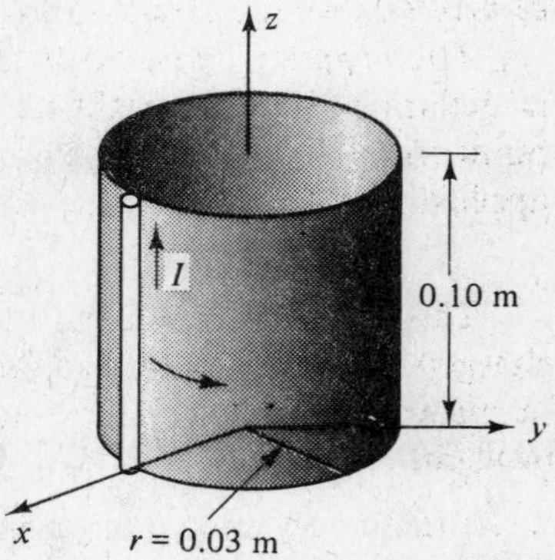

Fig. 10-4

10.5 TORQUE

The *moment of a force* or *torque* about a specified point is the cross product of the *lever arm* about that point and the force. The lever arm, **r**, is directed from the point about which the torque is to be obtained to the point of application of the force. In Fig. 10-5 the force at P has a torque about O given by

$$\mathbf{T} = \mathbf{r} \times \mathbf{F}$$

where **T** has the units N · m. (The units N · m/rad have been suggested, in order to distinguish torque from energy.)

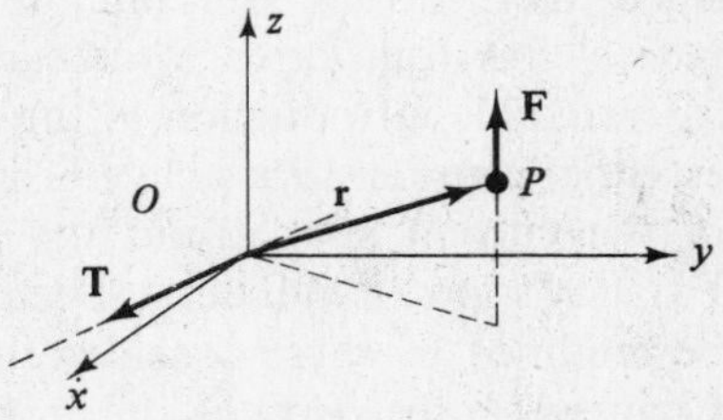

Fig. 10-5

In Fig. 10-5, **T** lies along an axis (in the xy plane) through O. If P were joined to O by a rigid rod freely pivoted at O, then the applied force would tend to rotate P about that axis. The torque **T** would then be said to be *about the axis*, rather than *about point O*.

10.6 MAGNETIC MOMENT OF A PLANAR COIL

Consider the single-turn coil in the $z = 0$ plane shown in Fig. 10-6, of width w in the x direction and length ℓ along y. The field **B** is uniform and in the $+x$ direction. The only forces result from the coil sides ℓ. For the side on the left,

$$\mathbf{F} = I(\ell \mathbf{a}_y \times B\mathbf{a}_x) = -BI\ell \mathbf{a}_z$$

and for the side on the right,

$$\mathbf{F} = BI\ell \mathbf{a}_z$$

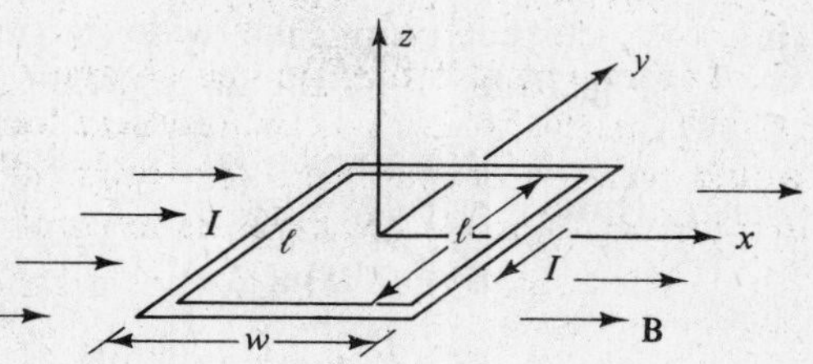

Fig. 10-6

The torque about the y axis from the left current element requires a lever arm $\mathbf{r} = -(w/2)\mathbf{a}_x$; the sign will change for the lever arm to the right current element. The torque from both elements is

$$\mathbf{T} = (-w/2)\mathbf{a}_x \times (-BI\ell)\mathbf{a}_z + (w/2)\mathbf{a}_x \times BI\ell \mathbf{a}_z = BI\ell w(-\mathbf{a}_y) = BI\,A(-\mathbf{a}_y)$$

where A is the area of the coil. It can be shown that this expression for the torque holds for a coil of arbitrary shape (and for any axis parallel to the y axis).

The *magnetic moment* $\mathbf{m}$ of a planar current loop is defined as $IA\mathbf{a}_n$, where the unit normal $\mathbf{a}_n$ is determined by the right-hand rule. (The right thumb gives the direction of $\mathbf{a}_n$ when the fingers point in the direction of the current.) It is seen that the torque on a planar coil is related to the applied field by

$$\mathbf{T} = \mathbf{m} \times \mathbf{B}$$

This concept of magnetic moment is essential to an understanding of the behavior of orbiting charged particles. For example, a positive charge Q moving in a circular orbit at a velocity U, or an angular velocity ω, is equivalent to a current $I = (\omega/2\pi)Q$, and so gives rise to a magnetic moment

$$\mathbf{m} = \frac{\omega}{2\pi} QA\mathbf{a}_n$$

as shown in Fig. 10-7. More important to the present discussion is the fact that in the presence of a magnetic field $\mathbf{B}$ there will be a torque $\mathbf{T} = \mathbf{m} \times \mathbf{B}$ which tends to turn the current loop until $\mathbf{m}$ and $\mathbf{B}$ are in the same direction, in which orientation the torque will be zero.

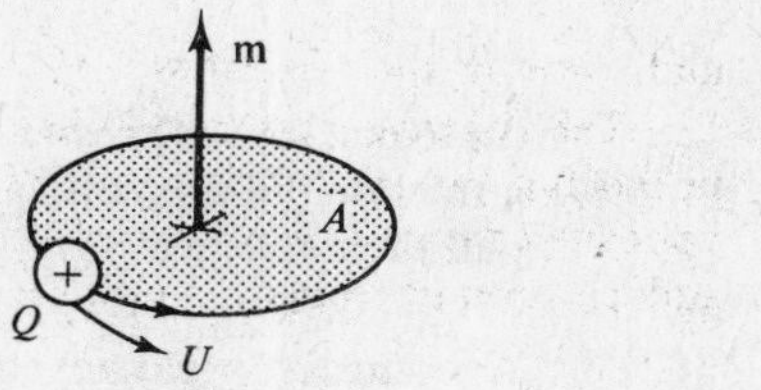

Fig. 10-7

Solved Problems

10.1. A conductor 4 m long lies along the y axis with a current of 10.0 A in the $\mathbf{a}_y$ direction. Find the force on the conductor if the field in the region is $\mathbf{B} = 0.05\mathbf{a}_x$ T.

$$\mathbf{F} = I\mathbf{L} \times \mathbf{B} = 10.0(4\mathbf{a}_y \times 0.05\,\mathbf{a}_x) = -2.0\,\mathbf{a}_z \text{ N}$$

10.2. A conductor of length 2.5 m located at $z = 0$, $x = 4$ m carries a current of 12.0 A in the $-\mathbf{a}_y$ direction. Find the uniform $\mathbf{B}$ in the region if the force on the conductor is 1.20×10^{-2} N in the direction $(-\mathbf{a}_x + \mathbf{a}_z)/\sqrt{2}$.

From $\mathbf{F} = I\mathbf{L} \times \mathbf{B}$,

$$(1.20 \times 10^{-2})\left(\frac{-\mathbf{a}_x + \mathbf{a}_z}{\sqrt{2}}\right) = \begin{vmatrix} \mathbf{a}_x & \mathbf{a}_y & \mathbf{a}_z \\ 0 & -(12.0)(2.5) & 0 \\ B_x & B_y & B_z \end{vmatrix} = -30B_z\mathbf{a}_x + 30B_x\mathbf{a}_z$$

whence

$$B_z = B_x = \frac{4 \times 10^{-4}}{\sqrt{2}} \text{ T}$$

The y component of $\mathbf{B}$ may have any value.

10.3. A current strip 2 cm wide carries a current of 15.0 A in the $\mathbf{a}_x$ direction, as shown in Fig. 10-8. Find the force on the strip per unit length if the uniform field is $\mathbf{B} = 0.20\,\mathbf{a}_y$ T.

In the expression for $d\mathbf{F}$, $I\,d\mathbf{l}$ may be replaced by $\mathbf{K}\,dS$.

$$d\mathbf{F} = (\mathbf{K}dS) \times \mathbf{B}$$

$$= \left(\frac{15.0}{0.02}\right) dx\,dy\,(0.20)\mathbf{a}_z$$

$$\mathbf{F} = \int_{-0.01}^{0.01} \int_0^L 150.0\,dx\,dy\,\mathbf{a}_z$$

$$\mathbf{F}/L = 3.0\mathbf{a}_z \text{ N/m}$$

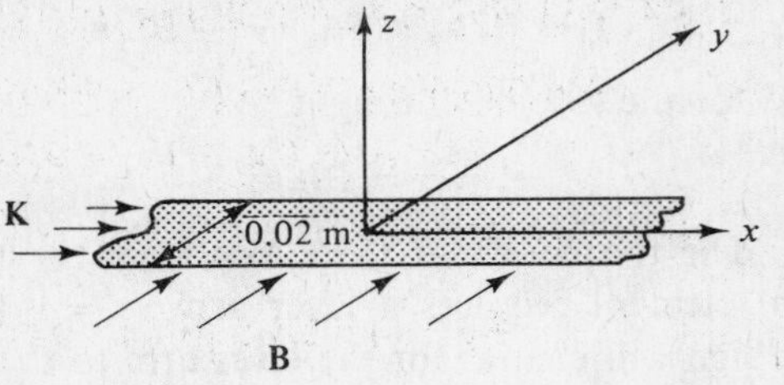

Fig. 10-8

10.4. Find the forces per unit length on two long, straight, parallel conductors if each carries a current of 10.0 A in the same direction and the separation distance is 0.20 m.

Consider the arrangement in cartesian coordinates shown in Fig. 10-9. The conductor on the left creates a field whose magnitude at the right-hand conductor is

$$B = \frac{\mu_0 I}{2\pi r} = \frac{(4\pi \times 10^{-7})(10.0)}{2\pi(0.20)} = 10^{-5}\text{ T}$$

and whose direction is $-\mathbf{a}_z$. Then the force on the right conductor is

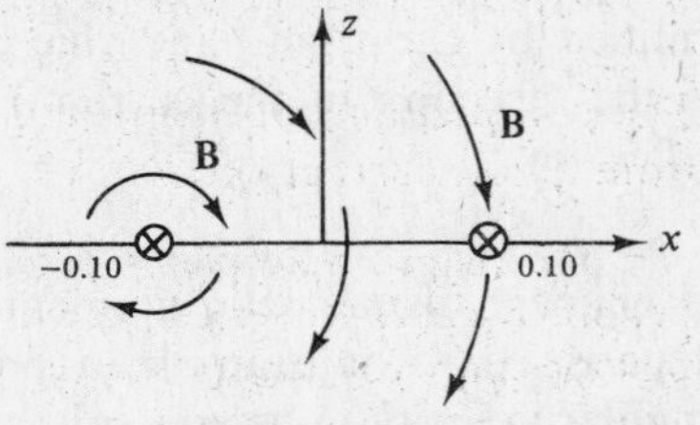

Fig. 10-9

$$\mathbf{F} = IL\mathbf{a}_y \times B(-\mathbf{a}_z) = ILB(-\mathbf{a}_x)$$

and

$$\mathbf{F}/L = 10^{-4}(-\mathbf{a}_x)\text{ N/m}$$

An equal but opposite force acts on the left-hand conductor. The force is seen to be one of attraction. Two parallel conductors carrying current in the same direction will have forces tending to pull them together.

10.5. A conductor carries current I parallel to a current strip of density K_0 and width w, as shown in Fig. 10-10. Find an expression for the force per unit length on the conductor. What is the result when the width w approaches infinity?

From Problem 10.4, the filament $K_0\,dx$ shown in Fig. 10-10 exerts an attractive force

$$d\mathbf{F}/L = IB\mathbf{a}_r = I\frac{\mu_0(K_0\,dx)}{2\pi r}\mathbf{a}_r$$

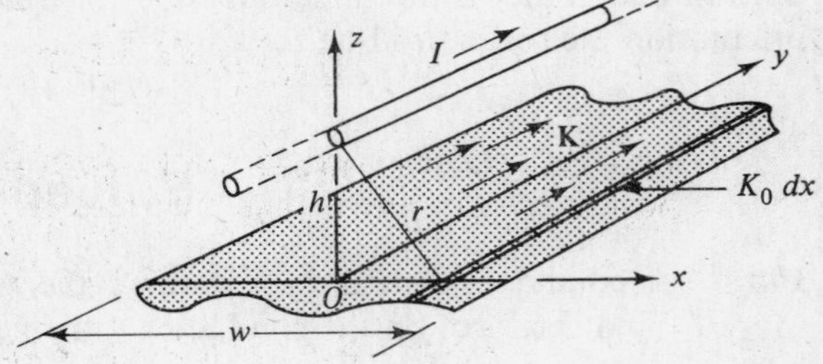

Fig. 10-10

on the conductor. Adding to this the force due to the similar filament at $-x$, the components in the x direction cancel, giving a resultant

$$d\mathbf{F}/L = I\frac{\mu_0(K_0\,dx)}{2\pi r}\left(2\frac{h}{r}\right)(-\mathbf{a}_z) = \frac{\mu_0 IK_0 h}{\pi}\,\frac{dx}{h^2 + x^2}(-\mathbf{a}_z)$$

Integrating over the half-width of the strip,

$$\mathbf{F}/L = \frac{\mu_0 IK_0 h}{\pi}(-\mathbf{a}_z)\int_0^{w/2}\frac{dx}{h^2 + x^2} = \left(\frac{\mu_0 IK_0}{\pi}\arctan\frac{w}{2h}\right)(-\mathbf{a}_z)$$

The force is one of attraction, as expected.

As the strip width approaches infinity, $\mathbf{F}/L \to (\mu_0 IK_0/2)(-\mathbf{a}_z)$.

10.6. Find the torque about the y axis for the two conductors of length ℓ, separated by a fixed distance w, in the uniform field **B** shown in Fig. 10-11.

The conductor on the left experiences the force

$$\mathbf{F}_1 = I\ell\mathbf{a}_y \times B\mathbf{a}_x = BI\ell(-\mathbf{a}_z)$$

the torque of which is

$$\mathbf{T}_1 = \frac{w}{2}(-\mathbf{a}_x) \times BI\ell(-\mathbf{a}_z) = BI\ell\frac{w}{2}(-\mathbf{a}_y)$$

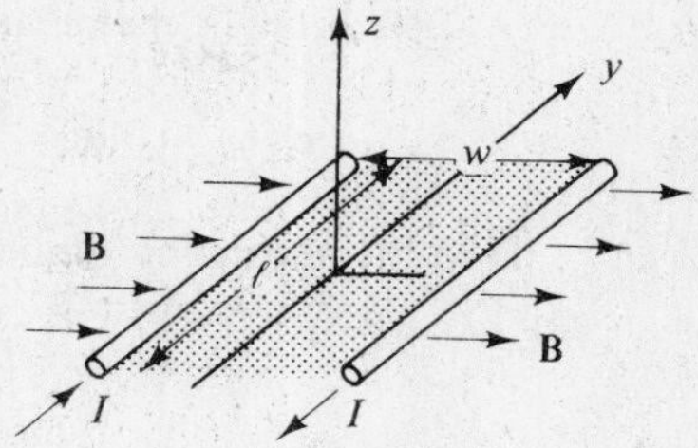

Fig. 10-11

The force on the conductor on the right results in the same torque. The sum is therefore

$$\mathbf{T} = BI\ell w(-\mathbf{a}_y)$$

10.7. A D'Arsonval meter movement has a uniform radial field of $B = 0.10\,\text{T}$ and a restoring spring with a torque $T = 5.87 \times 10^{-5}\theta$ (N · m), where the angle of rotation is in radians. The coil contains 35 turns and measures 23 mm by 17 mm. What angle of rotation results from a coil current of 15 mA?

The shaped pole pieces shown in Fig. 10-12 result in a uniform radial field over a limited range of deflection. Assuming that the entire coil length is in the field, the torque produced is

$$T = nBI\ell w = 35(0.10)(15 \times 10^{-3})(23 \times 10^{-3})(17 \times 10^{-3})$$
$$= 2.05 \times 10^{-5}\ \text{N} \cdot \text{m}$$

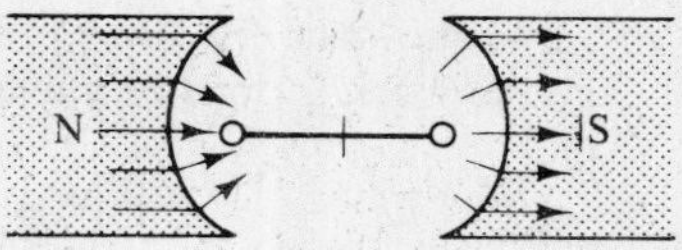

Fig. 10-12

This coil turns until this torque equals the spring torque.

$$2.05 \times 10^{-5} = 5.87 \times 10^{-5}\theta$$
$$\theta = 0.349\ \text{rad} \quad \text{or} \quad 20^\circ$$

10.8. The rectangular coil in Fig. 10-13 is in a field

$$\mathbf{B} = 0.05\,\frac{\mathbf{a}_x + \mathbf{a}_y}{\sqrt{2}}\ \text{T}$$

Find the torque about the z axis when the coil is in the position shown and carries a current of 5.0 A.

$$\mathbf{m} = IA\mathbf{a}_n = 1.60 \times 10^{-2}\mathbf{a}_x$$

$$\mathbf{T} = \mathbf{m} \times \mathbf{B} = 1.60 \times 10^{-2}\mathbf{a}_x \times 0.05\,\frac{\mathbf{a}_x + \mathbf{a}_y}{\sqrt{2}}$$

$$= 5.66 \times 10^{-4}\mathbf{a}_z\ \text{N} \cdot \text{m}$$

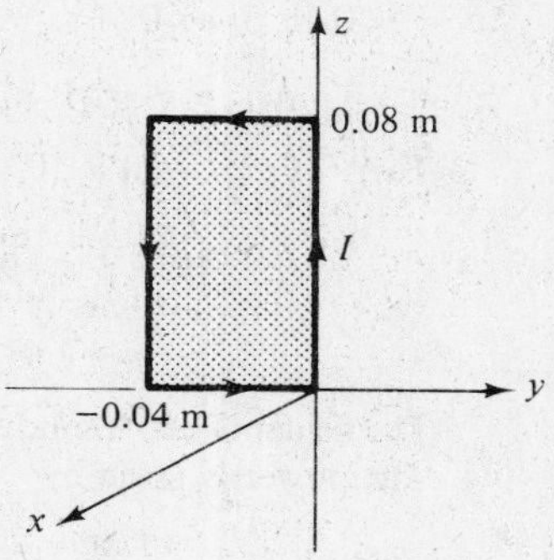

Fig. 10-13

If the coil turns through 45°, the direction of **m** will be $(\mathbf{a}_x + \mathbf{a}_y)/\sqrt{2}$ and the torque will be zero.

10.9. Find the maximum torque on an 85-turn, rectangular coil, 0.2 m by 0.3 m, carrying a current of 2.0 A in a field $B = 6.5$ T.

$$T_{\text{max}} = nBI\ell w = 85(6.5)(2.0)(0.2)(0.3) = 66.3\ \text{N} \cdot \text{m}$$

10.10. Find the maximum torque on an orbiting charged particle if the charge is 1.602×10^{-19} C, the circular path has a radius of 0.5×10^{-10} m, the angular velocity is 4.0×10^{16} rad/s, and $B = 0.4 \times 10^{-3}$ T.

The orbiting charge has a magnetic moment

$$\mathbf{m} = \frac{\omega}{2\pi} QA\mathbf{a}_n = \frac{4 \times 10^{16}}{2\pi}(1.602 \times 10^{-19})\pi(0.5 \times 10^{-10})^2\mathbf{a}_n = 8.01 \times 10^{-24}\mathbf{a}_n \text{ A} \cdot \text{m}^2$$

Then the maximum torque results when $\mathbf{a}_n$ is normal to $\mathbf{B}$.

$$T_{\max} = mB = 3.20 \times 10^{-27} \text{ N} \cdot \text{m}$$

10.11. A conductor of length 4 m, with current held at 10 A in the $\mathbf{a}_y$ direction, lies along the y axis between $y = \pm 2$ m. If the field is $\mathbf{B} = 0.05\mathbf{a}_x$ T, find the work done in moving the conductor parallel to itself at constant speed to $x = z = 2$ m.

For the entire motion,

$$\mathbf{F} = I\mathbf{L} \times \mathbf{B} = -2.0\,\mathbf{a}_z$$

The applied force is equal and opposite,

$$\mathbf{F}_a = 2.0\mathbf{a}_z$$

Because this force is constant, and therefore conservative, the conductor may be moved first along z, then in the x direction, as shown in Fig. 10-14. Since $\mathbf{F}_a$ is completely in the z direction, no work is done in moving along x. Then,

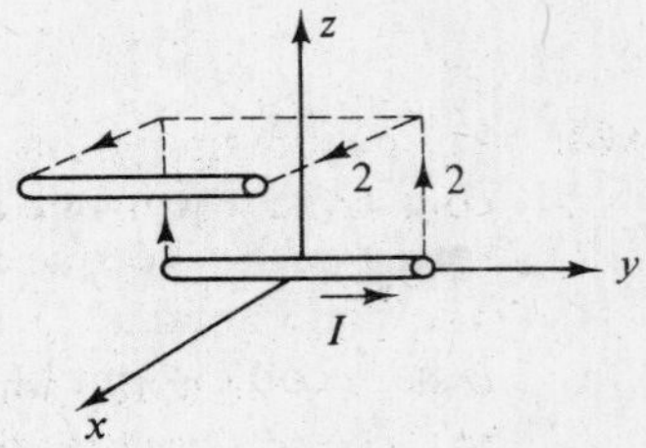

Fig. 10-14

$$W = \int_0^2 (2.0\mathbf{a}_z) \cdot dz\,\mathbf{a}_z = 4.0 \text{ J}$$

10.12. A conductor lies along the z axis at $-1.5 \leq z \leq 1.5$ m and carries a fixed current of 10.0 A in the $-\mathbf{a}_z$ direction. For a field

$$\mathbf{B} = 3.0 \times 10^{-4} e^{-0.2x}\mathbf{a}_y \quad \text{(T)}$$

find the work and power required to move the conductor at constant speed to $x = 2.0$ m, $y = 0$ in 5×10^{-3} s. Assume parallel motion along the x axis.

$$\mathbf{F} = I\mathbf{L} \times \mathbf{B} = 9.0 \times 10^{-3} e^{-0.2x}\mathbf{a}_x$$

Then $\mathbf{F}_a = -9.0 \times 10^{-3} e^{-0.2x}\mathbf{a}_x$ and

$$W = \int_0^2 (-9.0 \times 10^{-3} e^{-0.2x}\mathbf{a}_x) \cdot dx\,\mathbf{a}_x$$
$$= -1.48 \times 10^{-2} \text{ J}$$

The field moves the conductor, and therefore the work is negative. The power is given by

$$P = \frac{W}{t} = \frac{-1.48 \times 10^{-2}}{5 \times 10^{-3}} = -2.97 \text{ W}$$

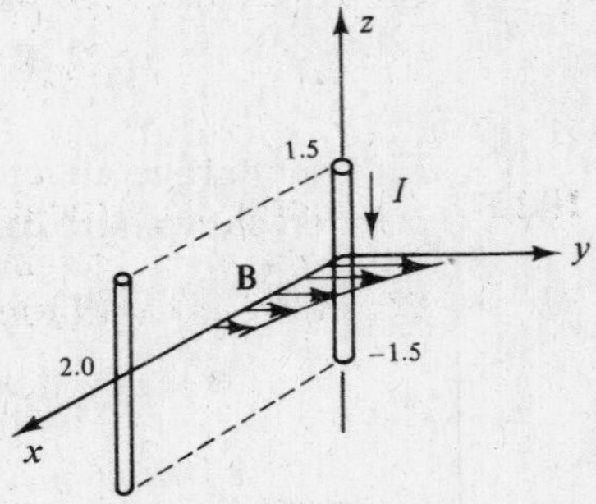

Fig. 10-15

10.13. Find the work and power required to move the conductor shown in Fig. 10-16 one full turn in the positive direction at a rotational frequency of N revolutions per minute, if $\mathbf{B} = B_0\,\mathbf{a}_r$ (B_0 a positive constant).

The force on the conductor is

$$\mathbf{F} = I\mathbf{L} \times \mathbf{B} = IL\mathbf{a}_z \times B_0\,\mathbf{a}_r = B_0\,IL\mathbf{a}_\phi$$

so that the applied force is

$$\mathbf{F}_a = B_0\,IL(-\mathbf{a}_\phi)$$

The conductor is to be turned in the $\mathbf{a}_\phi$ direction. Therefore, the work required for one full revolution is

$$W = \int_0^{2\pi} B_0 IL(-\mathbf{a}_\phi) \cdot r\,d\phi\,\mathbf{a}_\phi = -2\pi r B_0 IL$$

Since N revolutions per minute is $N/60$ per second, the power is

$$P = -\frac{2\pi r B_0 ILN}{60}$$

The negative signs on work and power indicate that the field does the work. The fact that work is done around a closed path shows that the force is nonconservative in this case.

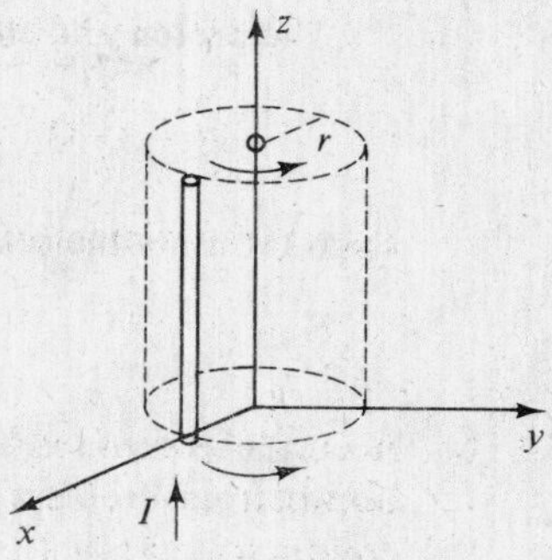

Fig. 10-16

10.14. In the configuration shown in Fig. 10-16 the conductor is 100 mm long and carries a constant 5.0 A in the $\mathbf{a}_z$ direction. If the field is

$$\mathbf{B} = -3.5 \sin\phi\,\mathbf{a}_r \quad \text{mT}$$

and $r = 25$ mm, find the work done in moving the conductor at constant speed from $\phi = 0$ to $\phi = \pi$, in the direction shown. If the current direction is reversed for $\pi < \phi < 2\pi$, what is the total work required for one full revolution?

$$\mathbf{F} = I\mathbf{L} \times \mathbf{B} = -1.75 \times 10^{-3} \sin\phi\,\mathbf{a}_\phi \quad \text{N}$$
$$\mathbf{F}_a = 1.75 \times 10^{-3} \sin\phi\,\mathbf{a}_\phi \quad \text{N}$$

Then
$$W = \int_0^{\pi} 1.75 \times 10^{-3} \sin\phi\,\mathbf{a}_\phi \cdot r\,d\phi\,\mathbf{a}_\phi = 87.5\ \mu\text{J}$$

If the current direction changes when the conductor is between π and 2π, the work will be the same. The total work is 175 μJ.

10.15. Compute the centripetal force necessary to hold an electron ($m_e = 9.107 \times 10^{-31}$ kg) in a circular orbit of radius 0.35×10^{-10} m with an angular velocity of 2×10^{16} rad/s.

$$F = m_e\omega^2 r = (9.107 \times 10^{-31})(2 \times 10^{16})^2(0.35 \times 10^{-10}) = 1.27 \times 10^{-8}\ \text{N}$$

10.16. A uniform magnetic field $\mathbf{B} = 85.3\,\mathbf{a}_z\ \mu$T exists in the region $x \geq 0$. If an electron enters this field at the origin with a velocity $\mathbf{U}_0 = 450\mathbf{a}_x$ km/s, find the position where it exits. Where would a proton with the same initial velocity exit?

$$r_e = \frac{m_e U_0}{|Q|B} = 3.00 \times 10^{-2}\ \text{m}$$

The electron experiences an initial force in the $\mathbf{a}_y$ direction and it exits the field at $x = z = 0$, $y = 6$ cm.

A proton would turn the other way, and part of the circular path is shown at P in Fig. 10-17. With $m_p = 1840 m_e$,

$$r_p = \frac{m_p}{m_e} r_e = 55\ \text{m}$$

and the proton exits at $x = z = 0$, $y = -110$ m.

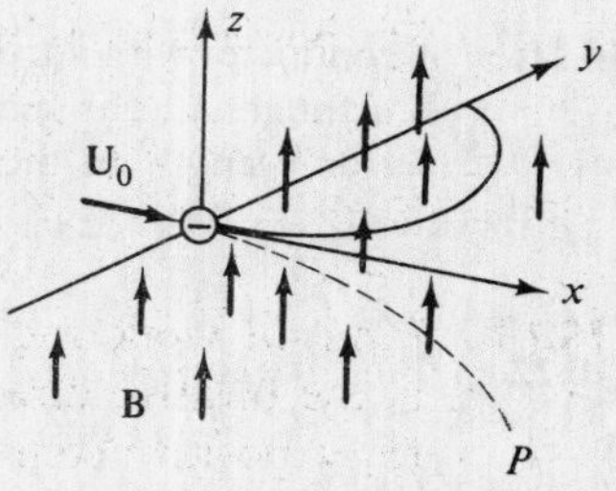

Fig. 10-17

10.17. If a proton is fixed in position and an electron revolves about it in a circular path of radius 0.35×10^{-10} m, what is the magnetic field at the proton?

The proton and electron are attracted by the coulomb force,

$$F = \frac{Q^2}{4\pi\epsilon_0 r^2}$$

which furnishes the centripetal force for the circular motion. Thus

$$\frac{Q^2}{4\pi\epsilon_0 r^2} = m_e \omega^2 r \qquad \text{or} \qquad \omega^2 = \frac{Q^2}{4\pi\epsilon_0 m_e r^3}$$

Now, the electron is equivalent to a current loop $I = (\omega/2\pi)Q$. The field at the center of such a loop is, from Problem 9.8,

$$B = \mu_0 H = \frac{\mu_0 I}{2r} = \frac{\mu_0 \omega Q}{4\pi r}$$

Substituting the value of ω found above,

$$B = \frac{(\mu_0/4\pi)Q^2}{r^2\sqrt{4\pi\epsilon_0 m_e r}} = \frac{(10^{-7})(1.6 \times 10^{-19})^2}{(0.35 \times 10^{-10})^2\sqrt{(\frac{1}{9} \times 10^{-9})(9.1 \times 10^{-31})(0.35 \times 10^{-10})}} = 35 \text{ T}$$

Supplementary Problems

10.18. A current element 2 m in length lies along the y axis centered at the origin. The current is 5.0 A in the $\mathbf{a}_y$ direction. If it experiences a force $1.50(\mathbf{a}_x + \mathbf{a}_z)/\sqrt{2}$ N due to a uniform field $\mathbf{B}$, determine $\mathbf{B}$.
Ans. $0.106(-\mathbf{a}_x + \mathbf{a}_z)$ T

10.19. A magnetic field, $\mathbf{B} = 3.5 \times 10^{-2}\mathbf{a}_z$ T, exerts a force on a 0.30 m conductor along the x axis. If the conductor current is 5.0 A in the $-\mathbf{a}_x$ direction, what force must be applied to hold the conductor in position? *Ans.* $-5.25 \times 10^{-2}\mathbf{a}_y$ N

10.20. A current sheet, $\mathbf{K} = 30.0\,\mathbf{a}_y$ A/m, lies in the plane $z = -5$ m and a filamentary conductor is on the y axis with a current of 5.0 A in the $\mathbf{a}_y$ direction. Find the force per unit length.
Ans. 94.2 μN/m (attraction)

10.21. A conductor with current I pierces a plane current sheet $\mathbf{K}$ orthogonally, as shown in Fig. 10-18. Find the force per unit length on the conductor above and below the sheet. *Ans.* $\pm\mu_0 KI/2$

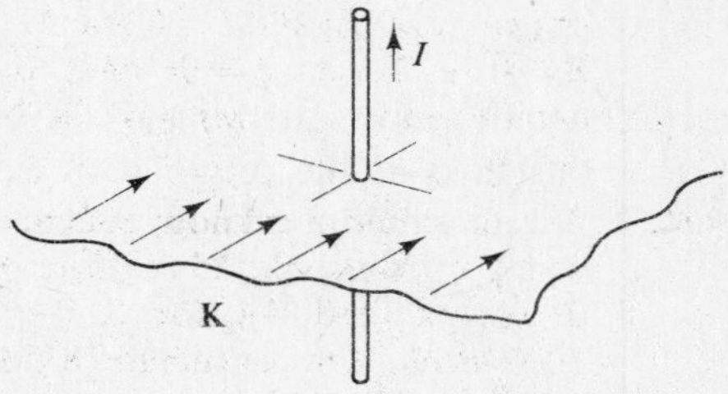

Fig. 10-18

10.22. Find the force on a 2 m conductor on the z axis with a current of 5.0 A in the $\mathbf{a}_z$ direction, if

$$\mathbf{B} = 2.0\mathbf{a}_x + 6.0\mathbf{a}_y \text{ T}$$

Ans. $-60\mathbf{a}_x + 20\mathbf{a}_y$ N

10.23. Two infinite current sheets, each of constant density K_0, are parallel and have their currents oppositely directed. Find the force per unit area on the sheets. Is the force one of repulsion or attraction?
Ans. $\mu_0 K_0^2/2$ (repulsion)

10.24. The circular current loop shown in Fig. 10-19 is in the plane $z = h$, parallel to a uniform current sheet, $\mathbf{K} = K_0\,\mathbf{a}_y$, at $z = 0$. Express the force on a differential length of the loop. Integrate and show that the total force is zero. *Ans.* $d\mathbf{F} = \frac{1}{2}Ia\mu_0 K_0 \cos\phi\, d\phi\,(-\mathbf{a}_z)$

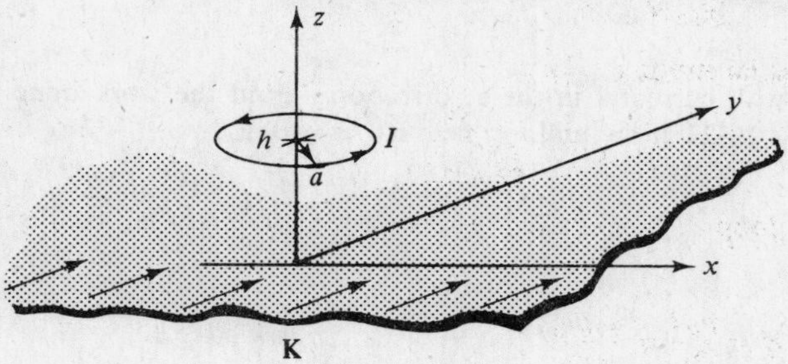

Fig. 10-19

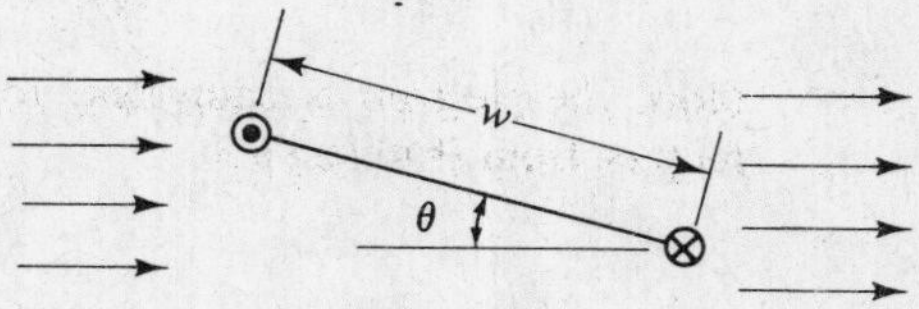

Fig. 10-20

10.25. Two conductors of length ℓ normal to $\mathbf{B}$ are shown in Fig. 10-20; they have a fixed separation w. Show that the torque about any axis parallel to the conductors is given by $BI\ell w \cos\theta$.

10.26. A circular current loop of radius r and current I lies in the $z = 0$ plane. Find the torque which results if the current is in the $\mathbf{a}_\phi$ direction and there is a uniform field $\mathbf{B} = B_0(\mathbf{a}_x + \mathbf{a}_z)/\sqrt{2}$.
Ans. $(\pi r^2 B_0 I/\sqrt{2})\mathbf{a}_y$

10.27. A current loop of radius $r = 0.35$ m is centered about the x axis in the plane $x = 0$ and at $(0, 0, 0.35)$ m the current is in the $-\mathbf{a}_y$ direction at a magnitude of 5.0 A. Find the torque if the uniform field is $\mathbf{B} = 88.4(\mathbf{a}_x + \mathbf{a}_z)$ μT. *Ans.* $1.70 \times 10^{-4}(-\mathbf{a}_y)$ N · m

10.28. A current of 2.5 A is directed generally in the $\mathbf{a}_\phi$ direction about a square conducting loop centered at the origin in the $z = 0$ plane with 0.60 m sides parallel to the x and y axes. Find the forces and the torque on the loop if $\mathbf{B} = 15\mathbf{a}_y$ mT. Would the torque be different if the loop were rotated through 45° in the $z = 0$ plane? *Ans.* $1.35 \times 10^{-2}(-\mathbf{a}_x)$ N · m; $\mathbf{T} = \mathbf{m} \times \mathbf{B}$

10.29. A 200-turn, rectangular coil, 0.30 m by 0.15 m with a current of 5.0 A, is in a uniform field $B = 0.2$ T. Find the magnetic moment m and the maximum torque. *Ans.* 45.0 A · m², 9.0 N · m

10.30. Two conductors of length 4.0 m are on a cylindrical shell of radius 2.0 m centered on the z axis, as shown in Fig. 10-21. Currents of 10.0 A are directed as shown and there is an external field $\mathbf{B} = 0.5\mathbf{a}_x$ T at $\phi = 0$ and $\mathbf{B} = -0.5\mathbf{a}_x$ T at $\phi = \pi$. Find the sum of the forces and the torque about the axis. *Ans.* $-40\mathbf{a}_y$ N, 0

10.31. A right circular cylinder contains 550 conductors on the curved surface and each has a current of constant magnitude 7.5 A. The magnetic field is $\mathbf{B} = 38 \sin\phi\, \mathbf{a}_r$ mT. The current direction is $\mathbf{a}_z$ for $0 < \phi < \pi$ and $-\mathbf{a}_z$ for $\pi < \phi < 2\pi$. Find the mechanical power required if the cylinder turns at 1600 revolutions per minute in the $-\mathbf{a}_\phi$ direction. *Ans.* 60.2 W

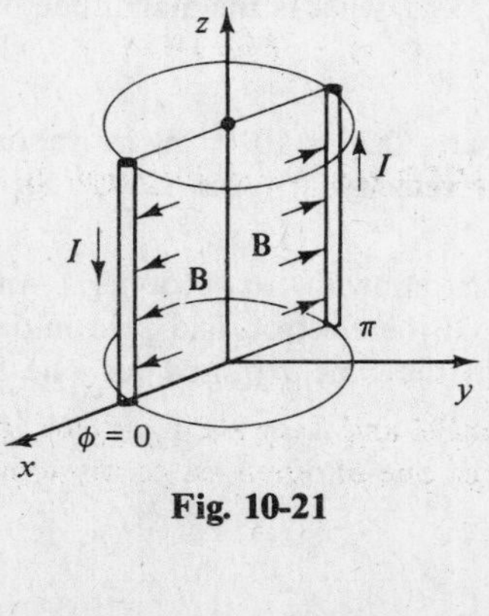

Fig. 10-21

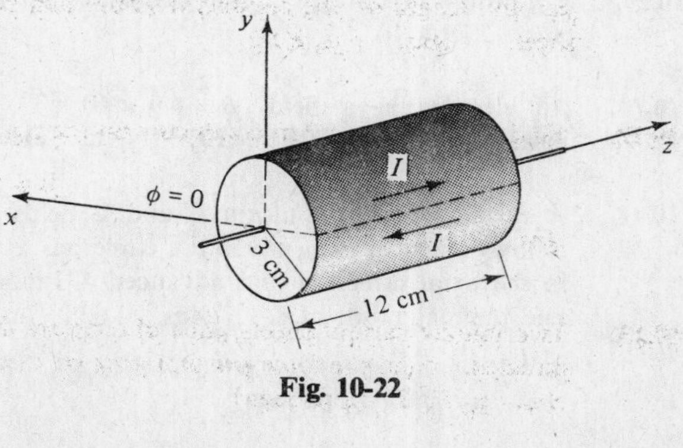

Fig. 10-22

10.32. Obtain an expression for the power required to turn a cylindrical set of n conductors (see Fig. 10-22) against the field at N revolutions per minute, if $\mathbf{B} = B_0 \sin 2\phi\, \mathbf{a}_r$ and the currents change direction in each quadrant where the sign of $\mathbf{B}$ changes. *Ans.* $\dfrac{B_0 n I \ell r N}{60}$ (W)

10.33. A conductor of length ℓ lies along the x axis with current I in the $\mathbf{a}_x$ direction. Find the work done in turning it at constant speed, as shown in Fig. 10-23, if the uniform field is $\mathbf{B} = B_0\, \mathbf{a}_z$.
Ans. $\pi B_0 \ell^2 I/4$

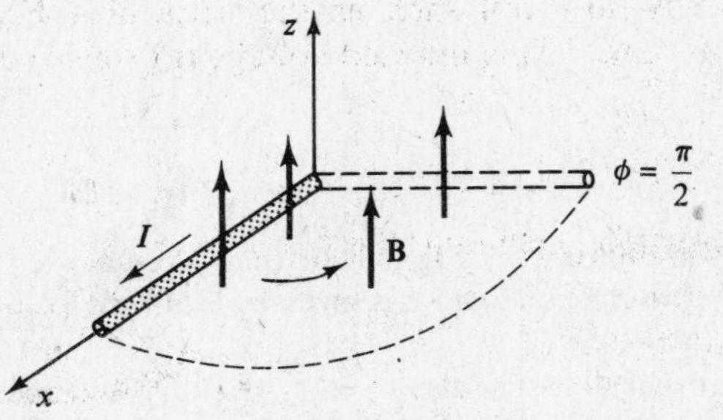

Fig. 10-23

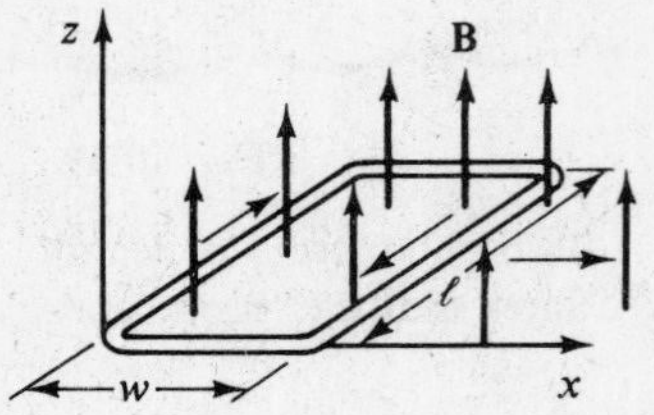

Fig. 10-24

10.34. A rectangular current loop, of length ℓ along the y axis, is in a uniform field $\mathbf{B} = B_0\, \mathbf{a}_z$, as shown in Fig. 10-24. Show that the work done in moving the loop along the x axis at constant speed is zero.

10.35. For the configuration shown in Fig. 10-24, the magnetic field is

$$\mathbf{B} = B_0 \left(\sin \frac{\pi x}{w} \right) \mathbf{a}_z$$

Find the work done in moving the coil a distance w along the x axis at constant speed, starting from the location shown. *Ans.* $-4B_0 I \ell w/\pi$

10.36. A conductor of length 0.25 m lies along the y axis and carries a current of 25.0 A in the $\mathbf{a}_y$ direction. Find the power needed for parallel translation of the conductor to $x = 5.0$ m at constant speed in 3.0 s if the uniform field is $\mathbf{B} = 0.06\mathbf{a}_z$ T. *Ans.* -0.625 W

10.37. Find the tangential velocity of a proton in a field $B = 30\ \mu$T if the circular path has a diameter of 1 cm.
Ans. 14.4 m/s

10.38. An alpha particle and a proton ($Q_\alpha = 2Q_p$) enter a magnetic field $B = 1\ \mu$T with an initial speed $U_0 = 8.5$ m/s. Given the masses 6.68×10^{-27} kg and 1.673×10^{-27} kg for the alpha particle and the proton, respectively, find the radii of the circular paths. *Ans.* 177 mm, 88.8 mm

10.39. If a proton in a magnetic field completes one circular orbit in 2.35 μs, what is the magnitude of $\mathbf{B}$?
Ans. 2.79×10^{-2} T

10.40. An electron in a field $B = 4.0 \times 10^{-2}$ T has a circular path 0.35×10^{-10} m in radius and a maximum torque of 7.85×10^{-26} N · m. Determine the angular velocity. *Ans.* 2.0×10^{16} rad/s

10.41. A region contains uniform $\mathbf{B}$ and $\mathbf{E}$ fields in the same direction, with $B = 650\ \mu$T. An electron follows a helical path, where the circle has a radius of 35 mm. If the electron had zero initial velocity in the axial direction and advanced 431 mm along the axis in the time required for one full circle, find the magnitude of $\mathbf{E}$. *Ans.* 1.62 kV/m

Chapter 11

Inductance and Magnetic Circuits

11.1 VOLTAGE OF SELF-INDUCTION

A voltage appears at the terminals of an N-turn coil such as that shown in Fig. 11-1, where the flux ϕ common to the turns is changing with time. This induced voltage is given by Faraday's law:

$$v = -N\frac{d\phi}{dt}$$

See Chapter 12 for a discussion of the polarity. Defining the *self-inductance* of the coil by

$$L = N\frac{d\phi}{di}$$

Faraday's law may be rewritten in the form

$$v = -L\frac{di}{dt}$$

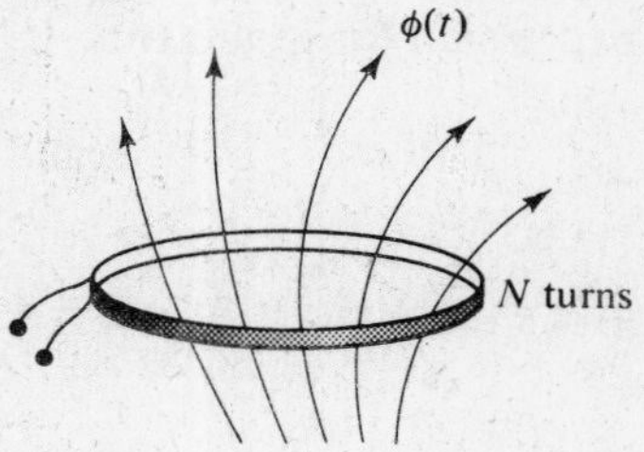

Fig. 11-1

The units on L are *henries*, where $1\ \text{H} = 1\ \text{Wb/A}$.

In any calculations of self-inductance, account must be taken of the presence of ferromagnetic materials. The flux, and hence the self-inductance, when free space is assumed will differ greatly from that which results when a portion of the region contains ferromagnetic materials.

11.2 INDUCTORS AND INDUCTANCE

An *inductor* (also called an *inductance*) is formed by two conductors separated by free space, where the conductor arrangement is such that magnetic flux from one links with the other. For static (or, at most, low-frequency) current I in the conductors, let the total flux linking the conductors be

$$\lambda = \begin{cases} N\Phi & \text{for coils} \\ \Phi & \text{for other arrangements} \end{cases}$$

Then the *inductance* of the inductor is defined to be

$$L = \frac{\lambda}{I}$$

It should be noted that L will always be the product of μ_0 and a geometrical factor having the dimensions of length. Compare with expressions for resistance R (Chapter 6) and capacitance C (Chapter 7).

EXAMPLE 1 Find the inductance per unit length of a coaxial conductor such as that shown in Fig. 11-2.

Between the conductors,

$$\mathbf{H} = \frac{I}{2\pi r}\mathbf{a}_\phi$$

$$\mathbf{B} = \frac{\mu_0 I}{2\pi r}\mathbf{a}_\phi$$

The currents in the two conductors are linked by the flux across the surface $\phi = \text{const.}$ For a length ℓ,

$$\lambda = \int_0^\ell \int_a^b \frac{\mu_0 I}{2\pi r} dr\, dz = \frac{\mu_0 I \ell}{2\pi} \ln\frac{b}{a}$$

and

$$\frac{L}{\ell} = \frac{\mu_0}{2\pi} \ln\frac{b}{a} \quad \text{(H/m)}$$

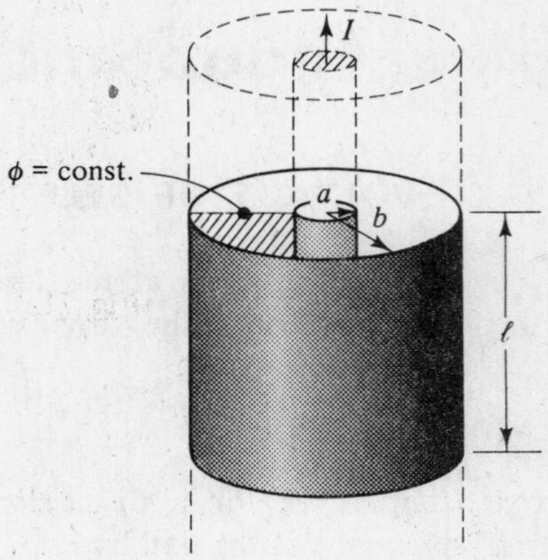

Fig. 11-2

The expressions for the stored energy in the magnetic field from circuit analysis and field theory provide another defining expression for inductance L.

$$W = \frac{1}{2} L I^2 \qquad \text{and} \qquad W = \frac{1}{2}\int_v (\mathbf{B}\cdot\mathbf{H})\, dv$$

give

$$L = \int_v \frac{(\mathbf{B}\cdot\mathbf{H})\, dv}{I^2}$$

EXAMPLE 2 The coaxial conductor of Example 1 can be treated from either approach. Using the **B** and **H** fields from Example 1,

$$L = \frac{\mu_0}{I^2} \int_0^\ell \int_0^{2\pi} \int_a^b \left(\frac{I^2}{4\pi^2 r^2}\right) r\, dr\, d\phi\, dz = \frac{\mu_0 \ell}{2\pi} \ln\frac{b}{a}$$

$$\frac{L}{\ell} = \frac{\mu_0}{2\pi} \ln\frac{b}{a} \quad \text{(H/m)}$$

11.3 STANDARD FORMS

In addition to coaxial conductors, several of the more common conductor configurations for which the inductance is required are shown in Figs. 11-3 through 11-9. The defining equations given above may be applied to develop the expressions for L, except in the case of the air-core coils, where they must be derived empirically. All dimensions are in meters.

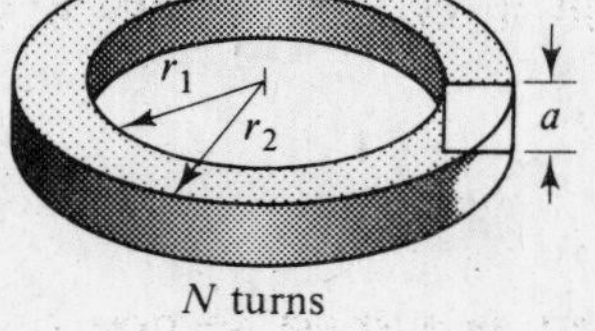

$$L = \frac{\mu_0 N^2 a}{2\pi} \ln\frac{r_2}{r_1} \quad \text{(H)}$$

Fig. 11-3. **Toroid, square cross section**

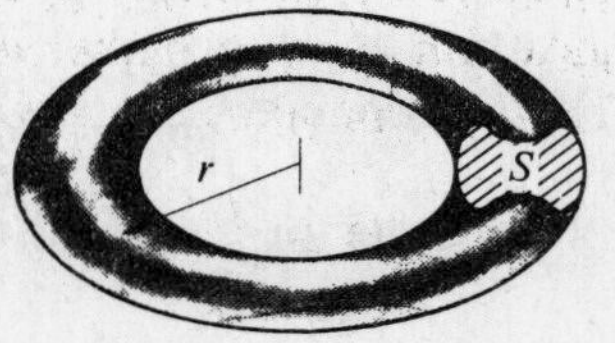

$$L \approx \frac{\mu_0 N^2 S}{2\pi r} \quad \text{(H)}$$

(assuming average flux density at average radius r)

Fig. 11-4. Toroid, general cross section S

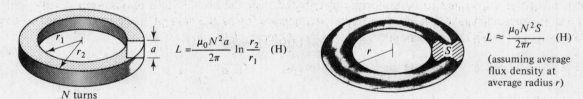

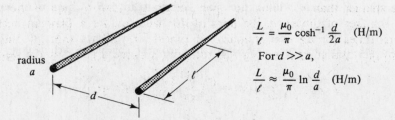

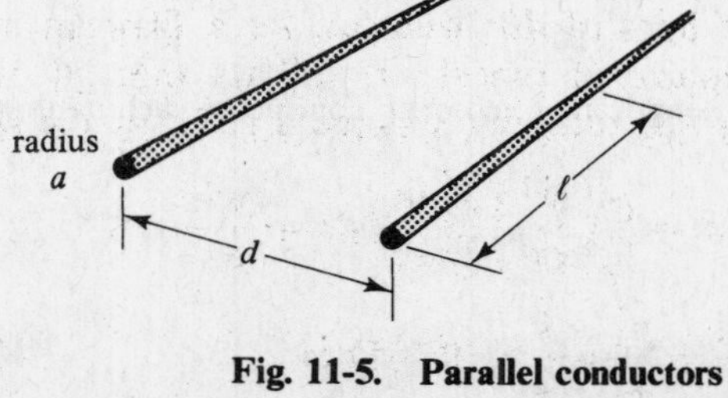

$$\frac{L}{\ell} = \frac{\mu_0}{\pi} \cosh^{-1}\frac{d}{2a} \quad \text{(H/m)}$$

For $d >> a$,

$$\frac{L}{\ell} \approx \frac{\mu_0}{\pi} \ln\frac{d}{a} \quad \text{(H/m)}$$

Fig. 11-5. Parallel conductors of radius a

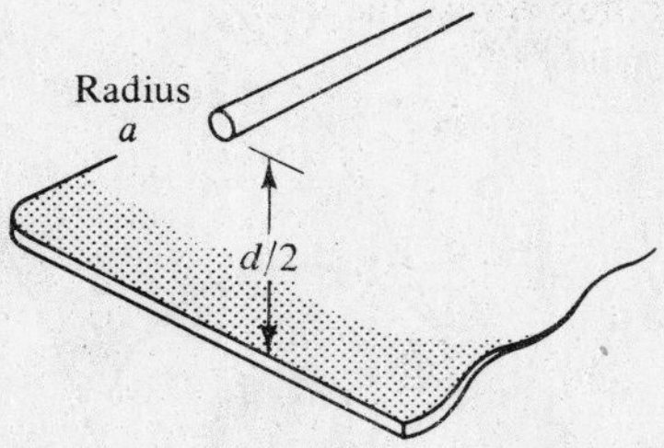

$$\frac{L}{\ell} = \frac{\mu_0}{2\pi}\cosh^{-1}\frac{d}{2a} \quad \text{(H/m)}$$

$$\approx \frac{\mu_0}{2\pi}\ln\frac{d}{a} \quad \text{(H/m)}$$

Fig. 11-6. Cylindrical conductor parallel to a ground plane

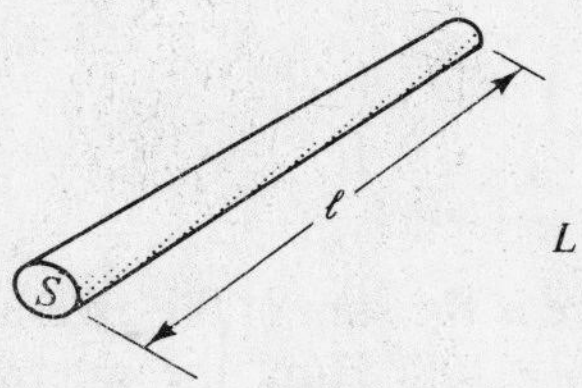

$$L = \frac{\mu_0 N^2 S}{\ell} \quad \text{(H)}$$

Fig. 11-7. Long solenoid of small cross-sectional area S

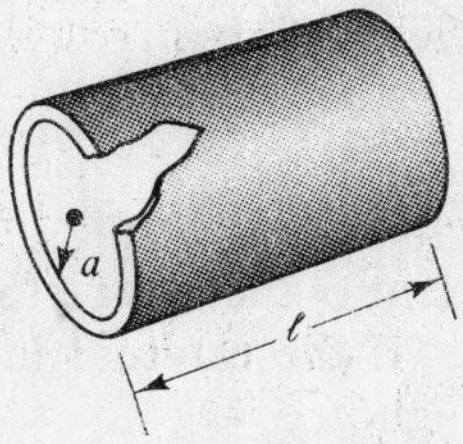

$$L = \frac{39.5\, N^2 a^2}{9a + 10\ell} \quad (\mu\text{H})$$

Fig. 11-8. Single-layer, air-core coil

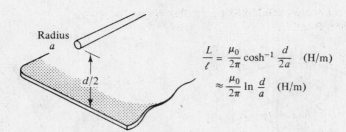

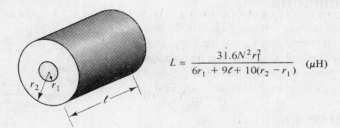

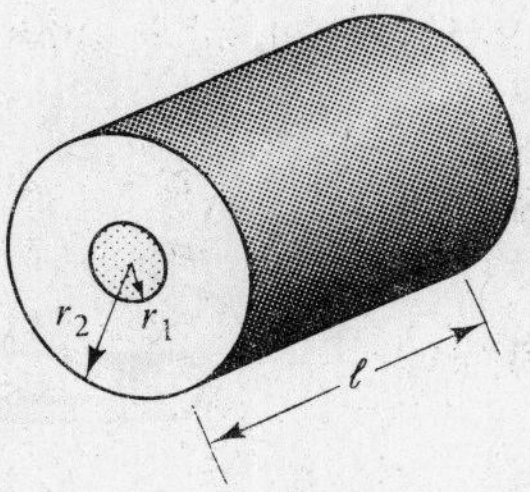

$$L = \frac{31.6 N^2 r_1^2}{6r_1 + 9\ell + 10(r_2 - r_1)} \quad (\mu\text{H})$$

Fig. 11-9. Multilayer, air-core coil

11.4 INTERNAL INDUCTANCE

Magnetic flux occurs within a conductor cross section as well as external to the conductor. This internal flux gives rise to an *internal inductance,* which is often small compared to the external inductance and frequently ignored. In Fig. 11-10(*a*) a conductor of circular cross section is shown, with a current I assumed to be uniformly distributed over the area. (This assumption is valid only at low frequencies, since *skin effect* at higher frequencies forces the current to be concentrated at the outer surface.) Within the conductor of radius a, Ampère's law gives

$$\mathbf{H} = \frac{Ir}{2\pi a^2}\mathbf{a}_\phi \qquad \text{and} \qquad \mathbf{B} = \frac{\mu_0 Ir}{2\pi a^2}\mathbf{a}_\phi$$

The straight piece of conductor shown in Fig. 11-10(*a*) must be imagined as a short section of an infinite torus, as suggested in Fig. 11-10(*b*). The current filaments become circles of infinite radius. The lines of flux $d\Phi$ through the strip $\ell\, dr$ encircle only those filaments whose distance from the conductor axis is smaller than r. Thus, an open surface bounded by one of those filaments is cut once (or an odd number of times) by the lines of $d\Phi$; whereas, for a filament such as *1* or *2*, the surface is cut zero times (or an even number of times). It follows that $d\Phi$ links only with the fraction $\pi r^2/\pi a^2$ of the total current, so that the total flux linkage is given by the weighted "sum"

$$\lambda = \int\left(\frac{\pi r^2}{\pi a^2}\right)d\Phi = \int_0^a\left(\frac{\pi r^2}{\pi a^2}\right)\frac{\mu_0 Ir}{2\pi a^2}\ell\, dr = \frac{\mu_0 I\ell}{8\pi}$$

and

$$\frac{L}{\ell} = \frac{\lambda/I}{\ell} = \frac{\mu_0}{8\pi} = \frac{1}{2}\times 10^{-7}\ \text{H/m}$$

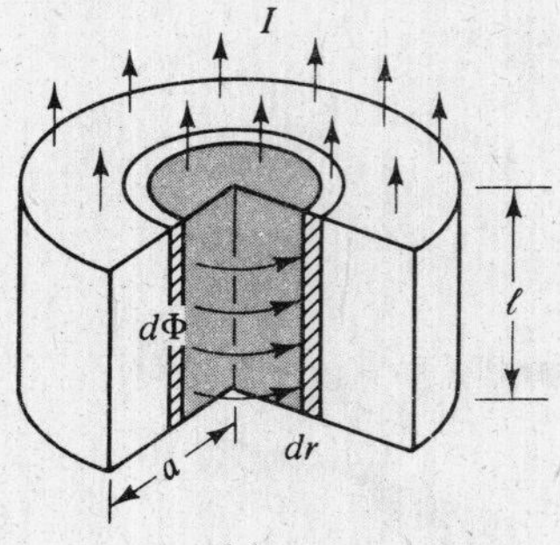

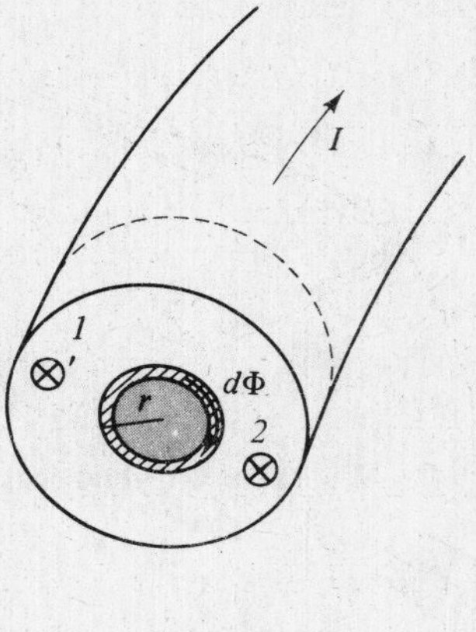

Fig. 11-10

This result is independent of the conductor radius. The total inductance is the sum of the external and internal inductances. If the external inductance is of the order of $\frac{1}{2} \times 10^{-7}$ H/m, the internal inductance should not be ignored.

11.5 MAGNETIC CIRCUITS

In Chapter 9, magnetic field intensity **H**, flux Φ, and magnetic flux density **B** were examined and various problems were solved where the medium was free space. For example, when Ampère's law is applied to the closed path *C* through the long, air-core coil shown in Fig. 11-11, the result is

$$\oint \mathbf{H} \cdot d\mathbf{l} = NI$$

But since the flux lines are widely spread outside of the coil, *B* is small there. The flux is effectively restricted to the inside of the coil, where

$$H \approx \frac{NI}{\ell}$$

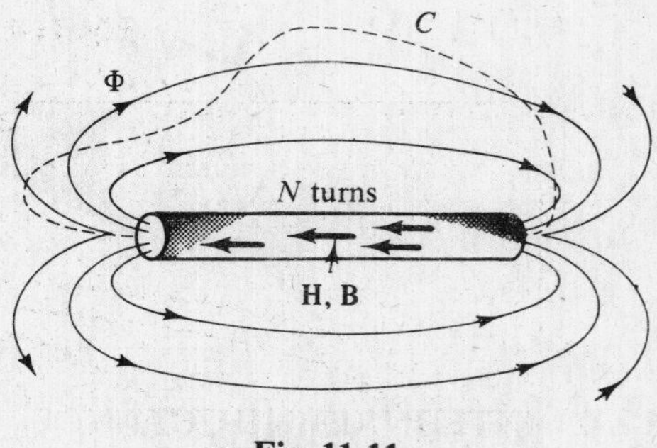

Fig. 11-11

Ferromagnetic materials have relative permeabilities μ_r in the order of thousands. Consequently, the flux density $B = \mu_0 \mu_r H$ is, for a given *H*, much greater than would result in free space. In Fig. 11-12, the coil is not distributed over the iron core. Even so, the *NI* of the coil causes a flux Φ which follows the core. It might be said that the flux prefers the core to the surrounding space by a ratio of several thousand to one. This is so different from the free-space magnetics of Chapter 9 that an entire subject area, known as *iron-core magnetics* or *magnetic circuits*, has developed. This brief introduction to the subject assumes that *all* of the flux is within the core. It is further assumed that the flux is uniformly distributed over the cross section of the core. Core lengths required for calculation of *NI* drops are mean lengths.

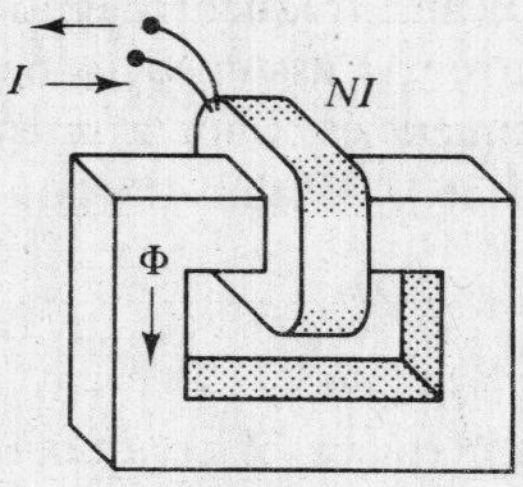

Fig. 11-12

11.6 NONLINEARITY OF THE *B-H* CURVE

A sample of ferromagnetic material could be tested by applying increasing values of *H* and measuring the corresponding values of flux density *B*. *Magnetization curves*, or simply *B-H curves*, for some common ferromagnetic materials are given in Figs. 11-13 and 11-14. The relative

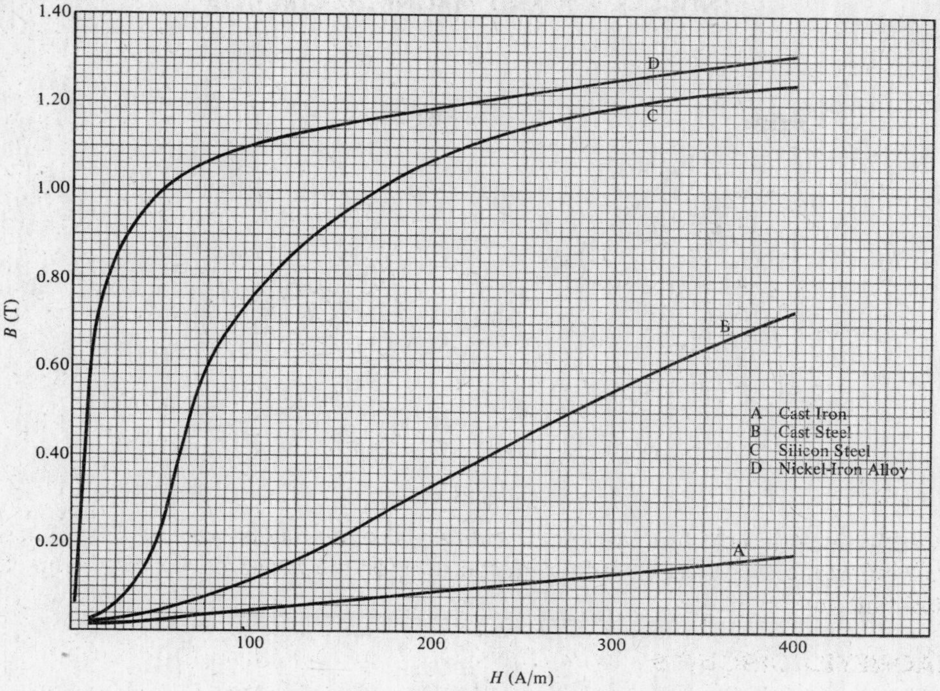

Fig. 11-13. ***B-H* curves,** $H < 400$ A/m

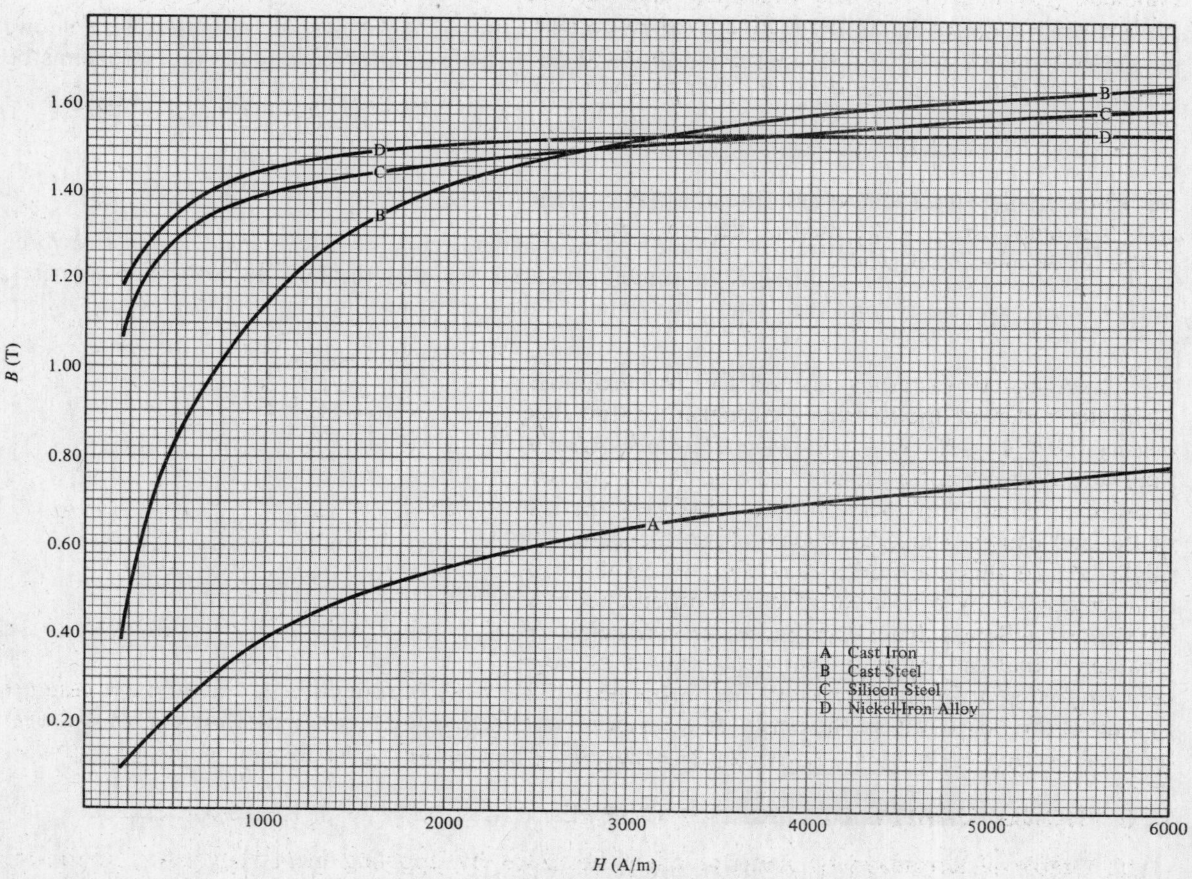

Fig. 11-14. ***B-H* curves,** $H > 400$ A/m

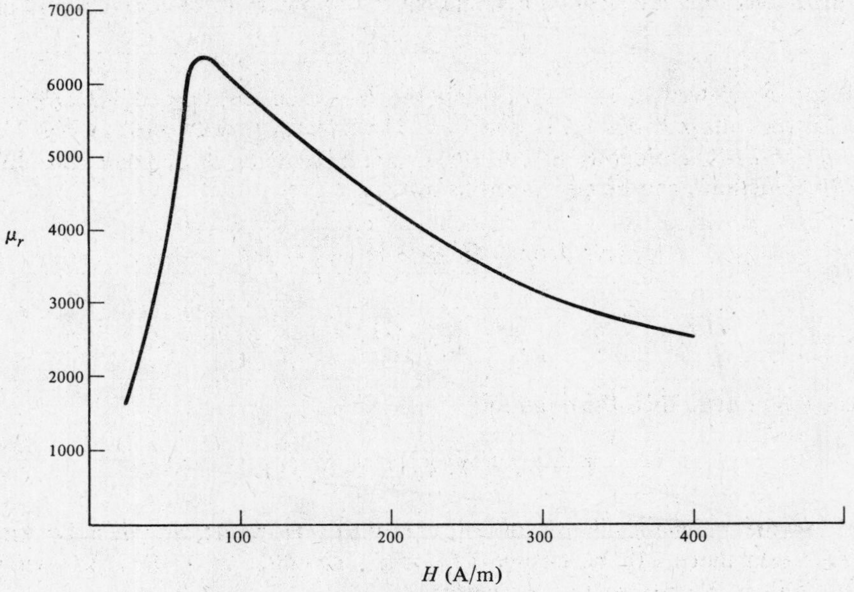

Fig. 11-15

permeability can be computed from the B-H curve by use of $\mu_r = B/\mu_0 H$. Figure 11-15 shows the extreme nonlinearity of μ_r versus H for silicon steel. This nonlinearity requires that problems be solved graphically.

11.7 AMPÈRE'S LAW FOR MAGNETIC CIRCUITS

A coil of N turns and current I around a ferromagnetic core produces a *magnetomotive force* (mmf) given by NI. The symbol F is sometimes used for this mmf. The units are amperes. Ampère's law, applied around the path in the center of the core shown in Fig. 11-16(*a*), gives

$$
\begin{aligned}
F = NI &= \oint \mathbf{H} \cdot d\mathbf{l} \\
&= \int_1 \mathbf{H} \cdot d\mathbf{l} + \int_2 \mathbf{H} \cdot d\mathbf{l} + \int_3 \mathbf{H} \cdot d\mathbf{l} \\
&= H_1 \ell_1 + H_2 \ell_2 + H_3 \ell_3
\end{aligned}
$$

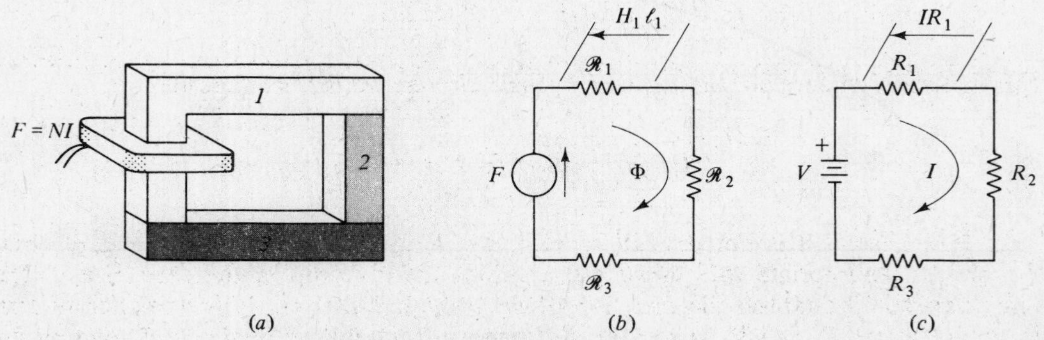

Fig. 11-16

Comparison with Kirchhoff's law around a single closed loop with three resistors and an emf V,

$$V = V_1 + V_2 + V_3$$

suggests that F can be viewed as an NI *rise* and the $H\ell$ terms considered NI *drops*, in analogy to the voltage rise V and voltage drops V_1, V_2 and V_3. The analogy is developed in Fig. 11-16(*b*) and (*c*). Flux Φ in Fig. 11-16(*b*) is analogous to current I, and *reluctance* $\mathscr{R}$ is analogous to resistance R. An expression for reluctance can be developed as follows.

$$NI \text{ drop} = H\ell = BA\left(\frac{\ell}{\mu A}\right) = \Phi\mathscr{R}$$

hence

$$\mathscr{R} = \frac{\ell}{\mu A} \quad (\text{H}^{-1})$$

If the reluctances are known, then the equation

$$F = NI = \Phi(\mathscr{R}_1 + \mathscr{R}_2 + \mathscr{R}_3)$$

can be written for the magnetic circuit of Fig. 11-16(*b*). However, μ_r must be known for each material before its reluctance can be calculated. And only after B or H is known will the value of μ_r be known. This is in contrast to the relation

$$R = \frac{\ell}{\sigma A}$$

(Section 6.7), in which the conductivity σ is independent of the current.

11.8 CORES WITH AIR GAPS

Magnetic circuits with small air gaps are very common. The gaps are generally kept as small as possible, since the NI drop of the air gap is often much greater than the drop in the core. The flux fringes outward at the gap, so that the area at the gap exceeds the area of the adjacent core. Provided that the gap length ℓ_a is less than 1/10 the smaller dimension of the core, an *apparent area*, S_a, of the air gap can be calculated. For a rectangular core of dimensions a and b,

$$S_a = (a + \ell_a)(b + \ell_a)$$

If the total flux in the air gap is known, H_a and $H_a\ell_a$ can be computed directly.

$$H_a = \frac{1}{\mu_0}\left(\frac{\Phi}{S_a}\right) \qquad H_a\ell_a = \frac{\ell_a\Phi}{\mu_0 S_a}$$

For a uniform iron core of length ℓ_i with a single air gap, Ampère's law reads

$$NI = H_i\ell_i + H_a\ell_a = H_i\ell_i + \frac{\ell_a\Phi}{\mu_0 S_a}$$

If the flux Φ is known, it is not difficult to compute the NI drop across the air gap, obtain B_i, take H_i from the appropriate B-H curve and compute the NI drop in the core, $H_i\ell_i$. The sum is the NI required to establish the flux Φ. However, with NI given, it is a matter of trial and error to obtain B_i and Φ, as will be seen in the problems. Graphical methods of solution are also available.

11.9 MULTIPLE COILS

Two or more coils on a core could be wound such that their mmfs either aid one another or oppose. Consequently, a method of indicating polarity is given in Fig. 11-17. An assumed direction for the resulting flux Φ could be incorrect, just as an assumed current in a dc circuit with two or more voltage sources may be incorrect. A negative result simply means that the flux is in the opposite direction.

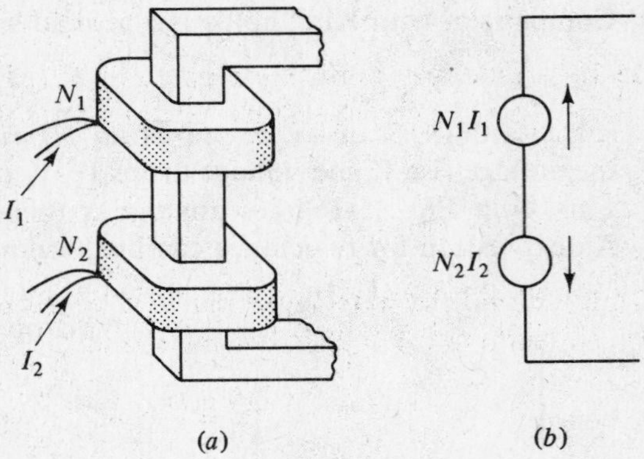

Fig. 11-17

11.10 PARALLEL MAGNETIC CIRCUITS

The method of solving a parallel magnetic circuit is suggested by the two-loop equivalent circuit shown in Fig. 11-18(*b*). The leg on the left contains an *NI* rise and an *NI* drop. The *NI* drop between the junctions *a* and *b* can be written for each leg as follows:

$$F - H_1\ell_1 = H_2\ell_2 = H_3\ell_3$$

and the fluxes satisfy

$$\Phi_1 = \Phi_2 + \Phi_3$$

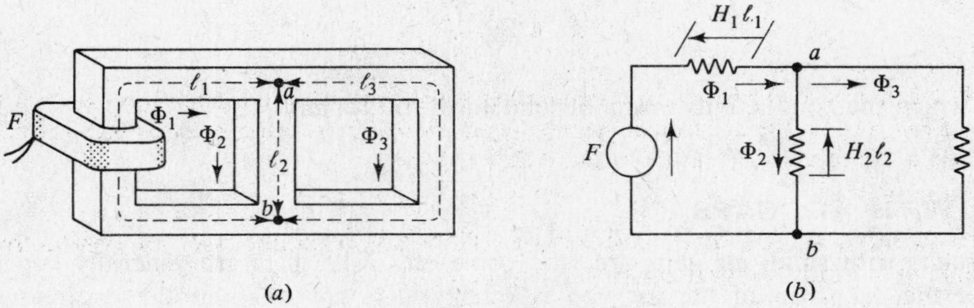

Fig. 11-18

Different materials for the core parts will necessitate working with several *B-H* curves. An air gap in one of the legs would lead to $H_i\ell_i + H_a\ell_a$ for the mmf between the junctions for that leg.

The equivalent magnetic circuit should be drawn for parallel magnetic circuit problems. It is good practice to mark the material types, cross-sectional areas and mean lengths directly on the diagram. In more complex problems a scheme like Table 11-1 can be helpful. The data are inserted directly into the table, and the remaining quantities are then calculated or taken from the appropriate *B-H* curve.

Table 11-1

Part	Material	Area	ℓ	Φ	B	H	$H\ell$
1							
2							
3							

Solved Problems

11.1. Find the inductance per unit length of the coaxial cable in Fig. 11-2 if $a = 1$ mm and $b = 3$ mm. Assume $\mu_r = 1$ and omit internal inductance.

$$\frac{L}{\ell} = \frac{\mu}{2\pi}\ln\frac{b}{a} = \frac{4\pi \times 10^{-7}}{2\pi}\ln 3 = 0.22\ \mu\text{H/m}$$

11.2. Find the inductance per unit length of the parallel cylindrical conductors shown in Fig. 11-5, where $d = 25$ ft, $a = 0.803$ in.

$$\frac{L}{\ell} = \frac{\mu_0}{\pi}\cosh^{-1}\frac{d}{2a} = (4 \times 10^{-7})\cosh^{-1}\frac{25(12)}{2(0.803)} = 2.37\ \mu\text{H/m}$$

The approximate formula gives

$$\frac{L}{\ell} = \frac{\mu_0}{\pi}\ln\frac{d}{a} = 2.37\ \mu\text{H/m}$$

When $d/a \geq 10$, the approximate formula may be used with an error of less than 0.5%.

11.3. A circular conductor with the same radius as in Problem 11.2 is 12.5 ft from an infinite conducting plane. Find the inductance.

$$\frac{L}{\ell} = \frac{\mu_0}{2\pi}\ln\frac{d}{a} = (2 \times 10^{-7})\ln\frac{25(12)}{0.803} = 1.18\ \mu\text{H/m}$$

This result is 1/2 that of Problem 11.2. A conducting plane may be inserted midway between the two conductors of Fig. 11-5. The inductance between each conductor and the plane is 1.18 μH/m. Since they are in series, the total inductance is the sum, 2.37 μH/m.

11.4. An air-core solenoid of 300 turns and length 0.50 m has a single layer of conductors at a radius of 0.02 m. Find the inductance L.

Using the empirical formula,

$$L = \frac{39.5\,N^2a^2}{9a + 10\ell} = 275\ \mu\text{H}$$

The equation for a long solenoid of small cross section results in

$$L = \frac{\mu_0 N^2 S}{\ell} = \frac{(4\pi \times 10^{-7})(300)^2\pi(0.02)^2}{0.50} = 284\ \mu\text{H}$$

This latter equation is an approximation based on the assumption that the magnetic field intensity H is constant throughout the interior of the coil.

11.5. Find the inductance of the coil shown in Fig. 11-9, where $N = 300$, $r_1 = 9$ mm, $r_2 = 25$ mm, $\ell = 20$ mm.

$$L = \frac{31.6(300)^2(9 \times 10^{-3})^2}{54 \times 10^{-3} + 180 \times 10^{-3} + 160 \times 10^{-3}} = 585\ \mu\text{H}$$

11.6. Assume that the air-core toroid shown in Fig. 11-4 has a circular cross section of radius 4 mm. Find the inductance if there are 2500 turns and the mean radius is $r = 20$ mm.

$$L = \frac{\mu N^2 S}{2\pi r} = \frac{(4\pi \times 10^{-7})(2500)^2\pi(0.004)^2}{2\pi(0.020)} = 3.14\ \text{mH}$$

11.7. Assume that the air-core toroid in Fig. 11-3 has 700 turns, an inner radius of 1 cm, an outer radius of 2 cm and height $a = 1.5$ cm. Find L using (*a*) the formula for square cross-section toroids; (*b*) the approximate formula for a general toroid, which assumes a uniform H at a mean radius.

(*a*)
$$L = \frac{\mu_0 N^2 a}{2\pi} \ln\frac{r_2}{r_1} = \frac{(4\pi \times 10^{-7})(700)^2(0.015)}{2\pi} \ln 2 = 1.02 \text{ mH}$$

(*b*)
$$L = \frac{\mu_0 N^2 S}{2\pi r} = \frac{(4\pi \times 10^{-7})(700)^2(0.01)(0.015)}{2\pi(0.015)} = 0.98 \text{ mH}$$

With a radius that is larger compared to the cross section, the two formulas yield the same result. See Problem 11.29.

11.8. Use the energy integral to find the internal inductance per unit length of a cylindrical conductor of radius a.

At a distance $r \le a$ from the conductor axis,

$$\mathbf{H} = \frac{Ir}{2\pi a^2}\mathbf{a}_\phi \qquad \mathbf{B} = \frac{\mu_0 Ir}{2\pi a^2}\mathbf{a}_\phi$$

whence
$$\mathbf{B}\cdot\mathbf{H} = \frac{\mu_0 I^2}{4\pi^2 a^4} r^2$$

The inductance corresponding to energy storage within a length ℓ of the conductor is then

$$L = \int \frac{(\mathbf{B}\cdot\mathbf{H})\,dv}{I^2} = \frac{\mu_0}{4\pi^2 a^4}\int_0^a r^2\, 2\pi r\ell\, dr = \frac{\mu_0 \ell}{8\pi}$$

or $L/\ell = \mu_0/8\pi$. This agrees with the result of Section 11.4.

11.9. The cast iron core shown in Fig. 11-19 has an inner radius of 7 cm and an outer radius of 9 cm. Find the flux Φ if the coil mmf is 500 A.

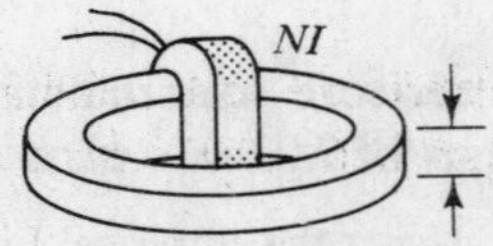

Fig. 11-19

$$\ell = 2\pi(0.08) = 0.503 \text{ m}$$

$$H = \frac{F}{\ell} = \frac{500}{0.503} = 995 \text{ A/m}$$

From the *B-H* curve for cast iron in Fig. 11-14, $B = 0.40$ T.

$$\Phi = BS = (0.40)(0.02)^2 = 0.16 \text{ mWb}$$

11.10. The magnetic circuit shown in Fig. 11-20 has a C-shaped cast steel part, *1*, and a cast iron part, *2*. Find the current required in the 150-turn coil if the flux density in the cast iron is $B_2 = 0.45$ T.

The calculated areas are $S_1 = 4 \times 10^{-4}\ \text{m}^2$ and $S_2 = 3.6 \times 10^{-4}\ \text{m}^2$. The mean lengths are

$$\ell_1 = 0.11 + 0.11 + 0.12 = 0.34 \text{ m}$$

$$\ell_2 = 0.12 + 0.009 + 0.009 = 0.138 \text{ m}$$

From the *B-H* curve for cast iron in Fig. 11-14, $H_2 = 1270$ A/m.

$$\Phi = B_2 S_2 = (0.45)(3.6 \times 10^{-4}) = 1.62 \times 10^{-4} \text{ Wb}$$

$$B_1 = \frac{\Phi}{S_1} = 0.41 \text{ T}$$

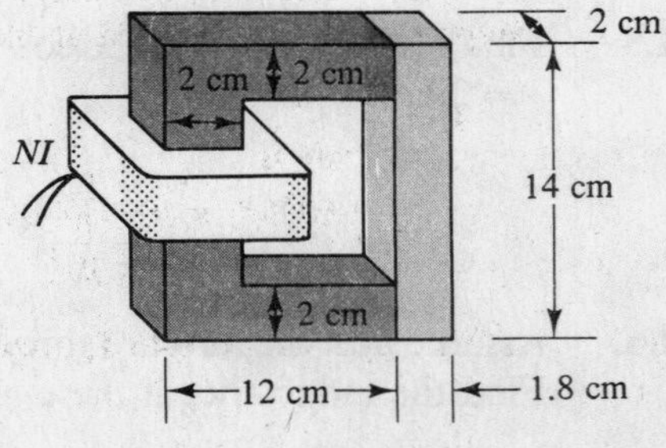

Fig. 11-20

Then, from the cast steel curve in Fig. 11-13, $H_1 = 233$ A/m.

The equivalent circuit, Fig. 11-21, suggests the equation

$$F = NI = H_1\ell_1 + H_2\ell_2$$
$$150I = 233(0.34) + 1270(0.138)$$
$$I = 1.70 \text{ A}$$

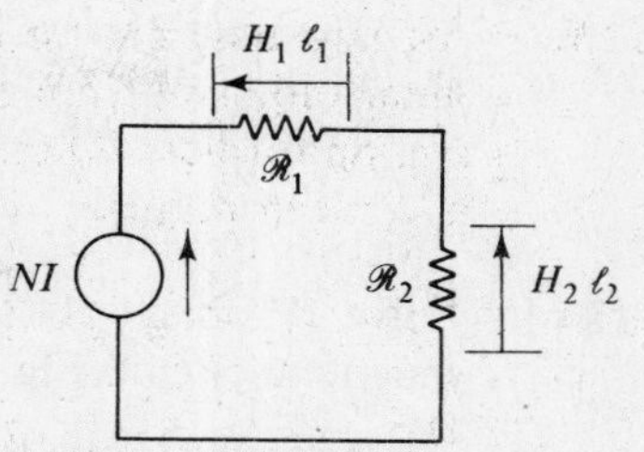

Fig. 11-21

11.11. The magnetic circuit shown in Fig. 11-22 is cast iron with a mean length $\ell_i = 0.44$ m and square cross section 0.02×0.02 m. The air gap length is $\ell_a = 2$ mm and the coil contains 400 turns. Find the current I required to establish an air gap flux of 0.141 mWb.

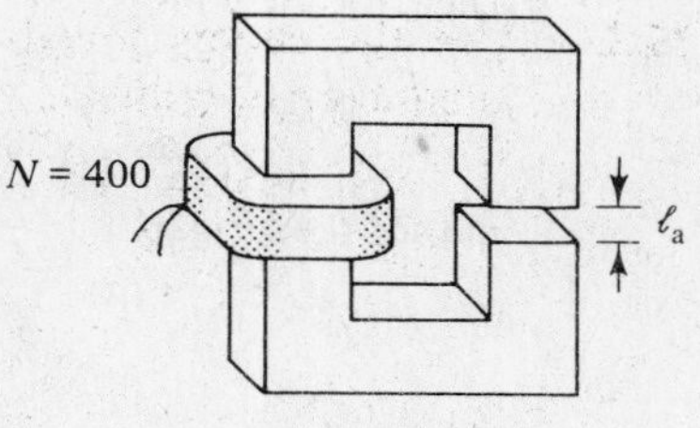

Fig. 11-22

The flux Φ in the air gap is also the flux in the core.

$$B_i = \frac{\Phi}{S_i} = \frac{0.141 \times 10^{-3}}{4 \times 10^{-4}} = 0.35 \text{ T}$$

From Fig. 11-14, $H_i = 850$ A/m. Then

$$H_i\ell_i = 850(0.44) = 374 \text{ A}$$

For the air gap, $S_a = (0.02 + 0.002)^2 = 4.84 \times 10^{-4}\ \text{m}^2$, and so

$$H_a\ell_a = \frac{\Phi}{\mu_0 S_a}\ell_a = \frac{0.141 \times 10^{-3}}{(4\pi \times 10^{-7})(4.84 \times 10^{-4})}(2 \times 10^{-3}) = 464 \text{ A}$$

Therefore, $F = H_i\ell_i + H_a\ell_a = 838$ A and

$$I = \frac{F}{N} = \frac{838}{400} = 2.09 \text{ A}$$

11.12. Determine the reluctance of an air gap in a dc machine where the apparent area is $S_a = 4.26 \times 10^{-2}\ \text{m}^2$ and the gap length $\ell_a = 5.6$ mm.

$$\mathcal{R} = \frac{\ell_a}{\mu_0 S_a} = \frac{5.6 \times 10^{-3}}{(4\pi \times 10^{-7})(4.26 \times 10^{-2})} = 1.05 \times 10^5\ \text{H}^{-1}$$

11.13. The cast iron magnetic core shown in Fig. 11-23 has an area $S_i = 4\ \text{cm}^2$ and a mean length 0.438 m. The 2 mm air gap has an apparent area $S_a = 4.84\ \text{cm}^2$. Determine the air-gap flux Φ.

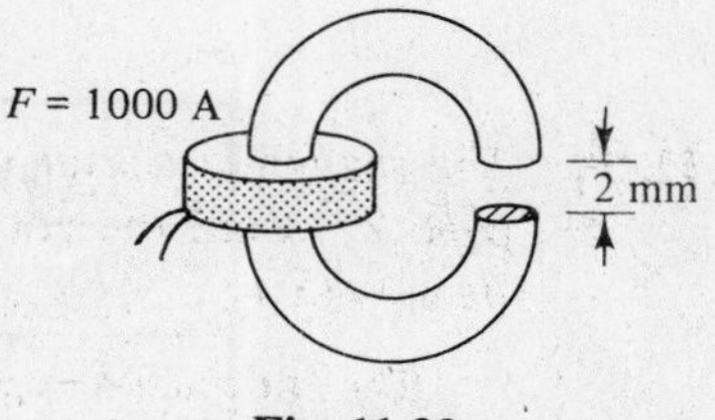

Fig. 11-23

The core is quite long compared to the length of the air gap, and cast iron is not a particularly good magnetic material. As a first estimate, therefore, assume that 600 of the total ampere turns are dropped at the air gap, i.e. $H_a\ell_a = 600$ A.

$$H_a\ell_a = \frac{\Phi}{\mu_0 S_a}\ell_a$$

$$\Phi = \frac{600(4\pi \times 10^{-7})(4.84 \times 10^{-4})}{2 \times 10^{-3}} = 1.82 \times 10^{-4} \text{ Wb}$$

Then $B_i = \Phi/S_i = 0.46$ T, and from Fig. 11-14, $H_i = 1340$ A/m. The core drop is then

$$H_i\ell_i = 1340(0.438) = 587 \text{ A}$$

so that

$$H_i\ell_i + H_a\ell_a = 1187 \text{ A}$$

This sum exceeds the 1000 A mmf of the coil. Consequently, values of B_i lower than 0.46 T should be tried until the sum of $H_i \ell_i$ and $H_a \ell_a$ is 1000 A. The values $B_i = 0.41$ T and $\Phi = 1.64 \times 10^{-4}$ Wb will result in a sum very close to 1000 A.

11.14. Solve Problem 11.13 using reluctances and the equivalent magnetic circuit, Fig. 11-24.

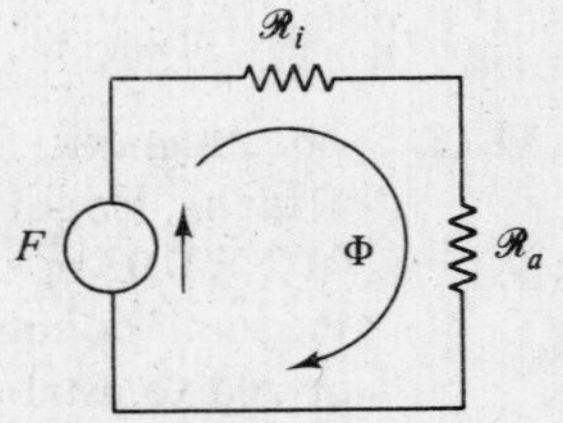

Fig. 11-24

From the values of B_i and H_i obtained in Problem 11.13,

$$\mu_0 \mu_r = \frac{B_i}{H_i} = 3.83 \times 10^{-4} \text{ H/m}$$

Then, for the core,

$$\mathscr{R}_i = \frac{\ell_i}{\mu_0 \mu_r S_i} = \frac{0.438}{(3.83 \times 10^{-4})(4 \times 10^{-4})} = 2.86 \times 10^6 \text{ H}^{-1}$$

and for the air gap,

$$\mathscr{R}_a = \frac{\ell_a}{\mu_0 S_a} = \frac{2 \times 10^{-3}}{(4\pi \times 10^{-7})(4.84 \times 10^{-4})} = 3.29 \times 10^6 \text{ H}^{-1}$$

The circuit equation,

$$F = \Phi(\mathscr{R}_i + \mathscr{R}_a)$$

gives

$$\Phi = \frac{1000}{(2.86 \times 10^6 + 3.29 \times 10^6)} = 1.63 \times 10^{-4} \text{ Wb}$$

The corresponding flux density in the iron is 0.41 T, in agreement with the results of Problem 11.13. While the air-gap reluctance can be calculated from the dimensions and μ_0, the same is not true for the reluctance of the iron. The reason is that μ_r for the iron depends on the values of B_i and H_i.

11.15. Solve Problem 11.13 graphically with a plot of Φ versus F.

Values of H_i from 700 through 1100 A/m are listed in the first column of Table 11-2; the corresponding values of B_i are found from the cast iron curve, Fig. 11-14. The values of Φ and $H_i \ell_i$ are computed, and $H_a \ell_a$ is obtained from $\Phi \ell_a / \mu_0 S_a$. Then F is given as the sum of $H_i \ell_i$ and $H_a \ell_a$. Since the air gap is linear, only two points are required.

Table 11-2

H_i (A/m)	B_i (T)	Φ (Wb)	$H_i \ell_i$ (A)	$H_a \ell_a$ (A)	F (A)
700	0.295	1.18×10^{-4}	307	388	695
800	0.335	1.34×10^{-4}	350	441	791
900	0.365	1.46×10^{-4}	395	480	874
1000	0.400	1.60×10^{-4}	438	526	964
1100	0.420	1.68×10^{-4}	482	552	1034

The flux Φ for $F = 1000$ A is seen from Fig. 11-25 to be approximately 1.65×10^{-4} Wb.

This method is simply a plot of the trial and error data used in Problem 11.13. However, it is helpful if several different coils or coil currents are to be examined.

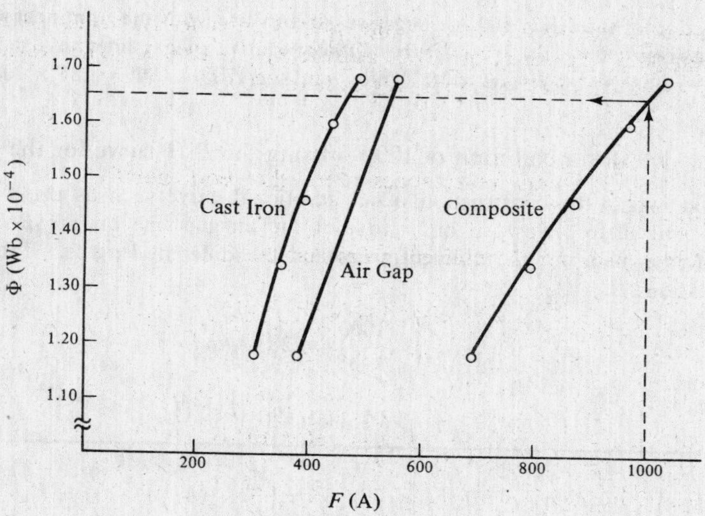

Fig. 11-25

11.16. Determine the fluxes Φ in the core of Problem 11.13 for coil mmfs of 800 and 1200 A. Use a graphical approach and the *negative air-gap line.*

The Φ versus $H_i \ell_i$ data for the cast iron core, developed in Problem 11.15, are plotted in Fig. 11-26. The air-gap Φ versus F is linear. One end of the negative air-gap line for the coil mmf of 800 A is at $\Phi = 0$, $F = 800$ A. The other end assumes $H_a \ell_a = 800$ A, from which

$$\Phi = \frac{\mu_0 S_a(H_a \ell_a)}{\ell_a} = 2.43 \times 10^{-4} \text{ Wb}$$

which locates this end at $\Phi = 2.43 \times 10^{-4}$ Wb, $F = 0$.

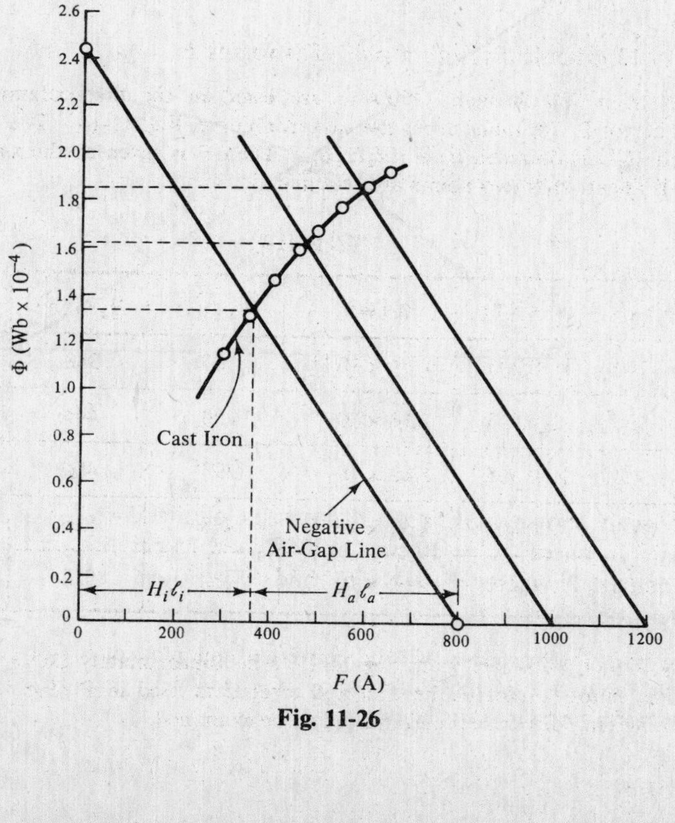

Fig. 11-26

The intersection of the $F = 800$ A negative air-gap line with the nonlinear Φ versus F curve for the cast iron core gives $\Phi = 1.34 \times 10^{-4}$ Wb. Other negative air-gap lines have the same negative slope. For a coil mmf of 1000 A, $\Phi = 1.62 \times 10^{-4}$ Wb and for 1200 A, $\Phi = 1.85 \times 10^{-4}$ Wb.

11.17. Solve Problem 11.13 for a coil mmf of 1000 A using the *B-H* curve for the cast iron.

This method avoids the construction of an additional curve such as the Φ versus F curves of Problems 11.15 and 11.16. Now, in order to plot the air-gap line on the *B-H* curve of the iron, adjustments must be made for the different areas and the different lengths. Table 11-3 suggests the necessary calculations.

$$\frac{F}{\ell_i} = \frac{1000}{0.438} = 2283 \text{ A/m}$$

Table 11-3

B_a (T)	H_a (A/m)	$B_a\left(\frac{S_a}{S_i}\right)$ (T)	$H_a\left(\frac{\ell_a}{\ell_i}\right)$ (A/m)	$\frac{F}{\ell_i} - H_a\left(\frac{\ell_a}{\ell_i}\right)$ (A/m)
0.10	0.80×10^5	0.12	363	1920
0.30	2.39×10^5	0.36	1091	1192
0.50	3.98×10^5	0.61	1817	466

The data from the third and fifth columns may be plotted directly on the cast iron *B-H* curve, as shown in Fig. 11-27. The air gap is linear and only two points are needed. The answer is seen to be $B_i = 0.41$ T. The method can be used with two nonlinear core parts, as well (see Problem 11.18).

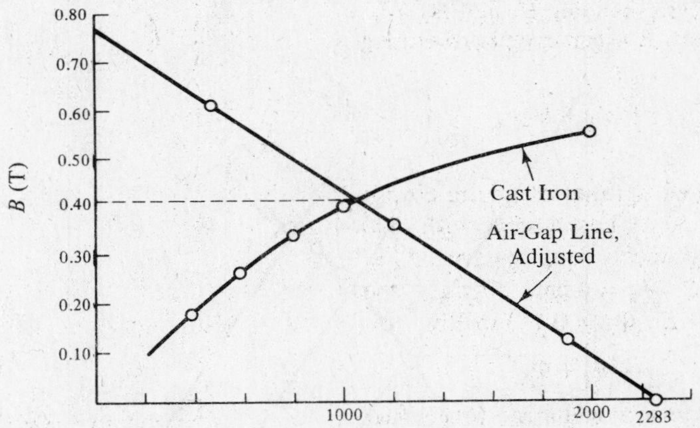

Fig. 11-27

11.18. The magnetic circuit shown in Fig. 11-28 consists of nickel-iron alloy in part *1*, where $\ell_1 = 10$ cm and $S_1 = 2.25$ cm^2, and cast steel for part *2*, where $\ell_2 = 8$ cm and $S_2 = 3$ cm^2. Find the flux densities B_1 and B_2.

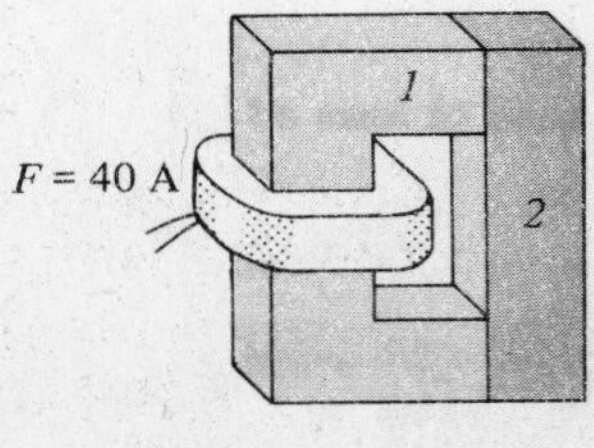

Fig. 11-28

The data for part *2* of cast steel will be converted and plotted on the *B-H* curve for part *1* of nickel-iron alloy ($F/\ell_1 = 400$ A/m). Table 11-4 suggests the necessary calculations.

Table 11-4

B_2 (T)	H_2 (A/m)	$B_2\left(\frac{S_2}{S_1}\right)$ (T)	$H_2\left(\frac{\ell_2}{\ell_1}\right)$ (A/m)	$\frac{F}{\ell_1} - H_2\left(\frac{\ell_2}{\ell_1}\right)$ (A/m)
0.33	200	0.44	160	240
0.44	250	0.59	200	200
0.55	300	0.73	240	160
0.65	350	0.87	280	120
0.73	400	0.97	320	80
0.78	450	1.04	360	40
0.83	500	1.11	400	0

From the graph, Fig. 11-29, $B_1 = 1.01$ T. Then, since $B_1S_1 = B_2S_2$,

$$B_2 = 1.01\left(\frac{2.25 \times 10^{-4}}{3 \times 10^{-4}}\right) = 0.76 \text{ T}$$

These values can be checked by obtaining the corresponding H_1 and H_2 from the appropriate *B-H* curves and substituting in

$$F = H_1\ell_1 + H_2\ell_2$$

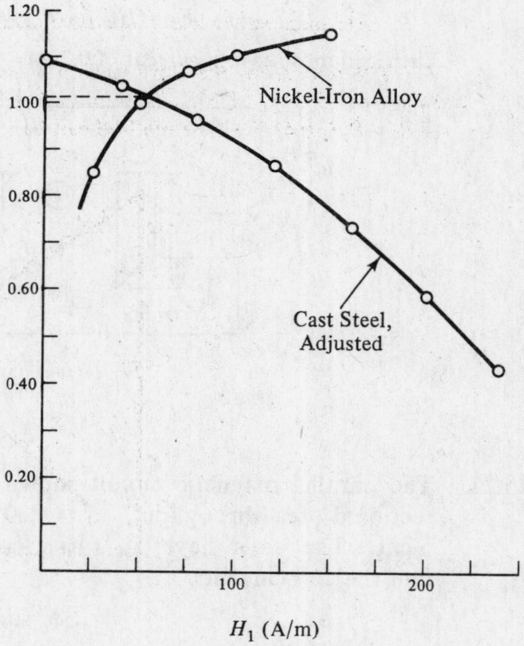

Fig. 11-29

11.19. The cast steel parallel magnetic circuit in Fig. 11.30(*a*) has a coil with 500 turns. The mean lengths are $\ell_2 = \ell_3 = 10$ cm, $\ell_1 = 4$ cm. Find the coil current if $\Phi_3 = 0.173$ mWb.

$$\Phi_1 = \Phi_2 + \Phi_3$$

Since the cross-sectional area of the center leg is twice that of the two side legs, the flux density is the same throughout the core, i.e.

$$B_1 = B_2 = B_3 = \frac{0.173 \times 10^{-3}}{1.5 \times 10^{-4}} = 1.15 \text{ T}$$

Corresponding to $B = 1.15$ T, Fig. 11-14 gives $H = 1030$ A/m. The NI drop between points a and b is now used to write the following equation [see Fig. 11-30(*b*)]:

$$F - H\ell_1 = H\ell_2 = H\ell_3 \quad \text{or} \quad F = H(\ell_1 + \ell_2) = 1030(0.14) = 144.2 \text{ A}$$

Then

$$I = \frac{F}{N} = \frac{144.2}{500} = 0.29 \text{ A}$$

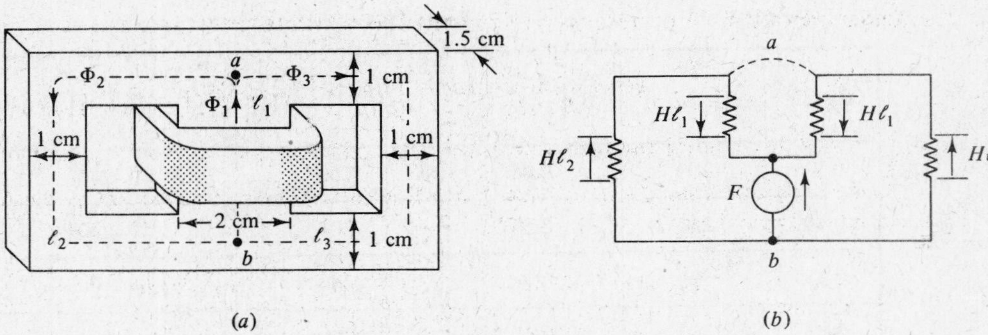

Fig. 11-30

11.20. The same cast steel core as in Problem 11.19 has identical 500-turn coils on the outer legs, with the winding sense as shown in Fig. 11-31(*a*). If again $\Phi_3 = 0.173$ mWb, find the coil currents.

The flux densities are the same throughout the core and consequently H is the same. The equivalent circuit in Fig. 11-31(*b*) suggests that the problem can be solved on a *per pole* basis.

$$B = \frac{\Phi_3}{S_3} = 1.15 \text{ T} \qquad \text{and} \qquad H = 1030 \text{ A/m} \quad \text{(From Fig. 11-14)}$$

$$F_3 = H(\ell_1 + \ell_3) = 1030(0.14) = 144.2 \text{ A} \qquad I = 0.29 \text{ A}$$

Each coil must have a current of 0.29 A.

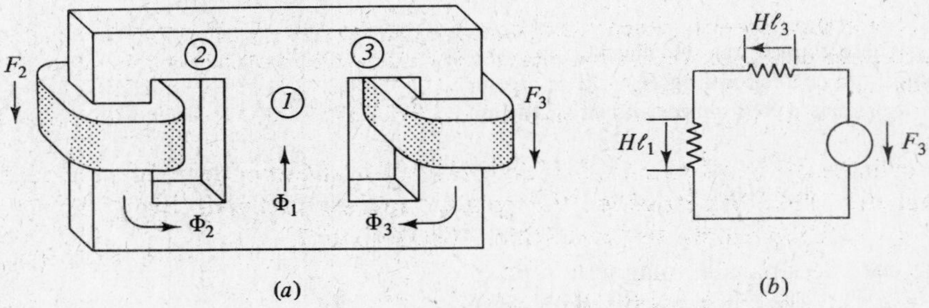

Fig. 11-31

11.21. The parallel magnetic circuit shown in Fig. 11-32(*a*) is silicon steel with the same cross-sectional area throughout, $S = 1.30 \text{ cm}^2$. The mean lengths are $\ell_1 = \ell_3 = 25$ cm, $\ell_2 = 5$ cm. The coils have 50 turns each. Given that $\Phi_1 = 90\ \mu$Wb and $\Phi_3 = 120\ \mu$Wb, find the coil currents.

$$\Phi_2 = \Phi_3 - \Phi_1 = 0.30 \times 10^{-4} \text{ Wb}$$

$$B_1 = \frac{90 \times 10^{-6}}{1.30 \times 10^{-4}} = 0.69 \text{ T}$$

From the silicon steel *B-H* curve, $H_1 = 87$ A/m. Then, $H_1\ell_1 = 21.8$ A. Similarly, $B_2 = 0.23$ T, $H_2 = 49$ A/m, $H_2\ell_2 = 2.5$ A; and $B_3 = 0.92$ T, $H_3 = 140$ A/m, $H_3\ell_3 = 35.0$ A. The equivalent circuit in Fig. 11-32(*b*) suggests the following equations for the *NI* drop between points *a* and *b*:

$$H_1\ell_1 - F_1 = H_2\ell_2 = F_3 - H_3\ell_3$$

$$21.8 - F_1 = 2.5 = F_3 - 35.0$$

from which $F_1 = 19.3$ A and $F_3 = 37.5$ A. The currents are $I_1 = 0.39$ A and $I_3 = 0.75$ A.

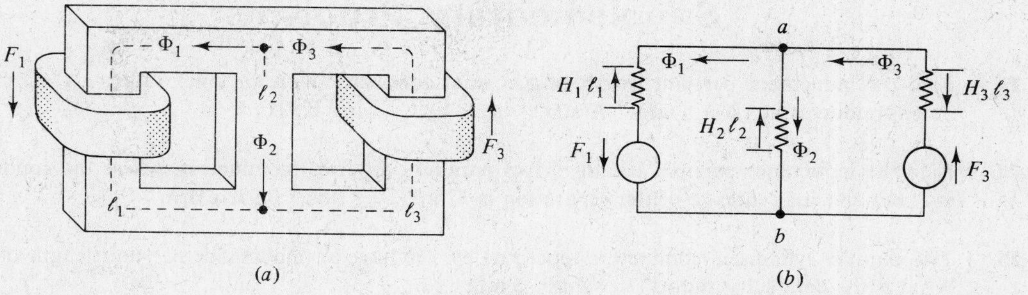

Fig. 11-32

11.22. Obtain the equivalent magnetic circuit for Problem 11.21 using reluctances for three legs, and calculate the flux in the core using $F_1 = 19.3$ A and $F_3 = 37.5$ A.

$$\mathscr{R} = \frac{\ell}{\mu_0 \mu_r S}$$

From the values of B and H found in Problem 11.21,

$$\mu_0 \mu_{r1} = 7.93 \times 10^{-3} \text{ H/m} \qquad \mu_0 \mu_{r2} = 4.69 \times 10^{-3} \text{ H/m} \qquad \mu_0 \mu_{r3} = 6.57 \times 10^{-3} \text{ H/m}$$

Now the reluctances are calculated:

$$\mathscr{R}_1 = \frac{\ell_1}{\mu_0 \mu_{r1} S_1} = 2.43 \times 10^5 \text{ H}^{-1}$$

$\mathscr{R}_2 = 8.20 \times 10^4 \text{ H}^{-1}$, $\mathscr{R}_3 = 2.93 \times 10^5 \text{ H}^{-1}$. From Fig. 11-33,

$$F_3 = \Phi_3 \mathscr{R}_3 + \Phi_2 \mathscr{R}_2 \qquad (1)$$

$$F_1 = \Phi_1 \mathscr{R}_1 - \Phi_2 \mathscr{R}_2 \qquad (2)$$

$$\Phi_1 + \Phi_2 = \Phi_3 \qquad (3)$$

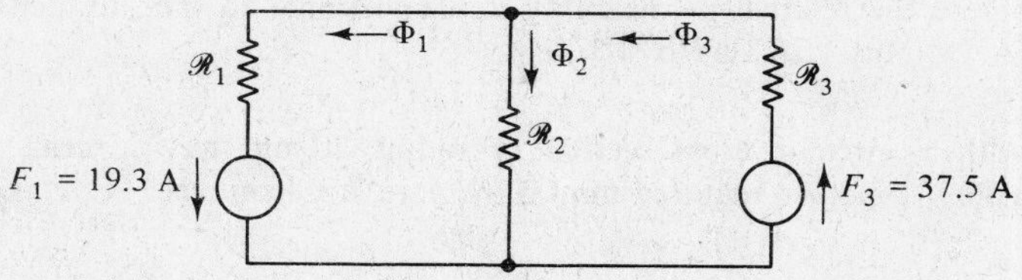

Fig. 11-33

Substituting Φ_2 from (3) into (1) and (2) results in the following set of simultaneous equations in Φ_1 and Φ_3:

$$\begin{aligned} F_1 &= \Phi_1(\mathscr{R}_1 + \mathscr{R}_2) - \Phi_3 \mathscr{R}_2 \\ F_3 &= -\Phi_1 \mathscr{R}_2 + \Phi_3(\mathscr{R}_2 + \mathscr{R}_3) \end{aligned} \qquad \text{or} \qquad \begin{aligned} 19.3 &= \Phi_1(3.25 \times 10^5) - \Phi_3(0.82 \times 10^5) \\ 37.5 &= -\Phi_1(0.82 \times 10^5) + \Phi_3(3.75 \times 10^5) \end{aligned}$$

Solving, $\Phi_1 = 89.7\ \mu$Wb, $\Phi_2 = 30.3\ \mu$Wb, $\Phi_3 = 120\ \mu$Wb.

Although the simultaneous equations above and the similarity to a two-mesh circuit problem may be interesting, it should be noted that the flux densities B_1, B_2 and B_3 had to be known before the relative permeabilities and reluctances could be computed. But if B is known, why not find the flux directly from $\Phi = BS$? Reluctance is simply not of much help in solving problems of this type.

Supplementary Problems

11.23. Find the inductance per unit length of a coaxial conductor with an inner radius $a = 2$ mm and an outer conductor at $b = 9$ mm. Assume $\mu_r = 1$. *Ans.* 0.301 μH/m

11.24. Find the inductance per unit length of two parallel cylindrical conductors, where the conductor radius is 1 mm and the center-to-center separation is 12 mm. *Ans.* 0.992 μH/m

11.25. Two parallel cylindrical conductors separated by 1 m have an inductance per unit length of 2.12 μH/m. What is the conductor radius? *Ans.* 5 mm

11.26. An air-core solenoid with 2500 evenly spaced turns has a length of 1.5 m and radius 2×10^{-2} m. Find the inductance L. *Ans.* 6.58 mH

11.27. Find the inductance of the coil of Fig. 11-9 if $r_1 = 1$ cm, $r_2 = 2$ cm, $\ell = 3$ cm and $N = 800$. *Ans.* 4.70 mH

11.28. A square-cross-section, air-core toroid such as that in Fig. 11-3 has inner radius 5 cm, outer radius 7 cm and height 1.5 cm. If the inductance is 495 μH, how many turns are there in the toroid? Examine the approximate formula and compare the result. *Ans.* 700, 704

11.29. A square-cross-section toroid such as that in Fig. 11-3 has $r_1 = 80$ cm, $r_2 = 82$ cm, $a = 1.5$ cm and 700 turns. Find L using both formulas and compare the results. (See Problem 11.7.) *Ans.* 36.3 μH (both formulas)

11.30. Determine the relative permeabilities of cast iron, cast steel, silicon steel and nickel-iron alloy at a flux density of 0.4 T. Use Figs. 11-13 and 11-14. *Ans.* 318, 1384, 5305, 42 440

11.31. An air gap of length $\ell_a = 2$ mm has a flux density of 0.4 T. Determine the length of a magnetic core with the same NI drop if the core is of (*a*) cast iron, (*b*) cast steel, (*c*) silicon steel. *Ans.* (*a*) 0.64 m; (*b*) 2.77 m; (*c*) 10.6 m

11.32. A magnetic circuit consists of two parts of the same ferromagnetic material ($\mu_r = 4000$). Part *1* has $\ell_1 = 50$ mm, $S_1 = 104\ \text{mm}^2$; part *2* has $\ell_2 = 30$ mm, $S_2 = 120\ \text{mm}^2$. The material is at a part of the curve where the relative permeability is proportional to the flux density. Find the flux Φ if the mmf is 4.0 A. *Ans.* 26.3 μWb

11.33. A toroid with a circular cross section of radius 20 mm has a mean length 280 mm and a flux $\Phi = 1.50$ mWb. Find the required mmf if the core is silicon steel. *Ans.* 83.2 A

11.34. Both parts of the magnetic circuit in Fig. 11-34 are cast steel. Part *1* has $\ell_1 = 34$ cm and $S_1 = 6\ \text{cm}^2$; part *2* has $\ell_2 = 16$ cm and $S_2 = 4\ \text{cm}^2$. Determine the coil current I_1, if $I_2 = 0.5$ A, $N_1 = 200$ turns, $N_2 = 100$ turns and $\Phi = 120\ \mu$Wb. *Ans.* 0.65 A

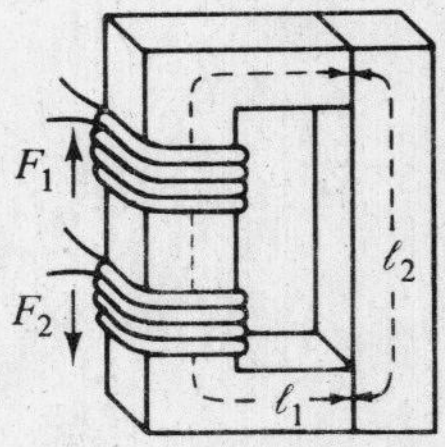

Fig. 11-34

11.35. The silicon steel core shown in Fig. 11-35 has a rectangular cross section 10 mm by 8 mm and a mean length 150 mm. The air-gap length is 0.8 mm and the air-gap flux is 80 μWb. Find the mmf. *Ans.* 561.2 A

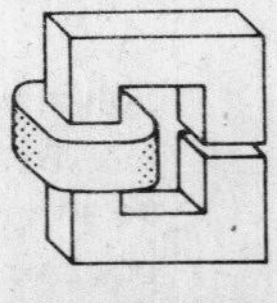
Fig. 11-35

11.36. Solve Problem 11.35 in reverse: the coil mmf is known to be 561.2 A and the air-gap flux is to be determined. Use the trial and error method, starting with the assumption that 90% of the NI drop is across the air gap.

11.37. The silicon steel magnetic circuit of Problem 11.35 has an mmf of 600 A. Determine the air-gap flux. *Ans.* 85.2 μWb

11.38. For the silicon steel magnetic circuit of Problem 11.35, calculate the reluctance of the iron, $\mathcal{R}_i$, and the reluctance of the air gap, $\mathcal{R}_a$. Assume the flux $\Phi = 80\ \mu\text{Wb}$ and solve for F. See Fig. 11-36.
Ans. $\mathcal{R}_i = 0.313\ \mu\text{H}^{-1}$, $\mathcal{R}_a = 6.70\ \mu\text{H}^{-1}$, $F = 561$ A

11.39. A silicon steel core such as shown in Fig. 11-35 has a rectangular cross section of area $S_i = 80\ \text{mm}^2$ and an air gap of length $\ell_a = 0.8$ mm with area $S_a = 95\ \text{mm}^2$. The mean length of the core is 150 mm and the mmf is 600 A. Solve graphically for the flux by plotting Φ versus F in the manner of Problem 11.15. *Ans.* 85 μWb

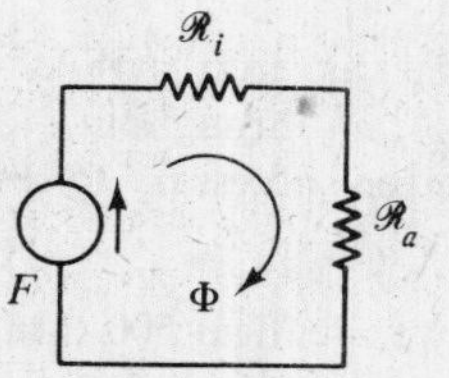

Fig. 11-36

11.40. Solve Problem 11.39 graphically using the negative air-gap line for an mmf of 600 A. *Ans.* 85 μWb

11.41. Solve Problem 11.39 graphically in the manner of Problem 11.17, obtaining the flux density in the core. *Ans.* 1.06 T

11.42. A rectangular ferromagnetic core 40 × 60 mm has a flux $\Phi = 1.44$ mWb. An air gap in the core is of length $\ell_a = 2.5$ mm. Find the NI drop across the air gap. *Ans.* 1079 A

11.43. A toroid with cross section of radius 2 cm has a silicon steel core of mean length 28 cm and an air gap of length 1 mm. Assume the air-gap area, S_a, is 10% greater than the adjacent core and find the mmf required to establish an air-gap flux of 1.5 mWb. *Ans.* 952 A

11.44. The magnetic circuit shown in Fig. 11-37 has an mmf of 500 A. Part *1* is cast steel with $\ell_1 = 340$ mm and $S_1 = 400\ \text{mm}^2$; part *2* is cast iron with $\ell_2 = 138$ mm and $S_2 = 360\ \text{mm}^2$. Determine the flux Φ.
Ans. 229 μWb

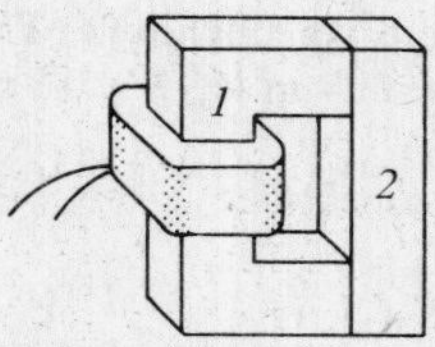

Fig. 11-37

11.45. Solve Problem 11.44 graphically in the manner of Problem 11.18. *Ans.* 229 μWb

11.46. A toroid of square cross section, with $r_1 = 2$ cm, $r_2 = 3$ cm and height $a = 1$ cm, has a two-part core. Part *1* is silicon steel of mean length 7.9 cm; part *2* is nickel-iron alloy of mean length 7.9 cm. Find the flux that results from an mmf of 17.38 A. *Ans.* 10^{-4} Wb

11.47. Solve Problem 11.46 by the graphical method of Problem 11.17. Why is it that the plotting of the second reverse *B-H* curve on the first is not as difficult as might be expected?
Ans. 10^{-4} Wb. The mean lengths and cross-sectional areas are the same.

11.48. The cast steel parallel magnetic circuit in Fig. 11-38 has a 500-turn coil in the center leg, where the cross-sectional area is twice that of the remainder of the core. The dimensions are: $\ell_a = 1$ mm, $S_2 = S_3 = 150\ \text{mm}^2$, $S_1 = 300\ \text{mm}^2$, $\ell_1 = 40$ mm, $\ell_2 = 110$ mm and $\ell_3 = 109$ mm. Find the coil current required to produce an air-gap flux of 125 μWb. Assume that S_a exceeds S_3 by 17%.
Ans. 1.34 A

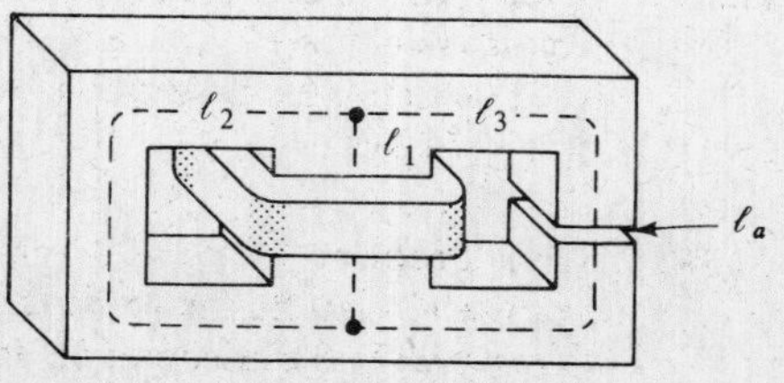

Fig. 11-38

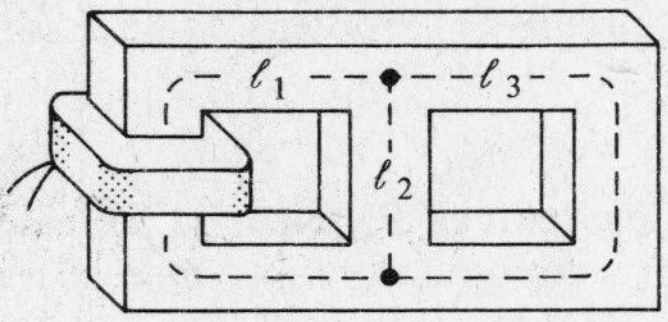

Fig. 11-39

11.49. The cast iron parallel circuit core in Fig. 11-39 has a 500-turn coil and a uniform cross section of $1.5\ \text{cm}^2$ throughout. The mean lengths are $\ell_1 = \ell_3 = 10$ cm and $\ell_2 = 4$ cm. Determine the coil current necessary to result in a flux density of 0.25 T in leg *3*. *Ans.* 1.05 A

11.50. Two identical 500-turn coils have equal currents and are wound as indicated in Fig. 11-40. The cast steel core has a flux in leg *3* of 120 μWb. Determine the coil currents and the flux in leg *1*.
Ans. 0.41 A, 0 Wb

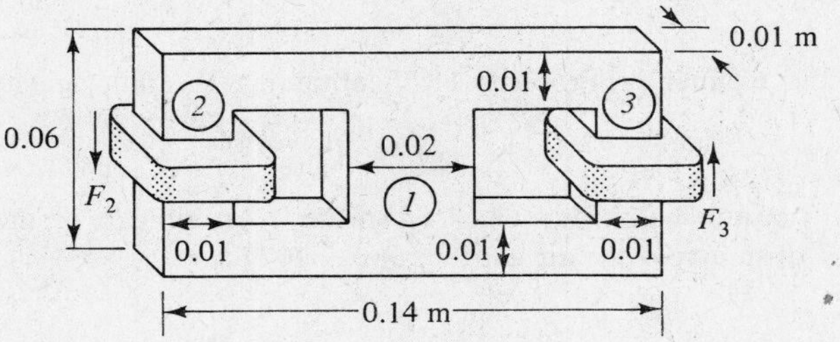

Fig. 11-40

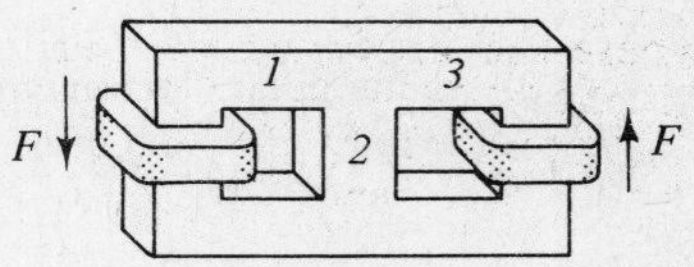

Fig. 11-41

11.51. Two identical coils are wound as indicated in Fig. 11-41. The silicon steel core has a cross section of $6\ \text{cm}^2$ throughout. The mean lengths are $\ell_1 = \ell_3 = 14$ cm and $\ell_2 = 4$ cm. Find the coil mmfs if the flux in leg *1* is 0.7 mWb. *Ans.* 38.5 A

Chapter 12

Displacement Current and Induced EMF

12.1 DISPLACEMENT CURRENT

For static fields, the curl of **H** is equal to **J**, the current density at the point (see Section 9.5). This current density, which is due to the motion of actual charges such as electrons, protons and ions, will now be called the *conduction current* density and will be denoted $\mathbf{J}_c$. It is clear that if $\nabla \times \mathbf{H} = \mathbf{J}_c$ also held for time-varying fields, then $\nabla \cdot \mathbf{J}_c = \nabla \cdot (\nabla \times \mathbf{H}) = 0$ (see Section 9.4.). But this would be incompatible with the continuity equation,

$$\nabla \cdot \mathbf{J}_c = -\frac{\partial \rho}{\partial t}$$

obtained in Section 6.9. Hence, James Clerk Maxwell postulated that

$$\nabla \times \mathbf{H} = \mathbf{J}_c + \mathbf{J}_D$$

where the *displacement current* density, $\mathbf{J}_D$, is defined by

$$\mathbf{J}_D = \frac{\partial \mathbf{D}}{\partial t}$$

With the displacement current added to the conduction current, the continuity equation is "saved." Indeed,

$$\nabla \cdot \mathbf{J}_c = \nabla \cdot (\nabla \times \mathbf{H}) - \nabla \cdot \mathbf{J}_D = 0 - \nabla \cdot \frac{\partial \mathbf{D}}{\partial t} = -\frac{\partial}{\partial t}(\nabla \cdot \mathbf{D}) = -\frac{\partial \rho}{\partial t}$$

where $\nabla \cdot \mathbf{D} = \rho$ (Section 4.3) has been used.

The displacement current (in amperes) through a specified open surface is obtained by integration in exactly the same way as is the conduction current. Thus

$$i_c = \int_S \mathbf{J}_c \cdot d\mathbf{S} \qquad i_D = \int_S \mathbf{J}_D \cdot d\mathbf{S} = \int_S \frac{\partial \mathbf{D}}{\partial t} \cdot d\mathbf{S}$$

The expression for i_D may be interpreted in terms of the motion of charges. If charge Q is moving with velocity **U**, the electric field surrounding Q also advances with velocity **U**. Therefore, even though Q may not be physically crossing a surface S (which would constitute a conduction current), the lines of **D** will be changing across S, resulting in a displacement current.

EXAMPLE 1 A time-varying voltage applied to a parallel-plate capacitor (Fig. 12-1) results in a (time-varying) current i_c in the connecting leads. Two open surfaces with a common contour C are shown. Because $\nabla \cdot (\nabla \times \mathbf{H}) = 0$, the divergence theorem (Section 4.5) gives

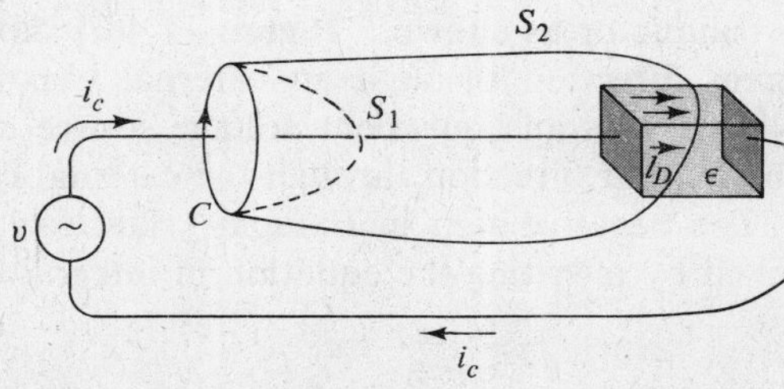

Fig. 12-1

$$\int_{S_1} (\nabla \times \mathbf{H}) \cdot d\mathbf{S} = \int_{S_2} (\nabla \times \mathbf{H}) \cdot d\mathbf{S}$$

or

$$\int_{S_1} \left(\mathbf{J}_c + \frac{\partial \mathbf{D}}{\partial t}\right) \cdot d\mathbf{S} = \int_{S_2} \left(\mathbf{J}_c + \frac{\partial \mathbf{D}}{\partial t}\right) \cdot d\mathbf{S}$$

In the absence of fringing, **D** will be changing only within the capacitor; moreover, there can be no moving charges ($\mathbf{J}_c = 0$) in the dielectric. Thus

$$\int_{S_1} \mathbf{J}_c \cdot d\mathbf{S} = \int_{S_3} \frac{\partial \mathbf{D}}{\partial t} \cdot d\mathbf{S}$$

where S_3 is that part of S_2 which lies within the dielectric. The integral on the left is simply i_c, the conduction current made up of the moving charges in the leads. The integral on the right is the displacement current i_D in the dielectric.

The equality of i_c and i_D for this case is verified in Problem 12.1.

12.2 RATIO OF J_c TO J_D

Some materials are neither good conductors nor perfect dielectrics, so that both conduction current and displacement current exist. A model for the poor conductor or lossy dielectric is shown in Fig. 12-2. Assuming the time dependence $e^{j\omega t}$ for **E**, the total current density is

$$\mathbf{J}_t = \mathbf{J}_c + \mathbf{J}_D = \sigma\mathbf{E} + \frac{\partial}{\partial t}(\epsilon\mathbf{E}) = \sigma\mathbf{E} + j\omega\epsilon\mathbf{E}$$

from which

$$\frac{J_c}{J_D} = \frac{\sigma}{\omega\epsilon}$$

As expected, the displacement current becomes increasingly important as the frequency increases.

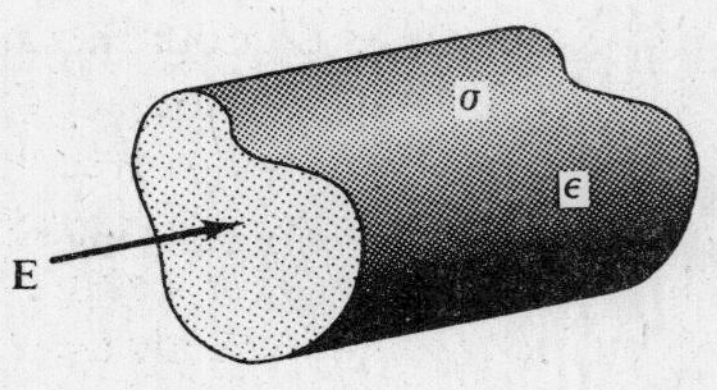

Fig. 12-2

12.3 FARADAY'S LAW

When a conductor moves through a magnetic field, cutting flux, a voltage is induced in the conductor. Similarly, when flux cuts across a stationary conductor, a voltage is induced. In either case, the voltage and the rate of cutting the flux are related by Faraday's law:

$$v = -\frac{d\phi}{dt}$$

The polarity of the induced voltage is sometimes illustrated by a distorted flux field as shown in Fig. 12-3. The lines appear to be pushed ahead of the moving conductor. A counterclockwise flux is shown around the conductor which would create the same distorted flux field. By the right-hand rule, the current, if a closed path were provided, would leave the conductor as shown. Figure 12-4(a) shows the current direction through an external circuit; Fig. 12-4(b) shows an equivalent voltage source and the same current direction through the external circuit.

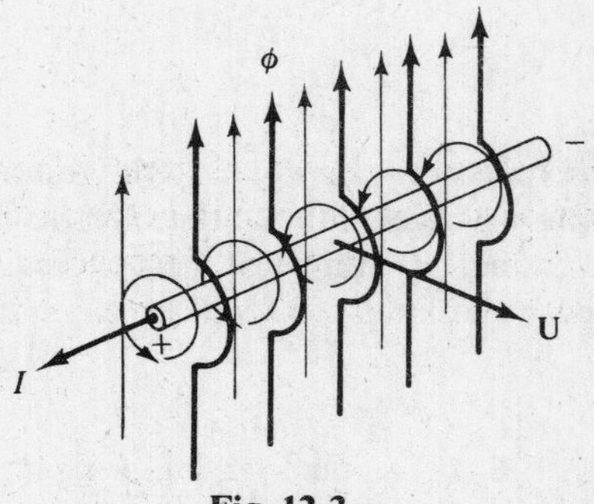

Fig. 12-3

The negative sign in Faraday's law can be explained by rewriting the equation in integral form.

$$v = -\frac{d\phi}{dt}$$

$$\oint_C \mathbf{E} \cdot d\mathbf{l} = -\frac{d}{dt}\int_S \mathbf{B} \cdot d\mathbf{S}$$

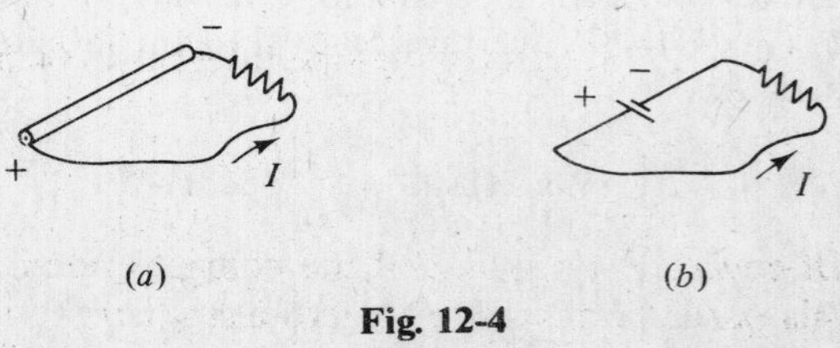

Fig. 12-4

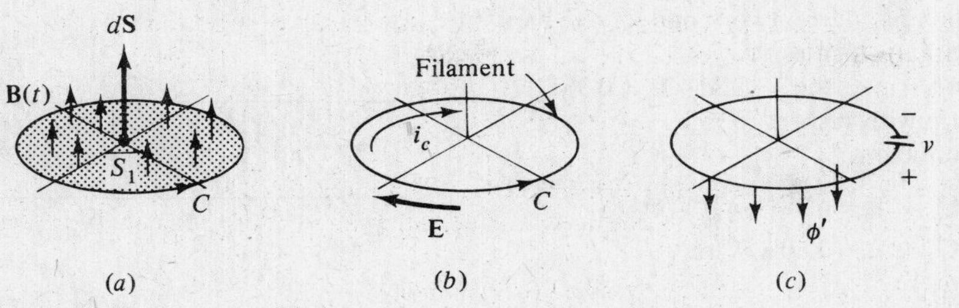

Fig. 12-5

The contour C around which the integral on the left is taken encloses the plane area S of the surface integral on the right. A positive direction is assigned to C and the direction of the normal, $d\mathbf{S}$, is then determined by the familiar right-hand rule. See Fig. 12-5(*a*). A changing magnetic flux density, $\mathbf{B}(t)$, is present within the contour. If $\mathbf{B}$ is increasing with time, the time derivative will be positive and, thus, the right side of the equation will be negative. In order for the left integral to be negative, the direction of $\mathbf{E}$ must be opposite to that of the contour, Fig. 12-5(*b*). A conducting filament in place of the contour would carry a current i_c, also in the direction of $\mathbf{E}$. As shown in Fig. 12-5(*c*), such a current loop generates a flux ϕ' which opposes the increase in $\mathbf{B}$. *Lenz's law* summarizes this somewhat involved discussion: *the voltage induced by a changing flux has a polarity such that the current established in a closed path gives rise to a flux which opposes the change in flux.*

In the special case of a conductor moving through a stationary magnetic field, the polarity predicted by Lenz's law will always be such that the conductor experiences magnetic forces which oppose its motion.

12.4 CONDUCTORS IN MOTION THROUGH TIME-INDEPENDENT FIELDS

The force $\mathbf{F}$ on a charge Q in a magnetic field $\mathbf{B}$, where the charge is moving with velocity $\mathbf{U}$, was examined in Chapter 10.

$$\mathbf{F} = Q(\mathbf{U} \times \mathbf{B})$$

A *motional* electric field intensity, $\mathbf{E}_m$, can be defined as the force per unit charge:

$$\mathbf{E}_m = \frac{\mathbf{F}}{Q} = \mathbf{U} \times \mathbf{B}$$

When a conductor with a great number of free charges moves through a field $\mathbf{B}$, the impressed $\mathbf{E}_m$ creates a voltage difference between the two ends of the conductor, the magnitude of which depends on how $\mathbf{E}_m$ is oriented with respect to the conductor. With conductor ends a and b, the voltage of a with respect to b is

$$v_{ab} = \int_b^a \mathbf{E}_m \cdot d\mathbf{l} = \int_b^a (\mathbf{U} \times \mathbf{B}) \cdot d\mathbf{l}$$

If the velocity $\mathbf{U}$ and the field $\mathbf{B}$ are at right angles, and the conductor is normal to both, then a conductor of length ℓ will have a voltage

$$v = B\ell U$$

For a closed loop the line integral must be taken around the entire loop:

$$v = \oint (\mathbf{U} \times \mathbf{B}) \cdot d\mathbf{l}$$

Of course, if only part of the complete loop is in motion, it is necessary only that the integral cover this part, since $\mathbf{E}_m$ will be zero elsewhere.

EXAMPLE 2 In Fig. 12-6, two conducting bars move outward with velocities $\mathbf{U}_1 = 12.5(-\mathbf{a}_y)$ m/s and $\mathbf{U}_2 = 8.0\mathbf{a}_y$ m/s in the field $\mathbf{B} = 0.35\mathbf{a}_z$ T. Find the voltage of b with respect to c.

At the two conductors,

$$\mathbf{E}_{m1} = \mathbf{U}_1 \times \mathbf{B} = 4.38(-\mathbf{a}_x) \text{ V/m}$$
$$\mathbf{E}_{m2} = \mathbf{U}_2 \times \mathbf{B} = 2.80\mathbf{a}_x \text{ V/m}$$

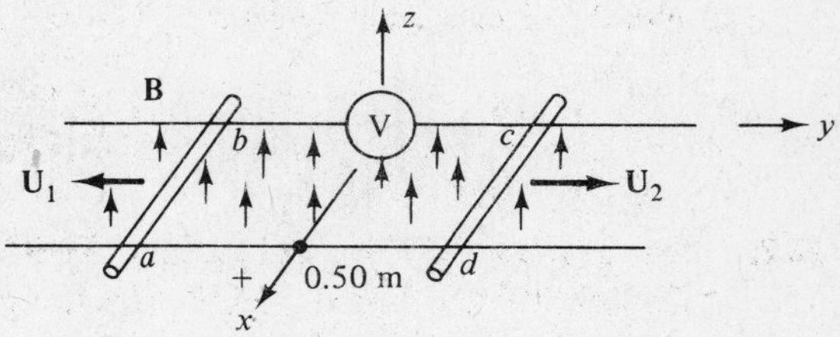

Fig. 12-6

and so

$$v_{ab} = \int_0^{0.50} 4.38(-\mathbf{a}_x) \cdot dx\,\mathbf{a}_x = -2.19 \text{ V} \qquad v_{dc} = \int_0^{0.50} 2.80\,\mathbf{a}_x \cdot dx\,\mathbf{a}_x = 1.40 \text{ V}$$
$$v_{bc} = v_{ba} + v_{ad} + v_{dc} = 2.19 + 0 + 1.40 = 3.59 \text{ V}$$

Since b is positive with respect to c, current through the meter will be in the $\mathbf{a}_y$ direction. This clockwise current in the circuit gives rise to flux in the $-\mathbf{a}_z$ direction, which, in accordance with Lenz's law, counters the increase in the flux in the $+\mathbf{a}_z$ direction due to the expansion of the circuit. Moreover, the forces that $\mathbf{B}$ exerts on the moving conductors are directed opposite to their velocities.

12.5 CONDUCTORS IN MOTION THROUGH TIME-DEPENDENT FIELDS

When a closed conducting loop is in motion (this includes changes in shape) and also the field $\mathbf{B}$ is a function of time (as well as of position), then the total induced voltage is made up of a contribution from each of the two sources of flux change. Faraday's law becomes

$$v = -\frac{d}{dt}\int_S \mathbf{B} \cdot d\mathbf{S} = -\int_S \frac{\partial \mathbf{B}}{\partial t} \cdot d\mathbf{S} + \oint (\mathbf{U} \times \mathbf{B}) \cdot d\mathbf{l}$$

The first term on the right is the voltage due to the change in $\mathbf{B}$, with the loop held fixed; the second term is the voltage arising from the motion of the loop, with $\mathbf{B}$ held fixed. The polarity of each term is found from the appropriate form of Lenz's law, and the two terms are then added with regard to those polarities. See Problem 12.17.

Solved Problems

12.1. Show that the displacement current in the dielectric of a parallel-plate capacitor is equal to the conduction current in the leads.

Refer to Fig. 12-1. The capacitance of the capacitor is

$$C = \frac{\epsilon A}{d}$$

where A is the plate area and d is the separation. The conduction current is then

$$i_c = C\frac{dv}{dt} = \frac{\epsilon A}{d}\frac{dv}{dt}$$

On the other hand, the electric field in the dielectric is, neglecting fringing, $E = v/d$. Hence

$$D = \epsilon E = \frac{\epsilon}{d} v \qquad \frac{\partial D}{\partial t} = \frac{\epsilon}{d}\frac{dv}{dt}$$

and the displacement current is (**D** is normal to the plates)

$$i_D = \int_A \frac{\partial \mathbf{D}}{\partial t} \cdot d\mathbf{S} = \int_A \frac{\epsilon}{d}\frac{dv}{dt}\, dS = \frac{\epsilon A}{d}\frac{dv}{dt} = i_c$$

12.2. In a material for which $\sigma = 5.0\,\text{S/m}$ and $\epsilon_r = 1$ the electric field intensity is $E = 250 \sin 10^{10}t$ (V/m). Find the conduction and displacement current densities, and the frequency at which they have equal magnitudes.

$$J_c = \sigma E = 1250 \sin 10^{10}t \quad (\text{A/m}^2)$$

On the assumption that the field direction does not vary with time,

$$J_D = \frac{\partial D}{\partial t} = \frac{\partial}{\partial t}(\epsilon_0 \epsilon_r 250 \sin 10^{10}t) = 22.1 \cos 10^{10}t \quad (\text{A/m}^2)$$

For $J_c = J_D$,

$$\sigma = \omega\epsilon \qquad \text{or} \qquad \omega = \frac{5.0}{8.854 \times 10^{-12}} = 5.65 \times 10^{11}\,\text{rad/s}$$

which is equivalent to a frequency $f = 8.99 \times 10^{10}$ Hz $= 89.9$ GHz.

12.3. A coaxial capacitor with inner radius 5 mm, outer radius 6 mm and length 500 mm has a dielectric for which $\epsilon_r = 6.7$ and an applied voltage $250 \sin 377t$ (V). Determine the displacement current i_D and compare with the conduction current i_c.

Assume the inner conductor to be at $v = 0$. Then, from Problem 8.7, the potential at $0.005 \leq r \leq 0.006$ m is

$$v = \left[\frac{250}{\ln(6/5)} \sin 377t\right]\left(\ln \frac{r}{0.005}\right) \quad (\text{V})$$

From this,

$$\mathbf{E} = -\nabla v = -\frac{1.37 \times 10^3}{r} \sin 377t\, \mathbf{a}_r \quad (\text{V/m})$$

$$\mathbf{D} = \epsilon_0 \epsilon_r \mathbf{E} = -\frac{8.13 \times 10^{-8}}{r} \sin 377t\, \mathbf{a}_r \quad (\text{C/m}^2)$$

$$\mathbf{J}_D = \frac{\partial \mathbf{D}}{\partial t} = -\frac{3.07 \times 10^{-5}}{r} \cos 377t\, \mathbf{a}_r \quad (\text{A/m}^2)$$

$$i_D = J_D(2\pi r L) = 9.63 \times 10^{-5} \cos 377t \quad (\text{A})$$

The circuit analysis method for i_c requires the capacitance,

$$C = \frac{2\pi\epsilon_0 \epsilon_r L}{\ln(6/5)} = 1.02 \times 10^{-9}\,\text{F}$$

Then
$$i_c = C\frac{dv}{dt} = (1.02 \times 10^{-9})(250)(377)(\cos 377t) = 9.63 \times 10^{-5} \cos 377t \quad (\text{A})$$

It is seen that $i_c = i_D$.

12.4. Moist soil has a conductivity of 10^{-3} S/m and $\epsilon_r = 2.5$. Find J_c and J_D where

$$E = 6.0 \times 10^{-6} \sin 9.0 \times 10^9 t \quad (\text{V/m})$$

First, $J_c = \sigma E = 6.0 \times 10^{-9} \sin 9.0 \times 10^9 t$ (A/m²). Then, since $D = \epsilon_0 \epsilon_r E$,

$$J_D = \frac{\partial D}{\partial t} = \epsilon_0 \epsilon_r \frac{\partial E}{\partial t} = 1.20 \times 10^{-6} \cos 9.0 \times 10^9 t \quad (\text{A/m}^2)$$

12.5. In Fig. 12-7 a three-meter-long conductor moves parallel to the x axis with velocity $\mathbf{U} = 2.50\mathbf{a}_y$ m/s in a uniform field $\mathbf{B} = 0.50\mathbf{a}_z$ T. Find the induced voltage.

Since the uniform field is at right angles to the velocity, and the conductor is normal to both,

$$v = B\ell U = 3.75 \text{ V}$$

Flux lines around the moving conductor, or Lenz's law, show that end a is positive.

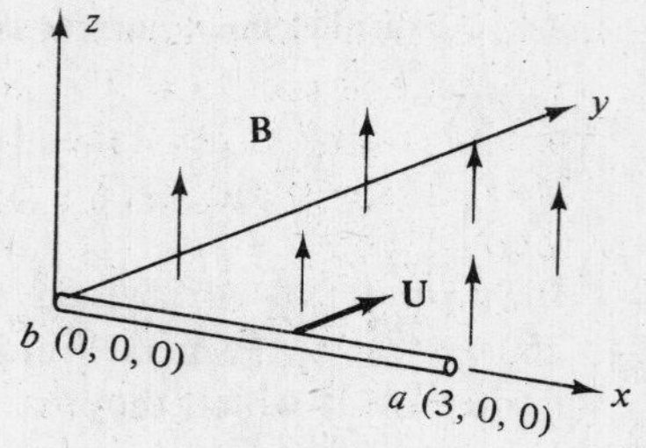

Fig. 12-7

12.6. With the same field and velocity as in Problem 12.5, find the voltage induced in a 3 m conductor lying along the y axis.

The motional **E** field is $\mathbf{U} \times \mathbf{B} = 1.25\mathbf{a}_x$ V/m, as before. Then, if the ends are at (0, 0, 0) and (0, 3, 0) m,

$$v = \int_0^3 1.25\mathbf{a}_x \cdot dy\,\mathbf{a}_y = 0$$

More generally, a conductor moving along its own length cuts no magnetic flux.

12.7. Find the induced voltage in the conductor of Fig. 12-8 where $\mathbf{B} = 0.04\mathbf{a}_y$ T and

$$\mathbf{U} = 2.5 \sin 10^3 t\,\mathbf{a}_z \quad \text{(m/s)}$$

$$\mathbf{E}_m = \mathbf{U} \times \mathbf{B} = 0.10 \sin 10^3 t\,(-\mathbf{a}_x) \quad \text{(V/m)}$$

$$v = \int_0^{0.20} 0.10 \sin 10^3 t(-\mathbf{a}_x) \cdot dx\,\mathbf{a}_x$$

$$= -0.02 \sin 10^3 t \quad \text{(V)}$$

The conductor first moves in the $\mathbf{a}_z$ direction. The $x = 0.20$ end is negative with respect to the end at the z axis for this half cycle.

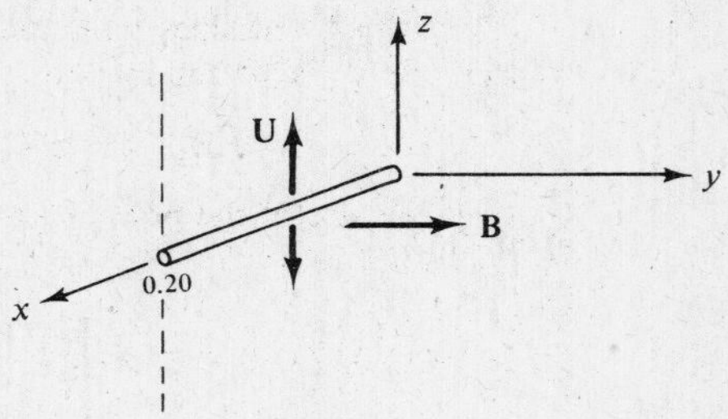

Fig. 12-8

12.8. An area of 0.65 m² in the $z = 0$ plane is enclosed by a filamentary conductor. Find the induced voltage, given that

$$\mathbf{B} = 0.05 \cos 10^3 t(\mathbf{a}_y + \mathbf{a}_z)/\sqrt{2} \quad \text{(T)}$$

See Fig. 12-9.

$$v = -\int_S \frac{\partial \mathbf{B}}{\partial t} \cdot dS\,\mathbf{a}_z$$

$$= \int_S 50 \sin 10^3 t \left(\frac{\mathbf{a}_y + \mathbf{a}_z}{\sqrt{2}}\right) \cdot dS\,\mathbf{a}_z$$

$$= 23.0 \sin 10^3 t \quad \text{(V)}$$

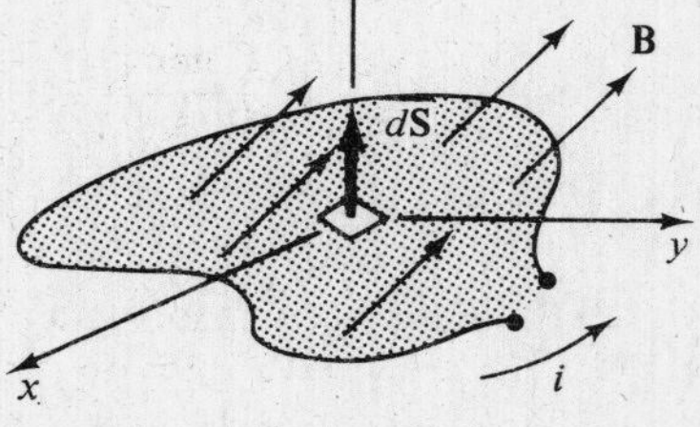

Fig. 12-9

The field is decreasing in the first half cycle of the cosine function. The direction of i in a closed circuit must be such as to oppose this decrease. Thus the conventional current must have the direction shown in Fig. 12-9.

12.9. The circular loop conductor shown in Fig. 12-10 lies in the $z = 0$ plane, has a radius of 0.10 m and a resistance of 5.0 Ω. Given $\mathbf{B} = 0.20 \sin 10^3 t\, \mathbf{a}_z$ (T), determine the current.

$$\phi = \mathbf{B} \cdot \mathbf{S} = 2 \times 10^{-3} \pi \sin 10^3 t \quad \text{(Wb)}$$

$$v = -\frac{d\phi}{dt} = -2\pi \cos 10^3 t \quad \text{(V)}$$

$$i = \frac{v}{R} = -0.4\pi \cos 10^3 t \quad \text{(A)}$$

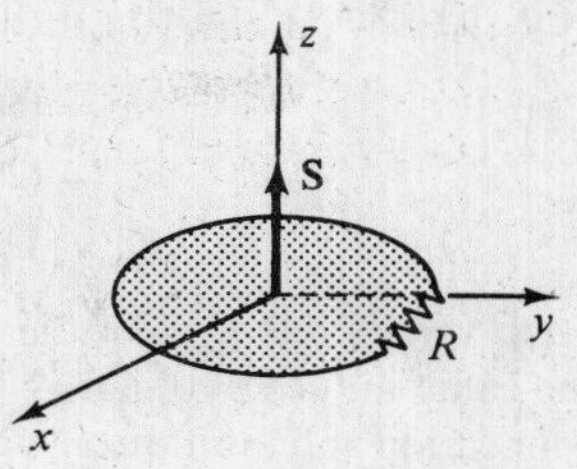

Fig. 12-10

At $t = 0+$ the flux is increasing. In order to oppose this increase, current in the loop must have an instantaneous direction $-\mathbf{a}_y$ where the loop crosses the positive x axis.

12.10. The rectangular loop shown in Fig. 12-11 moves toward the origin at a velocity $\mathbf{U} = -250\mathbf{a}_y$ m/s in a field

$$\mathbf{B} = 0.80 e^{-0.50\, y} \mathbf{a}_z \quad \text{(T)}$$

Find the current at the instant the coil sides are at $y = 0.50$ m and 0.60 m, if $R = 2.5$ Ω.

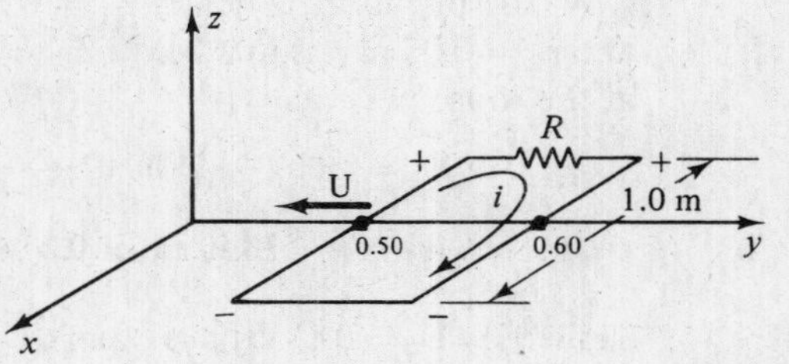

Fig. 12-11

Only the 1.0 m sides have induced voltages. Let the side at $y = 0.50$ m be *1*.

$$v_1 = B_1 \ell U = 0.80 e^{-0.25}(1)(250) = 155.8 \text{ V} \qquad v_2 = B_2 \ell U = 148.2 \text{ V}$$

The voltages are of the polarity shown. The instantaneous current is

$$i = \frac{155.8 - 148.2}{2.5} = 3.04 \text{ A}$$

12.11. A conductor 1 cm in length is parallel to the z axis and rotates at a radius of 25 cm at 1200 rev/min (see Fig. 12-12). Find the induced voltage if the radial field is given by $\mathbf{B} = 0.5\mathbf{a}_r$ T.

The angular velocity is

$$\left(1200 \frac{\text{rev}}{\text{min}}\right)\left(\frac{1 \text{ min}}{60 \text{ s}}\right)\left(2\pi \frac{\text{rad}}{\text{rev}}\right) = 40\pi \frac{\text{rad}}{\text{s}}$$

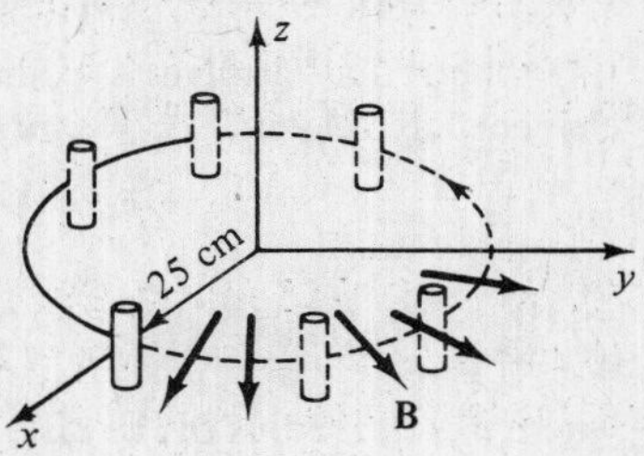

Fig. 12-12

Hence

$$U = r\omega = (0.25)(40\pi) \text{ m/s}$$

$$\mathbf{E}_m = 10\pi \mathbf{a}_\phi \times 0.5\,\mathbf{a}_r = 5.0\pi(-\mathbf{a}_z) \text{ V/m}$$

$$v = \int_0^{0.01} 5.0\pi(-\mathbf{a}_z) \cdot dz\, \mathbf{a}_z = -5.0 \times 10^{-2}\pi \text{ V}$$

The negative sign indicates that the lower end of the conductor is positive with respect to the upper end.

12.12. A conducting cylinder of radius 7 cm and height 15 cm rotates at 600 rev/min in a radial field $\mathbf{B} = 0.20\mathbf{a}_r$ T. Sliding contacts at the top and bottom connect to a voltmeter as shown in

Fig. 12-13. Find the induced voltage.

$$\omega = (600)(\tfrac{1}{60})(2\pi) = 20\pi \text{ rad/s}$$
$$\mathbf{U} = (20\pi)(0.07)\mathbf{a}_\phi \text{ m/s}$$
$$\mathbf{E}_m = \mathbf{U} \times \mathbf{B} = 0.88(-\mathbf{a}_z) \text{ V/m}$$

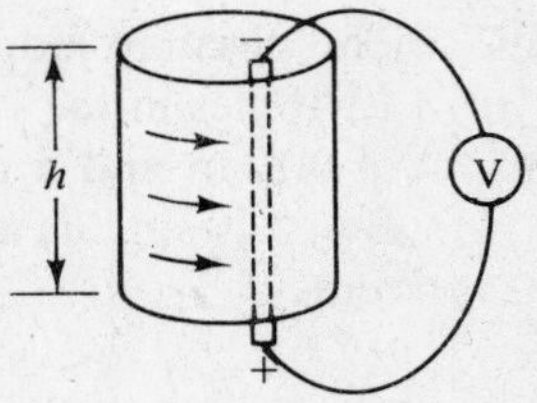

Fig. 12-13

Each vertical element of the curved surface cuts the same flux and has the same induced voltage. These elements are effectively in a parallel connection and the induced voltage of any element is the same as the total.

$$v = \int_0^{0.15} 0.88(-\mathbf{a}_z) \cdot dz\,\mathbf{a}_z = -0.13 \text{ V} \qquad (+ \text{ at the bottom})$$

12.13. In Fig. 12-14 a rectangular conducting loop with resistance $R = 0.20\ \Omega$ turns at 500 rev/min. The vertical conductor at $r_1 = 0.03$ m is in a field $\mathbf{B}_1 = 0.25\mathbf{a}_r$ T, and the conductor at $r_2 = 0.05$ m is in a field $\mathbf{B}_2 = 0.80\mathbf{a}_r$ T. Find the current in the loop.

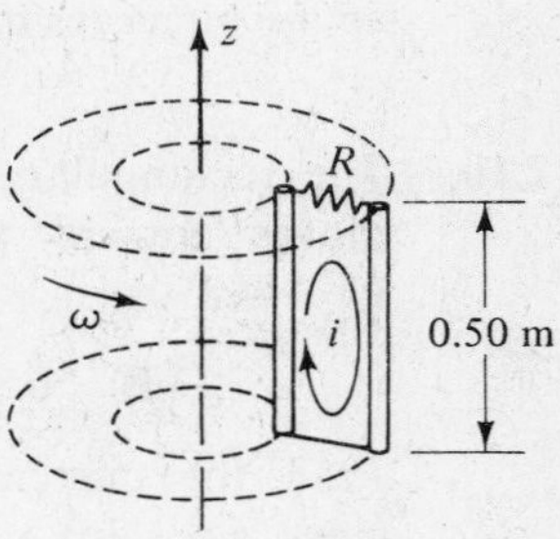

Fig. 12-14

$$\mathbf{U}_1 = (500)(\tfrac{1}{60})(2\pi)(0.03)\mathbf{a}_\phi = 0.50\pi\mathbf{a}_\phi \text{ m/s}$$
$$v_1 = \int_0^{0.50} (0.50\pi\mathbf{a}_\phi \times 0.25\mathbf{a}_r) \cdot dz\,\mathbf{a}_z = -0.20 \text{ V}$$

Similarly, $\mathbf{U}_2 = 0.83\pi\mathbf{a}_\phi$ m/s and $v_2 = -1.04$ V. Then

$$i = \frac{1.04 - 0.20}{0.20} = 4.20 \text{ A}$$

in the direction shown on the diagram.

12.14. The circular disk shown in Fig. 12-15 rotates at ω (rad/s) in a uniform flux density $\mathbf{B} = B\mathbf{a}_z$. Sliding contacts connect a voltmeter to the disk. What voltage is indicated on the meter from this *Faraday homopolar generator*?

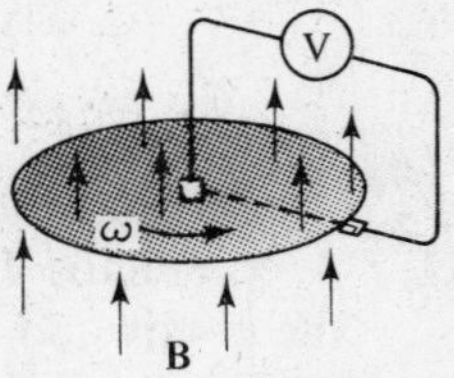

Fig. 12-15

One radial element is examined. A general point on this radial element has velocity $\mathbf{U} = \omega r\mathbf{a}_\phi$, so that

$$\mathbf{E}_m = \mathbf{U} \times \mathbf{B} = \omega r B\mathbf{a}_r$$

and
$$v = \int_0^a \omega r B\mathbf{a}_r \cdot dr\,\mathbf{a}_r = \frac{\omega a^2 B}{2}$$

where a is the radius of the disk. The positive result indicates that the outer point is positive with respect to the center for the directions of $\mathbf{B}$ and ω shown.

12.15. A square coil, 0.60 m on a side, rotates about the x axis at $\omega = 60\pi$ rad/s in a field $\mathbf{B} = 0.80\mathbf{a}_z$ T, as shown in Fig. 12-16(*a*). Find the induced voltage.

Assuming that the coil is initially in the xy plane,

$$\alpha = \omega t = 60\pi t \text{ (rad)}$$

The projected area on the xy plane becomes [see Fig. 12-16(*b*)]:

$$A = (0.6)(0.6 \cos 60\pi t) \text{ (m}^2\text{)}$$

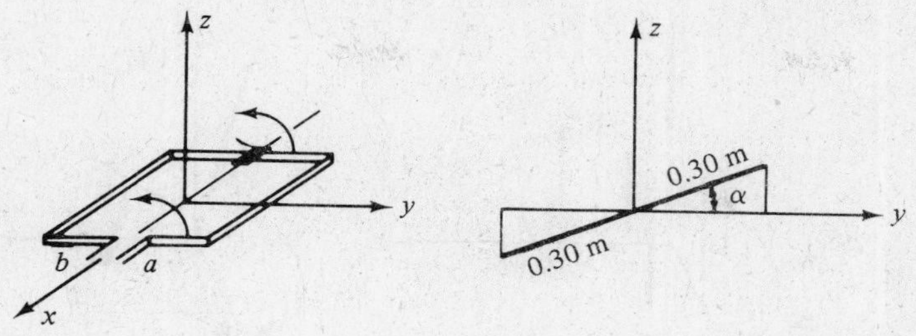

(a) (b)

Fig. 12-16

Then $\phi = BA = 0.288 \cos 60\pi t$ (Wb) and

$$v = -\frac{d\phi}{dt} = 54.3 \sin 60\pi t \quad \text{(V)}$$

Lenz's law shows that this is the voltage of a with respect to b.

Alternate Method

Each side parallel to the x axis has a y component of velocity whose magnitude is

$$|U_y| = |r\omega \sin\alpha| = |18.0\pi \sin 60\pi t| \quad \text{(m/s)}$$

The voltages $B\ell\,|U_y|$ for the two sides add, giving

$$|v| = 2(B\ell\,|U_y|) = |54.3 \sin 60\pi t| \quad \text{(V)}$$

Lenz's law again determines the proper sign.

12.16. Find the electrical power generated in the loop of Problem 12.13. Check the result by calculating the rate at which mechanical work is done on the loop.

The electrical power is the power loss in the resistor:

$$P_e = i^2R = (4.20)^2(0.20) = 3.53 \text{ W}$$

The forces exerted by the field on the two vertical conductors are

$$\mathbf{F}_1 = i(\mathbf{l}_1 \times \mathbf{B}_1) = (4.20)(0.50)(0.25)(\mathbf{a}_z \times \mathbf{a}_r) = 0.525\,\mathbf{a}_\phi \text{ N}$$
$$\mathbf{F}_2 = i(\mathbf{l}_2 \times \mathbf{B}_2) = (4.20)(0.50)(0.80)(-\mathbf{a}_z \times \mathbf{a}_r) = -1.68\,\mathbf{a}_\phi \text{ N}$$

To turn the loop, forces $-\mathbf{F}_1$ and $-\mathbf{F}_2$ must be applied; these do work at the rate

$$P = (-\mathbf{F}_1)\cdot\mathbf{U}_1 + (-\mathbf{F}_2)\cdot\mathbf{U}_2 = (-0.525)(0.50\pi) + (1.68)(0.83\pi) = 3.55 \text{ W}$$

To within rounding errors, $P = P_e$.

12.17. As shown in Fig. 12-17(a), a planar conducting loop rotates with angular velocity ω about the x axis; at $t = 0$ it is in the xy plane. A time-varying magnetic field, $\mathbf{B} = B(t)\mathbf{a}_z$, is present. Find the voltage induced in the loop (a) by using $v = -d\phi/dt$, (b) by using

$$v = -\int_S \frac{\partial \mathbf{B}}{\partial t}\cdot d\mathbf{S} + \oint (\mathbf{U} \times \mathbf{B})\cdot d\mathbf{l}$$

(a) Figure 12-17(b) shows the loop at time t. If the area of the loop is A, then the projected area normal to the field is $A\cos\omega t$, so that

$$\phi = B(t)(A\cos\omega t)$$

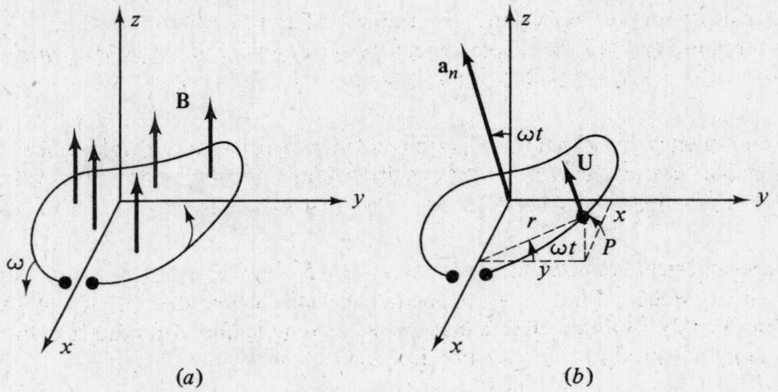

Fig. 12-17

and
$$v = -\frac{d\phi}{dt} = -\frac{dB}{dt}A\cos\omega t + BA\omega\sin\omega t$$

(b) The contribution to v due to the variation of $\mathbf{B}$ is

$$v_1 = -\int_S \frac{\partial \mathbf{B}}{\partial t}\cdot d\mathbf{S} = -\int_S \frac{dB}{dt}\mathbf{a}_z \cdot dS\,\mathbf{a}_n = -\frac{dB}{dt}A\cos\omega t$$

since $\mathbf{a}_z \cdot \mathbf{a}_n = \cos\omega t$.

To calculate the second, motional contribution to v, the velocity $\mathbf{U}$ of a point on the loop is needed. From Fig. 12-17(b) it is seen that

$$\mathbf{U} = r\omega\mathbf{a}_n = \frac{y}{\cos\omega t}\omega\mathbf{a}_n$$

so that

$$\mathbf{U}\times\mathbf{B} = \frac{y}{\cos\omega t}\omega\mathbf{a}_n \times B\mathbf{a}_z = \frac{y}{\cos\omega t}\omega B\sin\omega t\,(-\mathbf{a}_x)$$

since $\mathbf{a}_n \times \mathbf{a}_z = \sin\omega t\,(-\mathbf{a}_x)$. Consequently,

$$v_2 = \oint (\mathbf{U}\times\mathbf{B})\cdot d\mathbf{l} = -\frac{\omega B\sin\omega t}{\cos\omega t}\oint y\mathbf{a}_x \cdot d\mathbf{l}$$

Stokes' theorem (Section 9.8) can be used to evaluate the last integral. Since $\nabla \times y\mathbf{a}_x = -\mathbf{a}_z$,

$$\oint y\mathbf{a}_x \cdot d\boldsymbol{\ell} = \int_S (\nabla \times y\mathbf{a}_x)\cdot d\mathbf{S} = \int_S (-\mathbf{a}_z)\cdot dS\,\mathbf{a}_n = -A\cos\omega t$$

Therefore

$$v_2 = -\frac{\omega B\sin\omega t}{\cos\omega t}(-A\cos\omega t) = BA\omega\sin\omega t$$

It is seen that v_1 and v_2 are precisely the two terms found in part (a).

Supplementary Problems

12.18. Given the conduction current density in a lossy dielectric as $J_c = 0.02\sin 10^9 t$ (A/m²), find the displacement current density if $\sigma = 10^3$ S/m and $\epsilon_r = 6.5$. *Ans.* $1.15\times 10^{-6}\cos 10^9 t$ (A/m²)

12.19. A circular-cross-section conductor of radius 1.5 mm carries a current $i_c = 5.5 \sin 4 \times 10^{10} t$ (μA). What is the amplitude of the displacement current density, if $\sigma = 35$ MS/m and $\epsilon_r = 1$?
Ans. 7.87×10^{-3} μA/m^2

12.20. Find the frequency at which conduction current density and displacement current density are equal in (*a*) distilled water, where $\sigma = 2.0 \times 10^{-4}$ S/m and $\epsilon_r = 81$; (*b*) seawater, where $\sigma = 4.0$ S/m and $\epsilon_r = 1$. *Ans.* (a) 4.44×10^4 Hz; (*b*) 7.19×10^{10} Hz

12.21. Concentric spherical conducting shells at $r_1 = 0.5$ mm and $r_2 = 1$ mm are separated by a dielectric for which $\epsilon_r = 8.5$. Find the capacitance and calculate i_c, given an applied voltage $v = 150 \sin 5000t$ (V). Obtain the displacement current i_D and compare it with i_c.
Ans. $i_c = i_D = 7.09 \times 10^{-7} \cos 5000t$ (A)

12.22. Two parallel conducting plates of area 0.05 m^2 are separated by 2 mm of a lossy dielectric for which $\epsilon_r = 8.3$ and $\sigma = 8.0 \times 10^{-4}$ S/m. Given an applied voltage $v = 10 \sin 10^7 t$ (V), find the total rms current. *Ans.* 0.192 A

12.23. A parallel-plate capacitor of separation 0.6 mm and with a dielectric of $\epsilon_r = 15.3$ has an applied rms voltage of 25 V at a frequency of 15 GHz. Find the rms displacement current density. Neglect fringing. *Ans.* 5.32×10^5 A/m^2

12.24. A conductor on the x axis between $x = 0$ and $x = 0.2$ m has a velocity $\mathbf{U} = 6.0\mathbf{a}_z$ m/s in a field $\mathbf{B} = 0.04\mathbf{a}_y$ T. Find the induced voltage by using (*a*) the motional electric field intensity, (*b*) $d\phi/dt$ and (*c*) $B\ell U$. Determine the polarity and discuss Lenz's law if the conductor was connected to a closed loop. *Ans.* 0.048 V ($x = 0$ end is positive)

12.25. Repeat Problem 12.24 for $\mathbf{B} = 0.04 \sin kz\, \mathbf{a}_y$ (T). Discuss Lenz's law as the conductor moves from flux in one direction to the reverse direction. *Ans.* $0.048 \sin kz$ (V)

12.26. The bar conductor parallel to the y axis shown in Fig. 12-18 completes a loop by sliding contact with the conductors at $y = 0$ and $y = 0.05$ m. (*a*) Find the induced voltage when the bar is stationary at $x = 0.05$ m and $\mathbf{B} = 0.30 \sin 10^4 t\, \mathbf{a}_z$ (T). (*b*) Repeat for a velocity of the bar $\mathbf{U} = 150\mathbf{a}_x$ m/s. Discuss the polarity. *Ans.* (*a*) $-7.5 \cos 10^4 t$ (V); (*b*) $-7.5 \cos 10^4 t - 2.25 \sin 10^4 t$ (V)

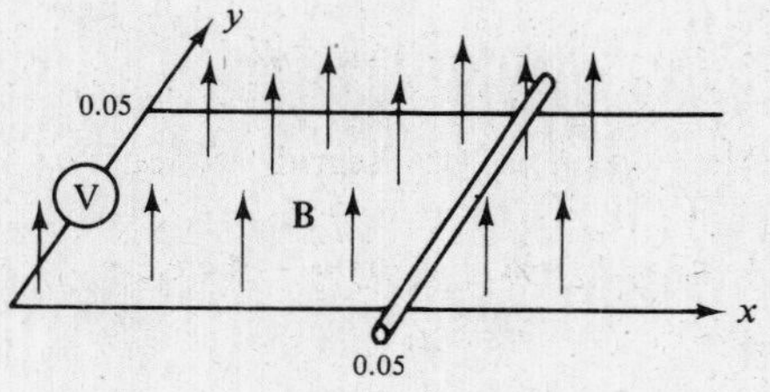

Fig. 12-18

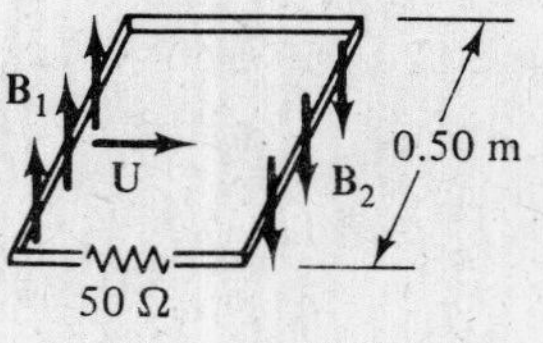

Fig. 12-19

12.27. The rectangular coil in Fig. 12-19 moves to the right at speed $U = 2.5$ m/s. The left side cuts flux at right angles, where $B_1 = 0.30$ T, while the right side cuts equal flux in the opposite direction. Find the instantaneous current in the coil and discuss its direction by use of Lenz's law.
Ans. 15 mA (counterclockwise)

12.28. A rectangular conducting loop in the $z = 0$ plane with sides parallel to the axes has y dimension 1 cm and x dimension 2 cm. Its resistance is 5.0 Ω. At a time when the coil sides are at $x = 20$ cm and $x = 22$ cm it is moving toward the origin at a velocity of 2.5 m/s along the x axis. Find the current if $\mathbf{B} = 5.0e^{-10x}\mathbf{a}_z$ (T). Repeat for the coil sides at $x = 5$ cm and $x = 7$ cm.
Ans. 0.613 mA, 2.75 mA

12.29. The 2.0 m conductor shown in Fig. 12-20 rotates at 1200 rev/min in the radial field $\mathbf{B} = 0.10 \sin \phi \, \mathbf{a}_r$ T. Find the current in the closed loop with a resistance of 100 Ω. Discuss the polarity and the current direction.
Ans. $5.03 \times 10^{-2} \sin 40\pi t$ (A)

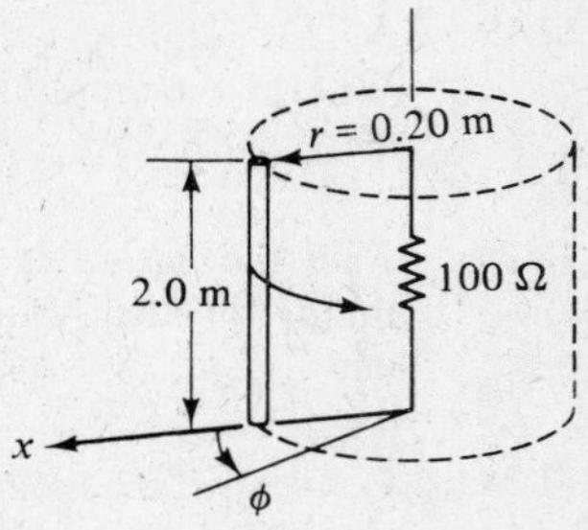

Fig. 12-20

12.30. In a radial field $\mathbf{B} = 0.50\mathbf{a}_r$ T, two conductors at $r = 0.23$ m and $r = 0.25$ m are parallel to the z axis and are 0.01 m in length. If both conductors are in the plane $\phi = 40\pi t$, what voltage is available to circulate a current when the two conductors are connected by radial conductors? *Ans.* 12.6 mV

12.31. In Fig. 12-21 a radial conductor, $3 \leq r \leq 6$ cm, is shown embedded in a rotating glass disk. Two 11.2 mΩ resistors complete two circuits. The disk turns at 12 rev/min. If the field at the disk is $\mathbf{B} = 0.30\mathbf{a}_n$ T, calculate the electrical power generated. What is the effect of this on the rotation? Discuss Lenz's law as it applies to this problem. *Ans.* 46.3 μW

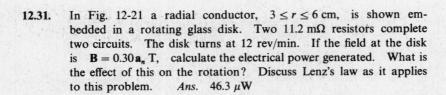

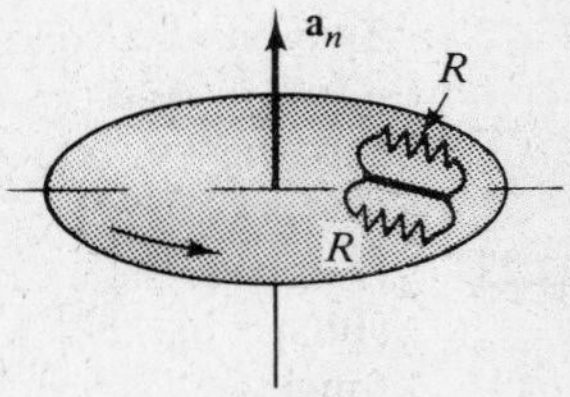

Fig. 12-21

12.32. What voltage is developed by a Faraday disk generator (Problem 12.14) with the meter connections at $r_1 = 1$ mm and $r_2 = 100$ mm when the disk turns at 500 rev/min in a flux density of 0.80 T?
Ans. 0.209 V

12.33. A coil such as that shown in Fig. 12-16(a) is 75 mm wide (y dimension) and 100 mm long (x dimension). What is the speed of rotation if an rms voltage of 0.25 V is developed in the uniform field $\mathbf{B} = 0.45\mathbf{a}_y$ T? *Ans.* 1000 rev/min

Chapter 13

Maxwell's Equations and Boundary Conditions

13.1 INTRODUCTION

The behavior of the electric field intensity **E** and the electric flux density **D** across the interface of two different materials was examined in Chapter 7, where the fields were static. A similar treatment will now be given for the magnetic field strength **H** and the magnetic flux density **B**, again with static fields. This will complete the study of the boundary conditions on the four principal vector fields.

In Chapter 12, where time-variable fields were treated, displacement current density $\mathbf{J}_D$ was introduced and Faraday's law was examined. In this chapter these same equations and others developed earlier are grouped together to form the set known as *Maxwell's equations.* These equations underlie all of electromagnetic field theory; they should be memorized.

13.2 BOUNDARY RELATIONS FOR MAGNETIC FIELDS

When **H** and **B** are examined at the interface between two different materials, abrupt changes can be expected, similar to those noted in **E** and **D** at the interface between two different dielectrics (see Section 7.4).

In Fig. 13-1 an interface is shown separating material *1*, with properties σ_1 and μ_{r1}, from *2*, with σ_2 and μ_{r2}. The behavior of **B** can be determined by use of a small right circular cylinder positioned across the interface as shown. Since magnetic flux lines are continuous,

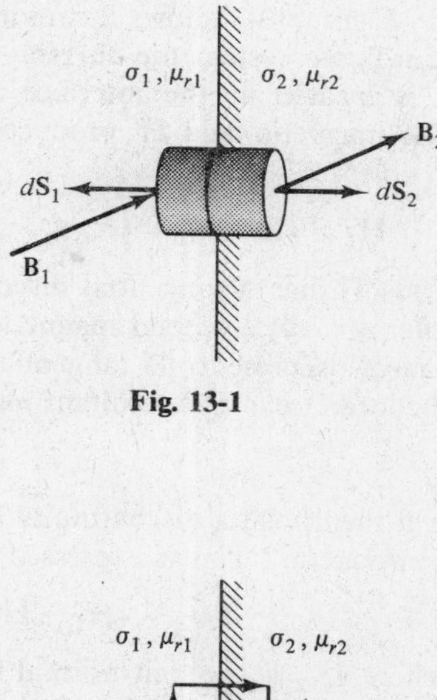

Fig. 13-1

$$\oint \mathbf{B}\cdot d\mathbf{S} = \int_{\text{end }1} \mathbf{B}_1\cdot d\mathbf{S}_1 + \int_{\text{cyl}} \mathbf{B}\cdot d\mathbf{S} + \int_{\text{end }2} \mathbf{B}_2\cdot d\mathbf{S}_2 = 0$$

Now if the two planes are allowed to approach one another, keeping the interface between them, the area of the curved surface will approach zero, giving

$$\int_{\text{end }1} \mathbf{B}_1\cdot d\mathbf{S}_1 + \int_{\text{end }2} \mathbf{B}_2\cdot d\mathbf{S}_2 = 0$$

or

$$-B_{n1}\int_{\text{end }1} dS_1 + B_{n2}\int_{\text{end }2} dS_2 = 0$$

from which

$$B_{n1} = B_{n2}$$

In words, *the normal component of* **B** *is continuous across an interface.* Note that *either* normal to the interface may be used in calculating B_{n1} and B_{n2}.

The variation in **H** across an interface is obtained by the application of Ampère's law around a closed rectangular path, as shown in Fig. 13-2. Assuming no current at the interface, and

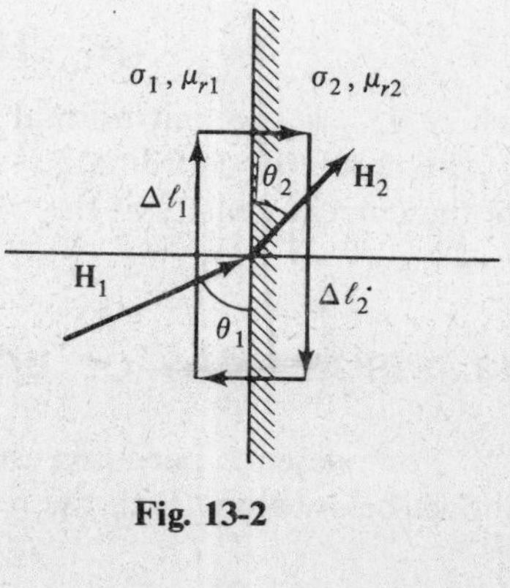

Fig. 13-2

letting the rectangle shrink to zero in the usual way,

$$0 = \oint \mathbf{H} \cdot d\mathbf{l} \to H_{\ell 1}\,\Delta\ell_1 - H_{\ell 2}\,\Delta\ell_2$$

whence

$$H_{\ell 1} = H_{\ell 2}$$

Thus tangential **H** has the same projection along the two sides of the rectangle. Since the rectangle can be rotated 90° and the argument repeated, it follows that

$$H_{t1} = H_{t2}$$

In words, *the tangential component of* **H** *is continuous across a current-free interface.*

The relation

$$\frac{\tan\theta_1}{\tan\theta_2} = \frac{\mu_{r2}}{\mu_{r1}}$$

between the angles made by $\mathbf{H}_1$ and $\mathbf{H}_2$ with a current-free interface (see Fig. 13-2) is obtained by analogy with Example 1, Section 7.4.

13.3 CURRENT SHEET AT THE BOUNDARY

If one material at the interface has a nonzero conductivity, a current may be present. This could be a current throughout the material; however, of more interest is the case of a current sheet at the interface.

Figure 13-3 shows a uniform current sheet. In the indicated coordinate system the current sheet has density $\mathbf{K} = K_0\,\mathbf{a}_y$ and it is located at the interface $x = 0$ between regions *1* and *2*. The magnetic field $\mathbf{H}'$ produced by this current sheet is given by Problem 9.3 as

$$\mathbf{H}'_1 = \tfrac{1}{2}\mathbf{K} \times \mathbf{a}_{n1} = \tfrac{1}{2}K_0\,\mathbf{a}_z \qquad \mathbf{H}'_2 = \tfrac{1}{2}\mathbf{K} \times \mathbf{a}_{n2} = \tfrac{1}{2}K_0(-\mathbf{a}_z)$$

Thus $\mathbf{H}'$ has a tangential discontinuity of magnitude $|K_0|$ at the interface. If a second magnetic field, $\mathbf{H}''$, arising from some other source, is present, its tangential component will be continuous at the interface. The resultant magnetic field,

$$\mathbf{H} = \mathbf{H}' + \mathbf{H}''$$

will then have a discontinuity of magnitude $|K_0|$ in its tangential component. This is expressed by the vector formula

$$(\mathbf{H}_1 - \mathbf{H}_2) \times \mathbf{a}_{n12} = \mathbf{K}$$

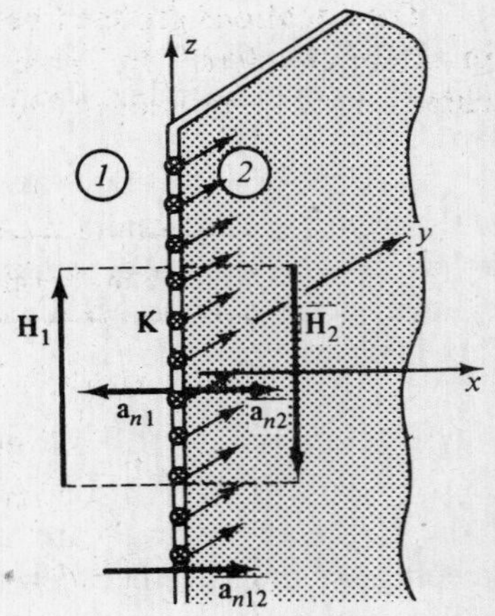

Fig. 13-3

where $\mathbf{a}_{n12}$ is the unit normal from region *1* to region *2*. The vector relation, which is independent of the choice of coordinate system, also holds for a nonuniform current sheet, where **K** is the value of the current density at the considered point of the interface.

13.4 SUMMARY OF BOUNDARY CONDITIONS

For reference purposes, the relationships for **E** and **D** across the interface of two dielectrics are shown below along with the relationships for **H** and **B**.

Magnetic Fields

$$B_{n1} = B_{n2}$$

$$\begin{cases} H_{t1} = H_{t2} & \text{(current-free)} \\ (\mathbf{H}_1 - \mathbf{H}_2) \times \mathbf{a}_{n12} = \mathbf{K} & \text{(with current sheet)} \end{cases}$$

$$\frac{\tan\theta_1}{\tan\theta_2} = \frac{\mu_{r2}}{\mu_{r1}} \quad \text{(current-free)}$$

Electric Fields

$$\begin{cases} D_{n1} = D_{n2} & \text{(charge-free)} \\ (\mathbf{D}_1 - \mathbf{D}_2) \cdot \mathbf{a}_{n12} = -\rho_s & \text{(with surface charge)} \end{cases}$$

$$E_{t1} = E_{t2}$$

$$\frac{\tan\theta_1}{\tan\theta_2} = \frac{\epsilon_{r2}}{\epsilon_{r1}} \quad \text{(charge-free)}$$

These relationships were obtained assuming static conditions. However, in Chapter 14 they will be found to apply equally well to time-variable fields.

13.5 MAXWELL'S EQUATIONS

A static **E** field can exist in the absence of a magnetic field **H**; a capacitor with a static charge Q furnishes an example. Likewise, a conductor with a constant current I has a magnetic field **H** without an **E** field. When fields are time-variable, however, **H** cannot exist without an **E** field nor can **E** exist without a corresponding **H** field. While much valuable information can be derived from static field theory, only with time-variable fields can the full value of electromagnetic field theory be demonstrated. The experiments of Faraday and Hertz and the theoretical analyses of Maxwell all involved time-variable fields.

The equations grouped below, called *Maxwell's equations*, were separately developed and examined in earlier chapters. In Table 13-1, the most general form is presented, where charges and conduction current may be present in the region.

Table 13-1. Maxwell's Equations, General Set

Point Form	Integral Form
$\nabla \times \mathbf{H} = \mathbf{J}_c + \dfrac{\partial \mathbf{D}}{\partial t}$	$\oint \mathbf{H} \cdot d\mathbf{l} = \int_S \left(\mathbf{J}_c + \dfrac{\partial \mathbf{D}}{\partial t}\right) \cdot d\mathbf{S}$ (Ampère's law)
$\nabla \times \mathbf{E} = -\dfrac{\partial \mathbf{B}}{\partial t}$	$\oint \mathbf{E} \cdot d\mathbf{l} = \int_S \left(-\dfrac{\partial \mathbf{B}}{\partial t}\right) \cdot d\mathbf{S}$ (Faraday's law; S fixed)
$\nabla \cdot \mathbf{D} = \rho$	$\oint_S \mathbf{D} \cdot d\mathbf{S} = \int_v \rho\, dv$ (Gauss's law)
$\nabla \cdot \mathbf{B} = 0$	$\oint_S \mathbf{B} \cdot d\mathbf{S} = 0$ (nonexistence of monopole)

Note that the point and integral forms of the first two equations are equivalent under Stokes' theorem, while the point and integral forms of the last two equations are equivalent under the divergence theorem.

For free space, where there are no charges ($\rho = 0$) and no conduction currents ($\mathbf{J}_c = 0$), Maxwell's equations take the form shown in Table 13-2.

The first and second point-form equations in the free-space set can be used to show that time-variable **E** and **H** fields cannot exist independently. For example, if **E** is a function of time, then $\mathbf{D} = \epsilon_0 \mathbf{E}$ will also be a function of time, so that $\partial \mathbf{D}/\partial t$ will be nonzero. Consequently, $\nabla \times \mathbf{H}$ is nonzero, and so a nonzero **H** must exist. In a similar way, the second equation can be used to show that if **H** is a function of time, then there must be an **E** field present.

The point form of Maxwell's equations is used most frequently in the problems. However, the integral form is important in that it displays the underlying physical laws.

Table 13-2. Maxwell's Equations, Free-Space Set

Point Form	Integral Form
$\nabla \times \mathbf{H} = \dfrac{\partial \mathbf{D}}{\partial t}$	$\oint \mathbf{H} \cdot d\mathbf{l} = \int_S \left(\dfrac{\partial \mathbf{D}}{\partial t}\right) \cdot d\mathbf{S}$
$\nabla \times \mathbf{E} = -\dfrac{\partial \mathbf{B}}{\partial t}$	$\oint \mathbf{E} \cdot d\mathbf{l} = \int_S \left(-\dfrac{\partial \mathbf{B}}{\partial t}\right) \cdot d\mathbf{S}$
$\nabla \cdot \mathbf{D} = 0$	$\oint_S \mathbf{D} \cdot d\mathbf{S} = 0$
$\nabla \cdot \mathbf{B} = 0$	$\oint_S \mathbf{B} \cdot d\mathbf{S} = 0$

Solved Problems

13.1. In region *1* of Fig. 13-4, $\mathbf{B}_1 = 1.2\,\mathbf{a}_x + 0.8\,\mathbf{a}_y + 0.4\,\mathbf{a}_z$ T. Find $\mathbf{H}_2$ (i.e. $\mathbf{H}$ at $z = +0$) and the angles between the field vectors and a tangent to the interface.

Write $\mathbf{H}_1$ directly below $\mathbf{B}_1$. Then write those components of $\mathbf{H}_2$ and $\mathbf{B}_2$ which follow directly from the two rules **B** *normal is continuous* and **H** *tangential is continuous* across a current-free interface.

$$\mathbf{B}_1 = 1.2\mathbf{a}_x + 0.8\,\mathbf{a}_y + 0.4\mathbf{a}_z \quad \text{T}$$

$$\mathbf{H}_1 = \frac{1}{\mu_0}(8.0\mathbf{a}_x + 5.33\,\mathbf{a}_y + 2.67\,\mathbf{a}_z)10^{-2} \quad \text{(A/m)}$$

$$\mathbf{H}_2 = \frac{1}{\mu_0}(8.0\mathbf{a}_x + 5.33\,\mathbf{a}_y + 10^2\mu_0 H_{z2}\,\mathbf{a}_z)10^{-2} \quad \text{(A/m)}$$

$$\mathbf{B}_2 = B_{x2}\,\mathbf{a}_x + B_{y2}\,\mathbf{a}_y + 0.4\mathbf{a}_z \quad \text{(T)}$$

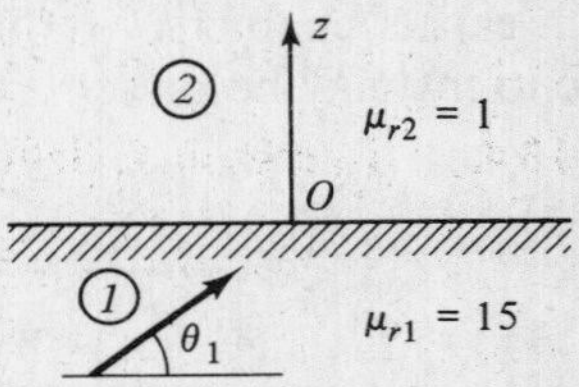

Fig. 13-4

Now the remaining terms follow directly:

$$B_{x2} = \mu_0 \mu_{r2} H_{x2} = 8.0 \times 10^{-2}\ \text{T} \qquad B_{y2} = 5.33 \times 10^{-2}\ \text{T} \qquad H_{z2} = \frac{B_{z2}}{\mu_0 \mu_{r2}} = \frac{0.4}{\mu_0}\ \text{(A/m)}$$

Angle θ_1 is $90° - \alpha_1$, where α_1 is the angle between $\mathbf{B}_1$ and the normal, $\mathbf{a}_z$.

$$\cos\alpha_1 = \frac{\mathbf{B}_1 \cdot \mathbf{a}_z}{|\mathbf{B}_1|} = 0.27$$

whence $\alpha_1 = 74.5°$ and $\theta_1 = 15.5°$. Similarly, $\theta_2 = 76.5°$.
Check: $(\tan\theta_1)/(\tan\theta_2) = \mu_{r2}/\mu_{r1}$.

13.2. Region *1*, for which $\mu_{r1} = 3$, is defined by $x < 0$ and region *2*, $x > 0$, has $\mu_{r2} = 5$. Given

$$\mathbf{H}_1 = 4.0\mathbf{a}_x + 3.0\mathbf{a}_y - 6.0\mathbf{a}_z \quad \text{A/m}$$

show that $\theta_2 = 19.7°$ and that $H_2 = 7.12$ A/m.

Proceed as in Problem 13.1.

$$\mathbf{H}_1 = 4.0\mathbf{a}_x + 3.0\mathbf{a}_y - 6.0\mathbf{a}_z \quad \text{A/m}$$
$$\mathbf{B}_1 = \mu_0(12.0\mathbf{a}_x + 9.0\mathbf{a}_y - 18.0\mathbf{a}_z) \quad \text{(T)}$$
$$\mathbf{B}_2 = \mu_0(12.0\mathbf{a}_x + 15.0\mathbf{a}_y - 30.0\mathbf{a}_z) \quad \text{(T)}$$
$$\mathbf{H}_2 = 2.40\mathbf{a}_x + 3.0\mathbf{a}_y - 6.0\mathbf{a}_z \quad \text{A/m}$$

Now

$$H_2 = \sqrt{(2.40)^2 + (3.0)^2 + (-6.0)^2} = 7.12 \text{ A/m}$$

The angle α_2 between $\mathbf{H}_2$ and the normal is given by

$$\cos\alpha_2 = \frac{H_{x2}}{H_2} = 0.34 \qquad \text{or} \qquad \alpha_2 = 70.3°$$

Then $\theta_2 = 90° - \alpha_2 = 19.7°$.

13.3. Region *1*, where $\mu_{r1} = 4$, is the side of the plane $y + z = 1$ containing the origin (see Fig. 13-5). In region *2*, $\mu_{r2} = 6$. Given $\mathbf{B}_1 = 2.0\mathbf{a}_x + 1.0\mathbf{a}_y$ T, find $\mathbf{B}_2$ and $\mathbf{H}_2$.

Choosing the unit normal $\mathbf{a}_n = (\mathbf{a}_y + \mathbf{a}_z)/\sqrt{2}$,

$$B_{n1} = (2.0\mathbf{a}_x + 1.0\mathbf{a}_y)\cdot(\mathbf{a}_y + \mathbf{a}_z)/\sqrt{2} = 1/\sqrt{2}$$

$$\mathbf{B}_{n1} = (1/\sqrt{2})\mathbf{a}_n = 0.5\mathbf{a}_y + 0.5\mathbf{a}_z = \mathbf{B}_{n2}$$

$$\mathbf{B}_{t1} = \mathbf{B}_1 - \mathbf{B}_{n1} = 2.0\mathbf{a}_x + 0.5\mathbf{a}_y - 0.5\mathbf{a}_z$$

$$\mathbf{H}_{t1} = \frac{1}{\mu_0}(0.5\mathbf{a}_x + 0.125\mathbf{a}_y - 0.125\mathbf{a}_z) = \mathbf{H}_{t2}$$

$$\mathbf{B}_{t2} = \mu_0\mu_{r2}\mathbf{H}_{t2} = 3.0\mathbf{a}_x + 0.75\mathbf{a}_y - 0.75\mathbf{a}_z$$

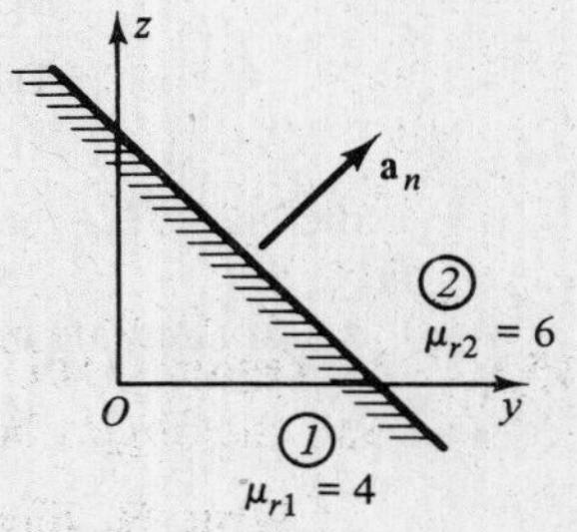

Fig. 13-5

Now the normal and tangential parts of $\mathbf{B}_2$ are combined.

$$\mathbf{B}_2 = 3.0\mathbf{a}_x + 1.25\mathbf{a}_y - 0.25\mathbf{a}_z \quad \text{T}$$

$$\mathbf{H}_2 = \frac{1}{\mu_0}(0.50\mathbf{a}_x + 0.21\mathbf{a}_y - 0.04\mathbf{a}_z) \quad \text{(A/m)}$$

13.4. In region *1*, defined by $z < 0$, $\mu_{r1} = 3$ and

$$\mathbf{H}_1 = \frac{1}{\mu_0}(0.2\mathbf{a}_x + 0.5\mathbf{a}_y + 1.0\mathbf{a}_z) \quad \text{(A/m)}$$

Find $\mathbf{H}_2$ if it is known that $\theta_2 = 45°$.

$$\cos\alpha_1 = \frac{\mathbf{H}_1\cdot\mathbf{a}_z}{|\mathbf{H}_1|} = 0.88 \qquad \text{or} \qquad \alpha_1 = 28.3°$$

Then, $\theta_1 = 61.7°$ and

$$\frac{\tan 61.7°}{\tan 45°} = \frac{\mu_{r2}}{3} \qquad \text{or} \qquad \mu_{r2} = 5.57$$

From the continuity of normal **B**, $\mu_{r1}H_{z1} = \mu_{r2}H_{z2}$, and so

$$\mathbf{H}_2 = \frac{1}{\mu_0}\left(0.2\mathbf{a}_x + 0.5\mathbf{a}_y + \frac{\mu_{r1}}{\mu_{r2}}1.0\mathbf{a}_z\right) = \frac{1}{\mu_0}(0.2\mathbf{a}_x + 0.5\mathbf{a}_y + 0.54\mathbf{a}_z) \quad \text{(A/m)}$$

13.5. A current sheet, $\mathbf{K} = 6.5\mathbf{a}_z$ A/m, at $x = 0$ separates region *1*, $x < 0$, where $\mathbf{H}_1 = 10\mathbf{a}_y$ A/m and region *2*, $x > 0$. Find $\mathbf{H}_2$ at $x = +0$.

Nothing is said about the permeabilities of the two regions; however, since $\mathbf{H}_1$ is entirely tangential, a change in permeability would have no effect. Since $B_{n1} = 0$, $B_{n2} = 0$ and therefore $H_{n2} = 0$.

$$(\mathbf{H}_1 - \mathbf{H}_2)\times\mathbf{a}_{n12} = \mathbf{K}$$

$$(10\mathbf{a}_y - H_{y2}\mathbf{a}_y)\times\mathbf{a}_x = 6.5\mathbf{a}_z$$

$$(10 - H_{y2})(-\mathbf{a}_z) = 6.5\mathbf{a}_z$$

$$H_{y2} = 16.5 \text{ A/m}$$

Thus, $\mathbf{H}_2 = 16.5\mathbf{a}_y$ A/m.

13.6. A current sheet, $\mathbf{K} = 9.0\mathbf{a}_y$ A/m, is located at $z = 0$, the interface between region *1*, $z < 0$, with $\mu_{r1} = 4$, and region *2*, $z > 0$, $\mu_{r2} = 3$. Given that $\mathbf{H}_2 = 14.5\mathbf{a}_x + 8.0\mathbf{a}_z$ A/m, find $\mathbf{H}_1$.

The current sheet shown in Fig. 13-6 is first examined alone.

$$\mathbf{H}'_1 = \tfrac{1}{2}(9.0)\mathbf{a}_y \times (-\mathbf{a}_z) = 4.5(-\mathbf{a}_x)$$
$$\mathbf{H}'_2 = \tfrac{1}{2}(9.0)\mathbf{a}_y \times \mathbf{a}_z = 4.5\mathbf{a}_x$$

From region *1* to region *2*, H_x will increase by 9.0 A/m due to the current sheet.

Now the complete **H** and **B** fields are examined.

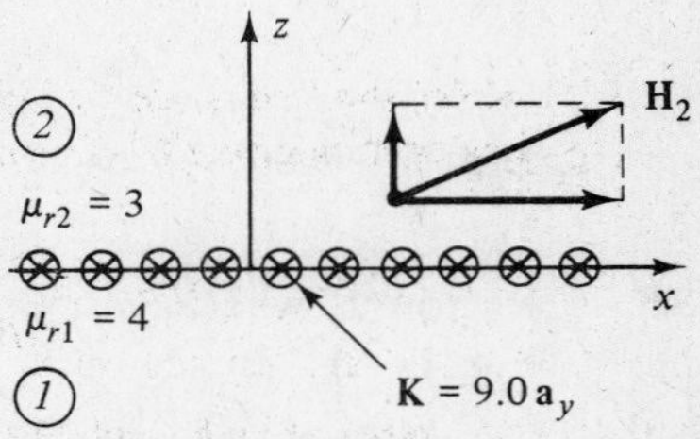

Fig. 13-6

$$\mathbf{H}_2 = 14.5\mathbf{a}_x + 8.0\mathbf{a}_z \quad \text{A/m}$$
$$\mathbf{B}_2 = \mu_0(43.5\mathbf{a}_x + 24.0\mathbf{a}_z) \quad \text{(T)}$$
$$\mathbf{B}_1 = \mu_0(22.0\mathbf{a}_x + 24.0\mathbf{a}_z) \quad \text{(T)}$$
$$\mathbf{H}_1 = 5.5\mathbf{a}_x + 6.0\mathbf{a}_z \quad \text{A/m}$$

Note that H_{x1} must be 9.0 A/m less than H_{x2} because of the current sheet. B_{x1} is obtained as $\mu_0 \mu_{r1} H_{x1}$.

An alternate method is to apply $(\mathbf{H}_1 - \mathbf{H}_2) \times \mathbf{a}_{n12} = \mathbf{K}$:

$$(H_{x1}\mathbf{a}_x + H_{y1}\mathbf{a}_y + H_{z1}\mathbf{a}_z) \times \mathbf{a}_z = \mathbf{K} + (14.5\mathbf{a}_x + 8.0\mathbf{a}_z) \times \mathbf{a}_z$$
$$-H_{x1}\mathbf{a}_y + H_{y1}\mathbf{a}_x = -5.5\mathbf{a}_y$$

from which $H_{x1} = 5.5$ A/m and $H_{y1} = 0$. This method deals exclusively with tangential **H**; any normal component must be determined by the previous methods.

13.7. Region *1*, $z < 0$, has $\mu_{r1} = 1.5$, while region *2*, $z > 0$, has $\mu_{r2} = 5$. Near (0,0,0),

$$\mathbf{B}_1 = 2.40\mathbf{a}_x + 10.0\mathbf{a}_z \quad \text{T} \qquad \mathbf{B}_2 = 25.75\mathbf{a}_x - 17.7\mathbf{a}_y + 10.0\mathbf{a}_z \quad \text{T}$$

If the interface carries a sheet current, what is its density at the origin?

Near the origin,

$$\mathbf{H}_1 = \frac{1}{\mu_0\mu_{r1}}\mathbf{B}_1 = \frac{1}{\mu_0}(1.60\mathbf{a}_x + 6.67\mathbf{a}_z) \quad \text{(A/m)}$$
$$\mathbf{H}_2 = \frac{1}{\mu_0}(5.15\mathbf{a}_x - 3.54\mathbf{a}_y + 2.0\mathbf{a}_z) \quad \text{(A/m)}$$

Then the local value of **K** is given by

$$\mathbf{K} = (\mathbf{H}_1 - \mathbf{H}_2) \times \mathbf{a}_{n12} = \frac{1}{\mu_0}(-3.55\mathbf{a}_x + 3.54\mathbf{a}_y + 4.67\mathbf{a}_z) \times \mathbf{a}_z = \frac{5.0}{\mu_0}\left(\frac{\mathbf{a}_x + \mathbf{a}_y}{\sqrt{2}}\right) \quad \text{(A/m)}$$

13.8. Given $\mathbf{E} = E_m \sin(\omega t - \beta z)\mathbf{a}_y$ in free space, find **D**, **B** and **H**. Sketch **E** and **H** at $t = 0$.

$$\mathbf{D} = \epsilon_0 \mathbf{E} = \epsilon_0 E_m \sin(\omega t - \beta z)\mathbf{a}_y$$

The Maxwell equation $\nabla \times \mathbf{E} = -\partial\mathbf{B}/\partial t$ gives

$$\begin{vmatrix} \mathbf{a}_x & \mathbf{a}_y & \mathbf{a}_z \\ \dfrac{\partial}{\partial x} & \dfrac{\partial}{\partial y} & \dfrac{\partial}{\partial z} \\ 0 & E_m \sin(\omega t - \beta z) & 0 \end{vmatrix} = -\frac{\partial \mathbf{B}}{\partial t}$$

or

$$-\frac{\partial \mathbf{B}}{\partial t} = \beta E_m \cos(\omega t - \beta z)\mathbf{a}_x$$

Integrating,

$$\mathbf{B} = -\frac{\beta E_m}{\omega}\sin(\omega t - \beta z)\mathbf{a}_x$$

where the "constant" of integration, which is a static field, has been neglected. Then,

$$\mathbf{H} = -\frac{\beta E_m}{\omega\mu_0}\sin(\omega t - \beta z)\mathbf{a}_x$$

Note that **E** and **H** are mutually perpendicular. At $t = 0$, $\sin(\omega t - \beta z) = -\sin\beta z$. Figure 13-7 shows the two fields along the z axis, on the assumption that E_m and β are positive.

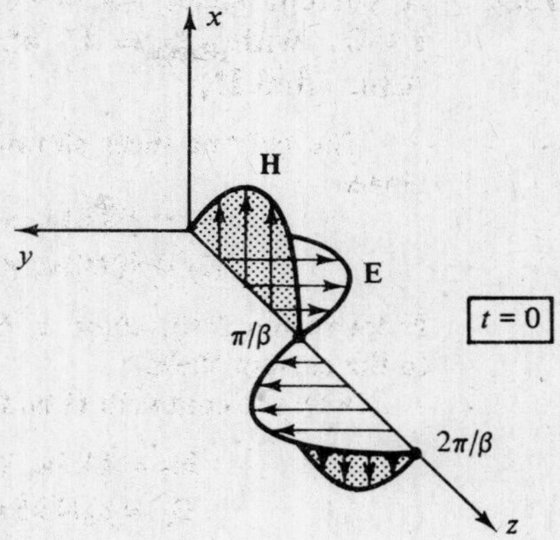

Fig. 13-7

13.9. Show that the **E** and **H** fields of Problem 13.8 constitute a wave traveling in the z direction. Verify that the wave speed and E/H depend only on the properties of free space.

E and **H** together vary as $\sin(\omega t - \beta z)$. A given state of **E** and **H** is then characterized by

$$\omega t - \beta z = \text{const.} = \omega t_0 \qquad \text{or} \qquad z = \frac{\omega}{\beta}(t - t_0)$$

But this is the equation of a plane moving with speed

$$c = \frac{\omega}{\beta}$$

in the direction of its normal, $\mathbf{a}_z$. (It is assumed that β, as well as ω, is positive; for β negative, the direction of motion would be $-\mathbf{a}_z$.) Thus, the entire pattern of Fig. 13-7 moves down the z axis with speed c.

The Maxwell equation $\nabla \times \mathbf{H} = \partial\mathbf{D}/\partial t$ gives

$$\begin{vmatrix} \mathbf{a}_x & \mathbf{a}_y & \mathbf{a}_z \\ \dfrac{\partial}{\partial x} & \dfrac{\partial}{\partial y} & \dfrac{\partial}{\partial z} \\ \dfrac{-\beta E_m}{\omega\mu_0}\sin(\omega t - \beta z) & 0 & 0 \end{vmatrix} = \frac{\partial}{\partial t}[\epsilon_0 E_m \sin(\omega t - \beta z)\mathbf{a}_y]$$

$$\frac{\beta^2 E_m}{\omega\mu_0}\cos(\omega t - \beta z)\mathbf{a}_y = \epsilon_0 E_m \omega\cos(\omega t - \beta z)\mathbf{a}_y$$

$$\frac{1}{\epsilon_0\mu_0} = \frac{\omega^2}{\beta^2}$$

Consequently,

$$c = \sqrt{\frac{1}{\epsilon_0\mu_0}} \approx \sqrt{\frac{1}{(10^{-9}/36\pi)(4\pi \times 10^{-7})}} = 3 \times 10^8 \text{ m/s}$$

Moreover,

$$\frac{E}{H} = \frac{\omega\mu_0}{\beta} = \sqrt{\frac{\mu_0}{\epsilon_0}} \approx 120\pi \text{ V/A} = 120\pi\ \Omega$$

13.10. Given $\mathbf{H} = H_m e^{j(\omega t + \beta z)}\mathbf{a}_x$ in free space, find **E**.

$$\nabla \times \mathbf{H} = \frac{\partial \mathbf{D}}{\partial t}$$

$$\frac{\partial}{\partial z} H_m e^{j(\omega t + \beta z)}\mathbf{a}_y = \frac{\partial \mathbf{D}}{\partial t}$$

$$j\beta H_m e^{j(\omega t + \beta z)}\mathbf{a}_y = \frac{\partial \mathbf{D}}{\partial t}$$

$$\mathbf{D} = \frac{\beta H_m}{\omega} e^{j(\omega t + \beta z)}\mathbf{a}_y$$

and $\mathbf{E} = \mathbf{D}/\epsilon_0$.

13.11. Given

$$\mathbf{E} = 30\pi e^{j(10^8 t + \beta z)}\mathbf{a}_x \quad \text{(V/m)} \qquad \mathbf{H} = H_m e^{j(10^8 t + \beta z)}\mathbf{a}_y \quad \text{(A/m)}$$

in free space, find H_m and β $(\beta > 0)$.

This is a plane wave, essentially the same as that in Problems 13.8 and 13.9 (except that, there, **E** was in the y direction and **H** in the x direction). The results of Problem 13.9 hold for any such wave in free space:

$$\frac{\omega}{\beta} = \frac{1}{\sqrt{\epsilon_0 \mu_0}} = 3 \times 10^8 \text{ m/s} \qquad \frac{E}{H} = \sqrt{\frac{\mu_0}{\epsilon_0}} = 120\pi\ \Omega$$

Thus, for the given wave,

$$\beta = \frac{10^8}{3 \times 10^8} = \frac{1}{3} \text{ rad/m} \qquad H_m = \pm\frac{30\pi}{120\pi} = \pm\frac{1}{4} \text{ A/m}$$

To fix the sign of H_m, apply $\nabla \times \mathbf{E} = -\partial \mathbf{B}/\partial t$:

$$j\beta\, 30\pi e^{j(10^8 t + \beta z)}\mathbf{a}_y = -j10^8 \mu_0 H_m e^{j(10^8 t + \beta z)}\mathbf{a}_y$$

which shows that H_m must be negative.

13.12. In a homogeneous nonconducting region where $\mu_r = 1$, find ϵ_r and ω if

$$\mathbf{E} = 30\pi e^{j[\omega t - (4/3)y]}\mathbf{a}_z \quad \text{(V/m)} \qquad \mathbf{H} = 1.0 e^{j[\omega t - (4/3)y]}\mathbf{a}_x \quad \text{(A/m)}$$

Here, by analogy to Problem 13.9,

$$\frac{\omega}{\beta} = \frac{1}{\sqrt{\epsilon\mu}} = \frac{3 \times 10^8}{\sqrt{\epsilon_r \mu_r}} \text{ (m/s)} \qquad \frac{E}{H} = \sqrt{\frac{\mu}{\epsilon}} = 120\pi\sqrt{\frac{\mu_r}{\epsilon_r}}\ (\Omega)$$

Thus, since $\mu_r = 1$,

$$\frac{\omega}{4/3} = \frac{3 \times 10^8}{\sqrt{\epsilon_r}} \qquad 30\pi = 120\pi\frac{1}{\sqrt{\epsilon_r}}$$

which yield $\epsilon_r = 16$, $\omega = 10^8$ rad/s. In this medium the speed of light is $c/4$.

Supplementary Problems

13.13. Region *1*, where $\mu_{r1} = 5$, is on the side of the plane $6x + 4y + 3z = 12$ that includes the origin. In region *2*, $\mu_{r2} = 3$. Given

$$\mathbf{H}_1 = \frac{1}{\mu_0}(3.0\mathbf{a}_x - 0.5\mathbf{a}_y) \quad \text{(A/m)}$$

find $\mathbf{B}_2$ and θ_2. *Ans.* $12.15\mathbf{a}_x + 0.60\mathbf{a}_y + 1.58\mathbf{a}_z$ T, 56.6°

13.14. The interface between two different regions is normal to one of the three cartesian axes. If

$$\mathbf{B}_1 = \mu_0(43.5\,\mathbf{a}_x + 24.0\,\mathbf{a}_z) \qquad \mathbf{B}_2 = \mu_0(22.0\,\mathbf{a}_x + 24.0\,\mathbf{a}_z)$$

what is the ratio $(\tan\theta_1)/(\tan\theta_2)$? *Ans.* 0.506

13.15. Inside a right circular cylinder, $\mu_{r1} = 1000$. The exterior is free space. If $\mathbf{B}_1 = 2.5\,\mathbf{a}_\phi$ T inside the cylinder, determine $\mathbf{B}_2$ just outside. *Ans.* $2.5\,\mathbf{a}_\phi$ mT

13.16. In spherical coordinates, region *1* is $r < a$, region *2* is $a < r < b$ and region *3* is $r > b$. Regions *1* and *3* are free space, while $\mu_{r2} = 500$. Given $\mathbf{B}_1 = 0.20\,\mathbf{a}_r$ T, find **H** in each region.
Ans. $\dfrac{0.20}{\mu_0}$ (A/m), $\dfrac{4 \times 10^{-4}}{\mu_0}$ (A/m), $\dfrac{0.20}{\mu_0}$ (A/m)

13.17. A current sheet, $\mathbf{K} = (8.0/\mu_0)\mathbf{a}_y$ (A/m), at $x = 0$ separates region *1*, $x < 0$ and $\mu_{r1} = 3$, from region *2*, $x > 0$ and $\mu_{r2} = 1$. Given $\mathbf{H}_1 = (10.0/\mu_0)(\mathbf{a}_y + \mathbf{a}_z)$ (A/m), find $\mathbf{H}_2$.
Ans. $\dfrac{1}{\mu_0}(10.0\,\mathbf{a}_y + 2.0\,\mathbf{a}_z)$ (A/m)

13.18. The $x = 0$ plane contains a current sheet of density **K** which separates region *1*, $x < 0$ and $\mu_{r1} = 2$, from region *2*, $x > 0$ and $\mu_{r2} = 7$. Given

$$\mathbf{B}_1 = 6.0\,\mathbf{a}_x + 4.0\,\mathbf{a}_y + 10.0\,\mathbf{a}_z \text{ T} \qquad \mathbf{B}_2 = 6.0\,\mathbf{a}_x - 50.96\,\mathbf{a}_y + 8.96\,\mathbf{a}_z \text{ T}$$

find **K**. *Ans.* $\dfrac{1}{\mu_0}(3.72\,\mathbf{a}_y - 9.28\,\mathbf{a}_z)$ (A/m)

13.19. In free space, $\mathbf{D} = D_m \sin(\omega t + \beta z)\mathbf{a}_x$. Using Maxwell's equations, show that

$$\mathbf{B} = \frac{-\omega\mu_0 D_m}{\beta}\sin(\omega t + \beta z)\,\mathbf{a}_y$$

Sketch the fields at $t = 0$ along the z axis, assuming that $D_m > 0$, $\beta > 0$. *Ans.* See Fig. 13-8.

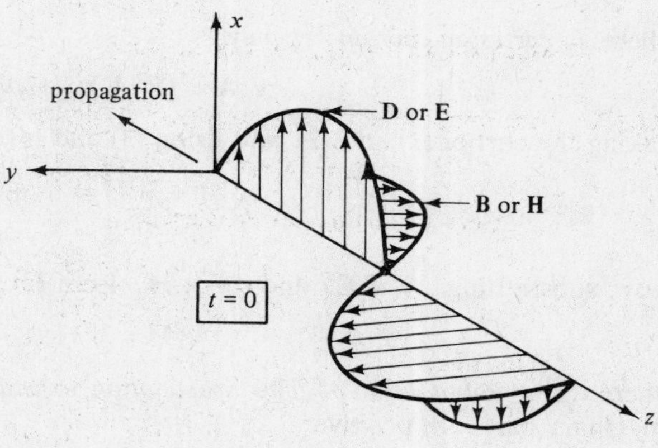

Fig. 13-8

13.20. In free space,

$$\mathbf{B} = B_m e^{j(\omega t + \beta z)}\mathbf{a}_y$$

Show that

$$\mathbf{E} = -\frac{\omega B_m}{\beta} e^{j(\omega t + \beta z)}\mathbf{a}_x$$

13.21. In a homogeneous region where $\mu_r = 1$ and $\epsilon_r = 50$,

$$\mathbf{E} = 20\pi e^{j(\omega t - \beta z)}\mathbf{a}_x \text{ (V/m)} \qquad \mathbf{B} = \mu_0 H_m e^{j(\omega t - \beta z)}\mathbf{a}_y \text{ (T)}$$

Find ω and H_m if the wavelength is 1.78 m. *Ans.* 1.5×10^8 rad/s, 1.18 A/m

Chapter 14

Electromagnetic Waves

14.1 INTRODUCTION

Some wave solutions to Maxwell's equations have already been encountered in the Solved Problems of Chapter 13. The present chapter will give a fuller treatment of electromagnetic waves. Since most regions of interest are free of charge, it will be assumed that $\rho = 0$. Moreover, linear isotropic materials will be assumed such that $\mathbf{D} = \epsilon \mathbf{E}$, $\mathbf{B} = \mu \mathbf{H}$ and $\mathbf{J}_c = \sigma \mathbf{E}$.

14.2 WAVE EQUATIONS

Under the above assumptions, and supposing that both **E** and **H** have the time-dependence $e^{j\omega t}$, Maxwell's equations (Table 13-1) give

$$\nabla \times \mathbf{H} = (\sigma + j\omega\epsilon)\mathbf{E} \qquad (1)$$

$$\nabla \times \mathbf{E} = -j\omega\mu\mathbf{H} \qquad (2)$$

$$\nabla \cdot \mathbf{E} = 0 \qquad (3)$$

$$\nabla \cdot \mathbf{H} = 0 \qquad (4)$$

Now apply the vector identity

$$\nabla \times (\nabla \times \mathbf{A}) \equiv \nabla(\nabla \cdot \mathbf{A}) - \nabla^2 \mathbf{A}$$

where, *in cartesian coordinates only*,

$$\nabla^2 \mathbf{A} = (\nabla^2 A_x)\mathbf{a}_x + (\nabla^2 A_y)\mathbf{a}_y + (\nabla^2 A_z)\mathbf{a}_z$$

Taking the curl of (*1*) and (*2*), and using (*3*) and (*4*),

$$-\nabla^2 \mathbf{H} = (\sigma + j\omega\epsilon)(\nabla \times \mathbf{E})$$

$$-\nabla^2 \mathbf{E} = -j\omega\mu(\nabla \times \mathbf{H})$$

Now, substituting $\nabla \times \mathbf{E}$ and $\nabla \times \mathbf{H}$ from (*2*) and (*1*), one obtains the vector *wave equations*

$$\nabla^2 \mathbf{H} = \gamma^2 \mathbf{H} \qquad \nabla^2 \mathbf{E} = \gamma^2 \mathbf{E}$$

where $\gamma^2 = j\omega\mu(\sigma + j\omega\epsilon)$. The *propagation constant*, γ, is that square root of γ^2 whose real and imaginary parts are positive:

$$\gamma = \alpha + j\beta$$

with

$$\alpha = \omega \sqrt{\frac{\mu\epsilon}{2}\left(\sqrt{1 + \left(\frac{\sigma}{\omega\epsilon}\right)^2} - 1\right)} \qquad (5)$$

$$\beta = \omega \sqrt{\frac{\mu\epsilon}{2}\left(\sqrt{1 + \left(\frac{\sigma}{\omega\epsilon}\right)^2} + 1\right)} \qquad (6)$$

The constant α is called the *attenuation factor*; β is called the *phase shift constant*. The reason for these names will become apparent in Section 14.4. From the wave equations it is seen that γ carries the units m^{-1}. However, it is customary to give α and β in Np/m and rad/m, respectively, where the *neper* (Np) is a dimensionless unit like the radian.

14.3 SOLUTIONS IN CARTESIAN COORDINATES

The familiar scalar wave equation in one dimension,

$$\frac{\partial^2 F}{\partial z^2} = \frac{1}{U^2}\frac{\partial^2 F}{\partial t^2}$$

has solutions of the form $F = f(z - Ut)$ and $F = g(z + Ut)$, where f and g are arbitrary functions. These represent waves traveling with speed U in the $+z$ and $-z$ directions, respectively. See Fig. 14-1. In particular, if harmonic time-variation $e^{j\omega t}$ is assumed, the wave equation becomes

$$\frac{d^2 F}{dz^2} = -\beta^2 F \qquad \left(\beta = \frac{\omega}{U}\right)$$

with solutions (including the time factor) of the form

$$F = Ce^{j(\omega t - \beta z)} \qquad F = De^{j(\omega t + \beta z)}$$

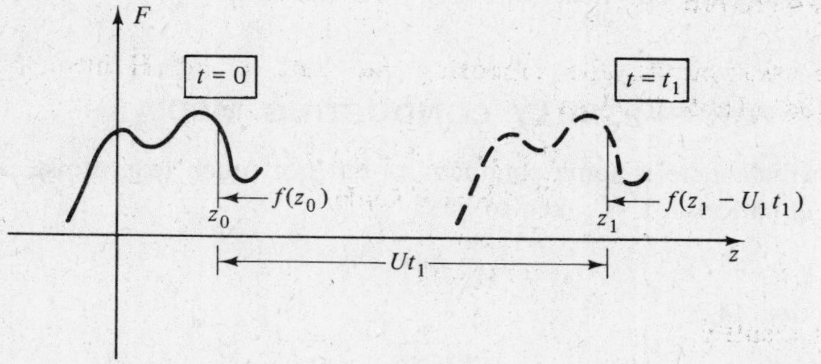

Fig. 14-1

or the real or imaginary parts of these. Figure 14-2 shows one of these solutions, $F = C\sin(\omega t - \beta z)$, at $t = 0$ and at $t = \pi/2\omega$; during this time interval the wave has moved a distance $d = U(\pi/2\omega) = \pi/2\beta$ to the right. At any fixed t, the waveform repeats itself when z changes by $2\pi/\beta$. The distance

$$\lambda = \frac{2\pi}{\beta}$$

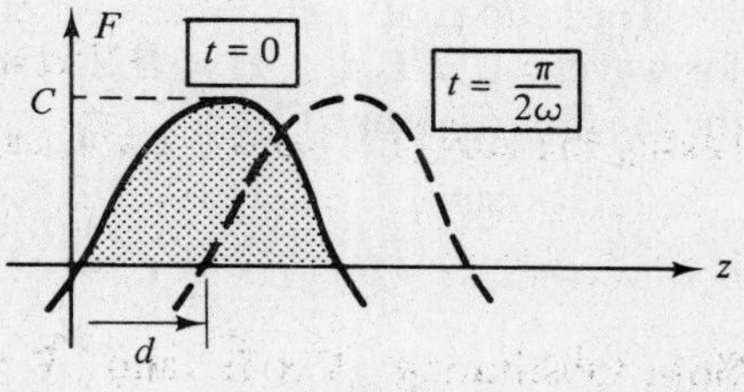

Fig. 14-2

is called the *wavelength*. Thus, in Fig. 14-2, the wave has advanced a quarter-wavelength to the right. The wavelength and the *frequency*, $f = \omega/2\pi$, have the well-known relation

$$\lambda f = U$$

Alternatively, $\lambda = TU$, where $T = 2\pi/\omega$ is the *period*.

The vector wave equations of Section 14.2 have solutions similar to those just discussed. Because the unit vectors $\mathbf{a}_x$, $\mathbf{a}_y$, and $\mathbf{a}_z$ in cartesian coordinates have fixed directions, the wave equation for $\mathbf{H}$ can be rewritten in the form

$$\frac{\partial^2 \mathbf{H}}{\partial x^2} + \frac{\partial^2 \mathbf{H}}{\partial y^2} + \frac{\partial^2 \mathbf{H}}{\partial z^2} = \gamma^2 \mathbf{H}$$

Of particular interest are solutions (plane waves) that depend on only one spatial coordinate, say z. Then the equation becomes

$$\frac{d^2 \mathbf{H}}{dz^2} = \gamma^2 \mathbf{H}$$

yielding

$$\mathbf{H} = H_0 e^{\pm \gamma z}\mathbf{a}_H \qquad \text{or} \qquad \mathbf{H}(z,t) = H_0 e^{\pm \gamma z} e^{j\omega t}\mathbf{a}_H$$

The corresponding solutions for the electric field are

$$\mathbf{E} = E_0 e^{\pm \gamma z}\mathbf{a}_E \qquad \text{or} \qquad \mathbf{E}(z,t) = E_0 e^{\pm \gamma z} e^{j\omega t}\mathbf{a}_E$$

Here, $\mathbf{a}_H$ and $\mathbf{a}_E$ are fixed unit vectors; the complex quantity γ is as given in Section 14.2. It is shown in Problem 14.2 that

$$\mathbf{a}_H \cdot \mathbf{a}_z = \mathbf{a}_E \cdot \mathbf{a}_z = 0$$

i.e. neither field has a component in the direction of propagation. This being the case, one can always rotate the axes to put one of the fields, say **E**, along the x axis. Then (*2*) of Section 14.2 shows that **H** lies along the y axis.

The plane wave solutions just obtained depend, via γ, on the properties of the medium (μ, ϵ, and σ). This dependence will be examined in the following three Sections.

14.4 SOLUTIONS FOR PARTIALLY CONDUCTING MEDIA

For a region in which there is some conductivity but not much (e.g. moist earth, seawater), the solution to the wave equation in **E** is taken to be

$$\mathbf{E} = E_0 e^{-\gamma z}\mathbf{a}_x$$

Then, from (*2*) of Section 14.2,

$$\mathbf{H} = \sqrt{\frac{\sigma + j\omega\epsilon}{j\omega\mu}}\, E_0 e^{-\gamma z}\mathbf{a}_y$$

The ratio E/H is characteristic of the medium (it is also frequency-dependent). More specifically, for waves $\mathbf{E} = E_x\mathbf{a}_x$, $\mathbf{H} = H_y\mathbf{a}_y$ which propagate in the $+z$ direction, the *intrinsic impedance*, η, of the medium is defined by

$$\eta = \frac{E_x}{H_y}$$

Thus

$$\eta = \sqrt{\frac{j\omega\mu}{\sigma + j\omega\epsilon}}$$

where the correct square root may be written in polar form, $|\eta|\angle\theta$, with

$$|\eta| = \frac{\sqrt{\mu/\epsilon}}{\sqrt[4]{1 + \left(\dfrac{\sigma}{\omega\epsilon}\right)^2}} \qquad \tan 2\theta = \frac{\sigma}{\omega\epsilon} \quad \text{and} \quad 0° < \theta < 45°$$

(If the wave propagate in the $-z$ direction, $E_x/H_y = -\eta$. In effect, γ is replaced by $-\gamma$ and the other square root used.)

Inserting the time factor $e^{j\omega t}$ and writing $\gamma = \alpha + j\beta$ results in the following equations for the fields in a partially conducting region:

$$\mathbf{E}(z,t) = E_0 e^{-\alpha z} e^{j(\omega t - \beta z)}\mathbf{a}_x$$

$$\mathbf{H}(z,t) = \frac{E_0}{|\eta|} e^{-\alpha z} e^{j(\omega t - \beta z - \theta)}\mathbf{a}_y$$

The factor $e^{-\alpha z}$ attenuates the magnitudes of both **E** and **H** as they propagate in the $+z$ direction. The expression for α, (*5*) of Section 14.2, shows that there will be some attenuation unless the conductivity σ is zero, which would be the case only for perfect dielectrics or free space. Likewise, the time phase difference θ between $\mathbf{E}(z, t)$ and $\mathbf{H}(z, t)$ vanishes only when σ is zero.

The velocity of propagation and the wavelength are given by

$$U = \frac{\omega}{\beta} = \frac{1}{\sqrt{\frac{\mu\epsilon}{2}\left(\sqrt{1 + \left(\frac{\sigma}{\omega\epsilon}\right)^2} + 1\right)}}$$

$$\lambda = \frac{2\pi}{\beta} = \frac{2\pi}{\omega\sqrt{\frac{\mu\epsilon}{2}\left(\sqrt{1 + \left(\frac{\sigma}{\omega\epsilon}\right)^2} + 1\right)}}$$

If the propagation velocity is known, $\lambda f = U$ may be used to determine the wavelength λ. The term $(\sigma/\omega\epsilon)^2$ has the effect of reducing both the velocity and the wavelength from what they would be in either free space or perfect dielectrics, where $\sigma = 0$. Observe that the medium is *dispersive*: waves with different frequencies ω have different velocities U.

14.5 SOLUTIONS FOR PERFECT DIELECTRICS

For a perfect dielectric, $\sigma = 0$, and so

$$\alpha = 0 \qquad \beta = \omega\sqrt{\mu\epsilon} \qquad \eta = \sqrt{\frac{\mu}{\epsilon}} \angle 0^\circ$$

Since $\alpha = 0$, there is no attenuation of the **E** and **H** waves. The zero angle on η results in **H** being in time phase with **E** at each fixed location. Assuming **E** in $\mathbf{a}_x$ and propagation in $\mathbf{a}_z$, the field equations may be obtained as limits of those in Section 14.4:

$$\mathbf{E}(z, t) = E_0\, e^{j(\omega t - \beta z)}\mathbf{a}_x$$

$$\mathbf{H}(z, t) = \frac{E_0}{\eta}\, e^{j(\omega t - \beta z)}\mathbf{a}_y$$

The velocity and the wavelength are

$$U = \frac{\omega}{\beta} = \frac{1}{\sqrt{\mu\epsilon}} \qquad \lambda = \frac{2\pi}{\beta} = \frac{2\pi}{\omega\sqrt{\mu\epsilon}}$$

Solutions in Free Space.

Free space is nothing more than the perfect dielectric for which

$$\mu = \mu_0 = 4\pi \times 10^{-7}\ \text{H/m} \qquad \epsilon = \epsilon_0 = 8.854 \times 10^{-12}\ \text{F/m} \approx \frac{10^{-9}}{36\pi}\ \text{F/m}$$

For free space, $\eta = \eta_0 \approx 120\pi\ \Omega$ and $U = c \approx 3 \times 10^8$ m/s.

14.6 SOLUTIONS FOR GOOD CONDUCTORS

In the ordinary materials classified as conductors, $\sigma \gg \omega\epsilon$ in the range of practical frequencies. For example, in copper, where $\sigma = 5.80 \times 10^7$ S/m and $\epsilon \approx \epsilon_0 = 8.854 \times 10^{-12}$ F/m, it would be necessary for frequencies to be of the order of 10^{16} Hz before $\omega\epsilon/\sigma$ would have to be retained in the expressions for the propagation constant and the intrinsic impedance. Therefore, letting $\omega\epsilon/\sigma \to 0$,

$$\alpha = \beta = \sqrt{\frac{\omega\mu\sigma}{2}} = \sqrt{\pi f \mu\sigma} \qquad \eta = \sqrt{\frac{\omega\mu}{\sigma}} \angle 45^\circ$$

It is seen that for all conductors the **E** and **H** waves are attenuated. Numerical examples will show that this is a very rapid attenuation. α will always be equal to β. At each fixed location **H** is out of time phase with **E** by 45° or $\pi/4$ rad. Once again assuming **E** in $\mathbf{a}_x$ and propagation in $\mathbf{a}_z$, the field equations are, from Section 14.4,

$$\mathbf{E}(z,t) = E_0\, e^{-\alpha z} e^{j(\omega t - \beta z)} \mathbf{a}_x \qquad \mathbf{H}(z,t) = \frac{E_0}{|\eta|}\, e^{-\alpha z} e^{j(\omega t - \beta z - \pi/4)} \mathbf{a}_y$$

Moreover,

$$U = \frac{\omega}{\beta} = \sqrt{\frac{2\omega}{\mu\sigma}} = \omega\delta \qquad \lambda = \frac{2\pi}{\beta} = \frac{2\pi}{\sqrt{\pi f \mu \sigma}} = 2\pi\delta$$

The velocity and wavelength in a conducting medium are written here in terms of the *skin depth* or *depth of penetration*,

$$\delta \equiv \frac{1}{\sqrt{\pi f \mu \sigma}}$$

The significance of this parameter is examined in Section 14.7.

14.7 SKIN DEPTH

The **E** and **H** traveling waves in a conducting medium are attenuated by the factor $e^{-\alpha z}$ as they advance along z. This attenuation is so rapid that often the waves may be considered to be zero after only a few millimeters of travel.

Consider that the region $z \geq 0$ is a conductor and just inside the conductor, at $z = +0$, **E** has magnitude 1.0 V/m. The *skin depth* δ is defined to be the distance after which $|\mathbf{E}|$ has decreased to $e^{-1} = 0.368$ V/m. Thus

$$\delta = \frac{1}{\alpha} = \frac{1}{\sqrt{\pi f \mu \sigma}} \text{ (m)}$$

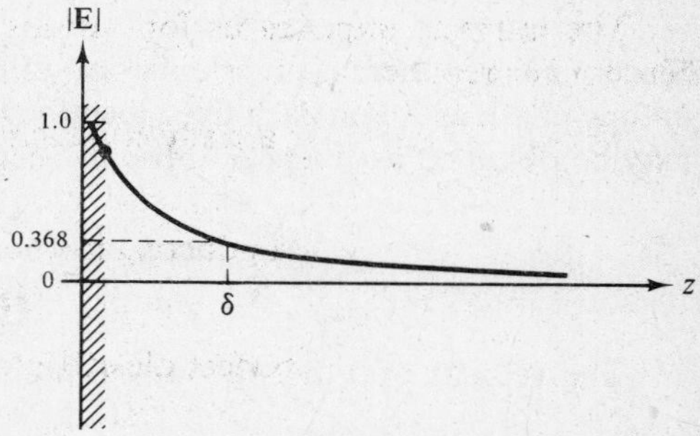

Fig. 14-3

For convenience, $z = 5\delta$ is often taken as the point where the function is zero, since its value there is 0.0067 or 0.67% of the initial value. At a frequency of 100 MHz in copper, the skin depth is 6.61 μm. The waves are attenuated to 0.67% in 5δ, or 33.0 μm. Consequently, the term *propagation* when used in conjunction with wave behavior within a conductor is somewhat misleading. The **E** and **H** waves hardly propagate at all. As will be determined shortly, the major part of an incident wave at a conductor surface is reflected. However, the portion which continues into the conductor and is rapidly attenuated cannot be completely ignored, since it gives rise to a conduction current density J_c and attendant ohmic power losses.

14.8 REFLECTED WAVES

When a traveling wave reaches an interface between two different regions, it is partly reflected and partly transmitted, with the magnitudes of the two parts determined by the constants of the two regions. In Fig. 14-4, a traveling **E** wave approaches the interface $z = 0$ from region *1*, $z < 0$. $\mathbf{E}^i$ and $\mathbf{E}^r$ are at $z = -0$, while $\mathbf{E}^t$ is at $z = +0$ (in region *2*). Here, *i* signifies "incident," *r* "reflected" and *t* "transmitted." Normal incidence is assumed. The equations for **E** and **H** can be written

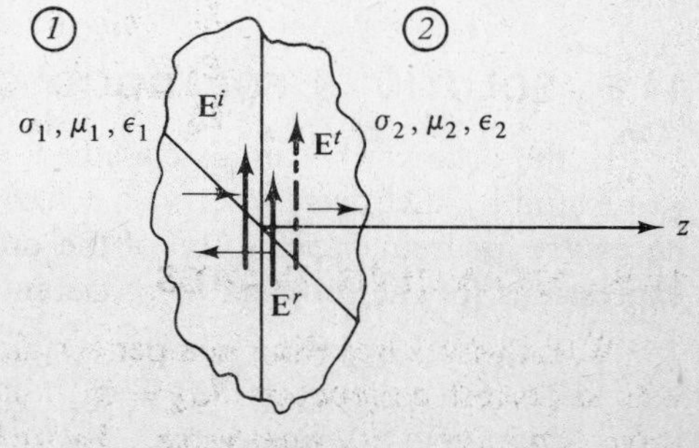

Fig. 14-4

$$\mathbf{E}^i(z,t) = E_0^i e^{-\gamma_1 z} e^{j\omega t}\mathbf{a}_x$$
$$\mathbf{E}^r(z,t) = E_0^r e^{\gamma_1 z} e^{j\omega t}\mathbf{a}_x$$
$$\mathbf{E}^t(z,t) = E_0^t e^{-\gamma_2 z} e^{j\omega t}\mathbf{a}_x$$
$$\mathbf{H}^i(z,t) = H_0^i e^{-\gamma_1 z} e^{j\omega t}\mathbf{a}_y$$
$$\mathbf{H}^r(z,t) = H_0^r e^{\gamma_1 z} e^{j\omega t}\mathbf{a}_y$$
$$\mathbf{H}^t(z,t) = H_0^t e^{-\gamma_2 z} e^{j\omega t}\mathbf{a}_y$$

With normal incidence, **E** and **H** are entirely tangential to the interface, and thus are continuous across it. At $z = 0$ this implies

$$E_0^i + E_0^r = E_0^t \qquad H_0^i + H_0^r = H_0^t$$

Furthermore, the intrinsic impedance in either region is equal to $\pm E_x/H_y$ (see Section 14.4).

$$\frac{E_0^i}{H_0^i} = \eta_1 \qquad \frac{E_0^r}{H_0^r} = -\eta_1 \qquad \frac{E_0^t}{H_0^t} = \eta_2$$

The five equations above can be combined to produce the following ratios in terms of the intrinsic impedances:

$$\frac{E_0^r}{E_0^i} = \frac{\eta_2 - \eta_1}{\eta_1 + \eta_2} \qquad \frac{H_0^r}{H_0^i} = \frac{\eta_1 - \eta_2}{\eta_1 + \eta_2}$$

$$\frac{E_0^t}{E_0^i} = \frac{2\eta_2}{\eta_1 + \eta_2} \qquad \frac{H_0^t}{H_0^i} = \frac{2\eta_1}{\eta_1 + \eta_2}$$

The intrinsic impedances for various materials have been examined earlier. They are repeated here for reference.

partially conducting medium: $\eta = \sqrt{\dfrac{j\omega\mu}{\sigma + j\omega\epsilon}}$

conducting medium: $\eta = \sqrt{\dfrac{\omega\mu}{\sigma}}\,\underline{/45^\circ}$

perfect dielectric: $\eta = \sqrt{\dfrac{\mu}{\epsilon}}$

free space: $\eta_0 = \sqrt{\dfrac{\mu_0}{\epsilon_0}} \approx 120\pi\ \Omega$

EXAMPLE 1 Traveling **E** and **H** waves in free space (region *1*) are normally incident on the interface with a perfect dielectric (region *2*) for which $\epsilon_r = 3.0$. Compare the magnitudes of the incident, reflected, and transmitted **E** and **H** waves at the interface.

$$\eta_1 = \eta_0 = 120\pi \qquad \eta_2 = \sqrt{\frac{\mu}{\epsilon}} = \frac{120\pi}{\sqrt{\epsilon_r}} = 217.7$$

$$\frac{E_0^r}{E_0^i} = \frac{\eta_2 - \eta_1}{\eta_1 + \eta_2} = -0.268 \qquad \frac{H_0^r}{H_0^i} = \frac{\eta_1 - \eta_2}{\eta_1 + \eta_2} = 0.268$$

$$\frac{E_0^t}{E_0^i} = \frac{2\eta_2}{\eta_1 + \eta_2} = 0.732 \qquad \frac{H_0^t}{H_0^i} = \frac{2\eta_1}{\eta_1 + \eta_2} = 1.268$$

14.9 STANDING WAVES

When waves traveling in a perfect dielectric $(\sigma_1 = \alpha_1 = 0)$ are normally incident on the interface with a perfect conductor $(\sigma_2 = \infty,\ \eta_2 = 0)$, the reflected wave in combination with the incident wave produces a *standing wave*. In such a wave, which is readily demonstrated on a clamped taut string, the oscillations at all points of a half-wavelength interval are in time phase. The combination of incident and reflected waves may be written

$$\mathbf{E}(z, t) = [E_0^i\, e^{j(\omega t - \beta z)} + E_0^r\, e^{j(\omega t + \beta z)}]\mathbf{a}_x = e^{j\omega t}(E_0^i\, e^{-j\beta z} + E_0^r\, e^{j\beta z})\mathbf{a}_x$$

Since $\eta_2 = 0$, $E_0^r/E_0^i = -1$ and

$$\mathbf{E}(z, t) = e^{j\omega t}(E_0^i\, e^{-j\beta z} - E_0^i\, e^{j\beta z})\mathbf{a}_x = -2jE_0^i \sin \beta z\, e^{j\omega t}\mathbf{a}_x$$

or, taking the real part,

$$\mathbf{E}(z, t) = 2E_0^i \sin \beta z \sin \omega t\, \mathbf{a}_x$$

The standing wave is shown in Fig. 14-5, at time intervals of $T/8$, where $T = 2\pi/\omega$ is the period. At $t = 0$, $\mathbf{E} = 0$ everywhere; at $t = 1(T/8)$, the endpoints of the $\mathbf{E}$ vectors lie on sine curve *1*; at $t = 2(T/8)$, they lie on sine curve *2*; and so forth. Sine curves *2* and *6* form an envelope for the oscillations; the amplitude of this envelope is twice the amplitude of the incident wave. Note that adjacent half-wavelength segments are 180° out of phase with each other.

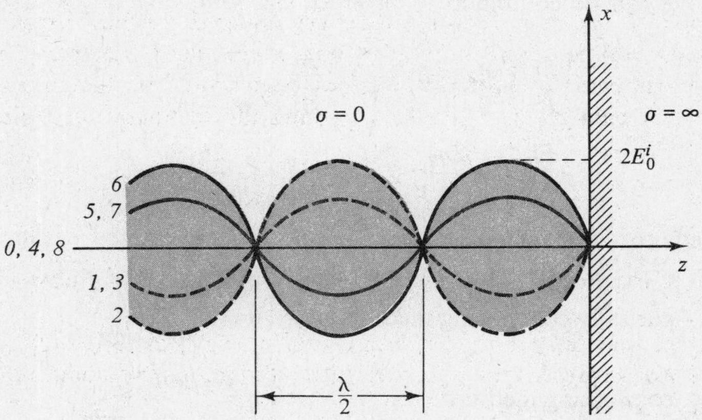

Fig. 14-5

14.10 POWER AND THE POYNTING VECTOR

Maxwell's first equation for a region with conductivity σ is written and then $\mathbf{E}$ is dotted with each term.

$$\nabla \times \mathbf{H} = \sigma \mathbf{E} + \epsilon \frac{\partial \mathbf{E}}{\partial t}$$

$$\mathbf{E} \cdot (\nabla \times \mathbf{H}) = \sigma E^2 + \mathbf{E} \cdot \epsilon \frac{\partial \mathbf{E}}{\partial t}$$

where, as usual, $E^2 = \mathbf{E} \cdot \mathbf{E}$. The vector identity $\nabla \cdot (\mathbf{A} \times \mathbf{B}) = \mathbf{B} \cdot (\nabla \times \mathbf{A}) - \mathbf{A} \cdot (\nabla \times \mathbf{B})$ is employed to change the left side of the equation.

$$\mathbf{H} \cdot (\nabla \times \mathbf{E}) - \nabla \cdot (\mathbf{E} \times \mathbf{H}) = \sigma E^2 + \mathbf{E} \cdot \epsilon \frac{\partial \mathbf{E}}{\partial t}$$

By Maxwell's second equation,

$$\mathbf{H} \cdot (\nabla \times \mathbf{E}) = \mathbf{H} \cdot \left(-\mu \frac{\partial \mathbf{H}}{\partial t}\right) = -\frac{\mu}{2} \frac{\partial H^2}{\partial t}$$

Similarly,

$$\mathbf{E}\cdot\epsilon\frac{\partial\mathbf{E}}{\partial t}=\frac{\epsilon}{2}\frac{\partial E^2}{\partial t}$$

Substituting, and rearranging terms,

$$\sigma E^2=-\frac{\epsilon}{2}\frac{\partial E^2}{\partial t}-\frac{\mu}{2}\frac{\partial H^2}{\partial t}-\nabla\cdot(\mathbf{E}\times\mathbf{H})$$

If this equality is valid, then the integration of its terms throughout a general volume v must also be valid.

$$\int_v \sigma E^2\,dv=-\int_v\left(\frac{\epsilon}{2}\frac{\partial E^2}{\partial t}+\frac{\mu}{2}\frac{\partial H^2}{\partial t}\right)dv-\oint_S(\mathbf{E}\times\mathbf{H})\cdot d\mathbf{S}$$

where the last term has been converted to an integral over the surface of v by use of the divergence theorem.

The integral on the left has the units of watts and is the usual ohmic term representing energy dissipated per unit time in heat. This dissipated energy has its source in the integrals on the right. Because $\epsilon E^2/2$ and $\mu H^2/2$ are the densities of energy stored in the electric and magnetic fields, respectively, the negative time derivatives can be viewed as a decrease in this stored energy. Consequently, the final integral (including the minus sign) must be the rate of energy entering the volume from outside. A change of sign then produces the *instantaneous rate of energy leaving the volume*:

$$P(t)=\oint_S(\mathbf{E}\times\mathbf{H})\cdot d\mathbf{S}=\oint_S\mathcal{P}\cdot d\mathbf{S}$$

where $\mathcal{P}=\mathbf{E}\times\mathbf{H}$ is the *Poynting vector*, the instantaneous rate of energy flow per unit area at a point.

In the cross product that defines the Poynting vector, the fields are supposed to be in real form. If, instead, $\mathbf{E}$ and $\mathbf{H}$ are expressed in complex form and have the common time-dependence $e^{j\omega t}$, then the time-average of $\mathcal{P}$ is given by

$$\mathcal{P}_{\text{avg}}=\tfrac{1}{2}\,\text{Re}\,(\mathbf{E}\times\mathbf{H}^*)$$

where $\mathbf{H}^*$ is the complex conjugate of $\mathbf{H}$. This follows the *complex power* of circuit analysis, $\mathbf{S}=\frac{1}{2}\mathbf{VI}^*$, of which the power is the real part, $P=\frac{1}{2}\,\text{Re}\,\mathbf{VI}^*$.

For plane waves, the direction of energy flow is the direction of propagation. Thus the Poynting vector offers a useful, coordinate-free way of specifying the direction of propagation, or of determining the directions of the fields if the direction of propagation is known. This can be particularly valuable where incident, transmitted and reflected waves are being examined.

Solved Problems

14.1. A traveling wave is described by $y=10\sin(\beta z-\omega t)$. Sketch the wave at $t=0$ and at $t=t_1$, when it has advanced $\lambda/8$, if the velocity is 3×10^8 m/s and the angular frequency $\omega=10^6$ rad/s. Repeat for $\omega=2\times10^6$ rad/s and the same time t_1.

The wave advances λ in one period, $T=2\pi/\omega$. Hence,

$$t_1=\frac{T}{8}=\frac{\pi}{4\omega}$$

$$\frac{\lambda}{8}=ct_1=(3\times10^8)\frac{\pi}{4(10^6)}=236\text{ m}$$

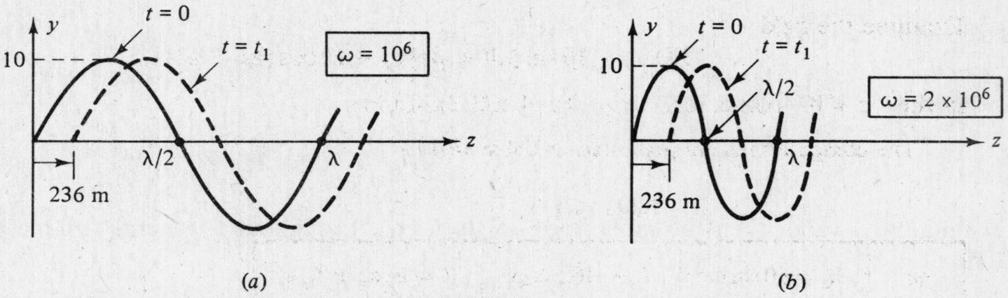

Fig. 14-6

The wave is shown at $t = 0$ and $t = t_1$ in Fig. 14-6(a). At twice the frequency, the wavelength λ is one-half, and the phase shift constant β is twice, the former value. See Fig. 14-6(b). At t_1 the wave has also advanced 236 m, but this distance is now $\lambda/4$.

14.2. Show that for the plane wave (time-dependence omitted)

$$\mathbf{H} = H_0 e^{\pm \gamma z} \mathbf{a}_H$$

where $\mathbf{a}_H$ is a fixed unit vector, $\mathbf{a}_H \cdot \mathbf{a}_z = 0$.

In terms of the cartesian unit vectors,

$$\mathbf{a}_H = (\mathbf{a}_H \cdot \mathbf{a}_x)\mathbf{a}_x + (\mathbf{a}_H \cdot \mathbf{a}_y)\mathbf{a}_y + (\mathbf{a}_H \cdot \mathbf{a}_z)\mathbf{a}_z$$

and so

$$\mathbf{H} = H_0 e^{\pm \gamma z}(\mathbf{a}_H \cdot \mathbf{a}_x)\mathbf{a}_x + H_0 e^{\pm \gamma z}(\mathbf{a}_H \cdot \mathbf{a}_y)\mathbf{a}_y + H_0 e^{\pm \gamma z}(\mathbf{a}_H \cdot \mathbf{a}_z)\mathbf{a}_z$$

The Maxwell equation $\nabla \cdot \mathbf{H} = 0$ then gives

$$\frac{\partial}{\partial z}[H_0 e^{\pm \gamma z}(\mathbf{a}_H \cdot \mathbf{a}_z)] = 0 \qquad \text{or} \qquad \pm \gamma H_0 e^{\pm \gamma z}(\mathbf{a}_H \cdot \mathbf{a}_z) = 0$$

which can hold only if $\mathbf{a}_H \cdot \mathbf{a}_z = 0$.

14.3. In free space, $\mathbf{E}(z, t) = 10^3 \sin(\omega t - \beta z)\mathbf{a}_y$ (V/m). Obtain $\mathbf{H}(z, t)$.

Examination of the phase, $\omega t - \beta z$, shows that the direction of propagation is $+z$. Since $\mathbf{E} \times \mathbf{H}$ must also be in $+z$, $\mathbf{H}$ must have the direction $-\mathbf{a}_x$. Consequently,

$$\frac{E_y}{-H_x} = \eta_0 = 120\pi\ \Omega \qquad \text{or} \qquad H_x = -\frac{10^3}{120\pi}\sin(\omega t - \beta z) \quad \text{(A/m)}$$

and

$$\mathbf{H}(z, t) = -\frac{10^3}{120\pi}\sin(\omega t - \beta z)\mathbf{a}_x \quad \text{(A/m)}$$

14.4. For the wave of Problem 14.3 determine the propagation constant γ, given that the frequency is $f = 95.5$ MHz.

In general, $\gamma = \sqrt{j\omega\mu(\sigma + j\omega\epsilon)}$. In free space, $\sigma = 0$, so that

$$\gamma = j\omega\sqrt{\mu_0 \epsilon_0} = j(2\pi f/c) = j\frac{2\pi(95.5 \times 10^6)}{3 \times 10^8} = j(2.0)\ \text{m}^{-1}$$

Note that this result shows that the attentuation factor is $\alpha = 0$ and the phase shift constant is $\beta = 2.0$ rad/m.

14.5. Examine the field

$$\mathbf{E}(z,t) = 10\sin(\omega t + \beta z)\mathbf{a}_x + 10\cos(\omega t + \beta z)\mathbf{a}_y$$

in the $z = 0$ plane, for $\omega t = 0, \pi/4, \pi/2, 3\pi/4$ and π.

The computations are presented in Table 14-1.

Table 14-1

ωt	$E_x = 10\sin\omega t$	$E_y = 10\cos\omega t$	$\mathbf{E} = E_x\mathbf{a}_x + E_y\mathbf{a}_y$
0	0	10	$10\mathbf{a}_y$
$\pi/4$	$10/\sqrt{2}$	$10/\sqrt{2}$	$10\left(\frac{\mathbf{a}_x + \mathbf{a}_y}{\sqrt{2}}\right)$
$\pi/2$	10	0	$10\mathbf{a}_x$
$3\pi/4$	$10/\sqrt{2}$	$-10/\sqrt{2}$	$10\left(\frac{\mathbf{a}_x + \mathbf{a}_y}{\sqrt{2}}\right)$
π	0	-10	$10(-\mathbf{a}_y)$

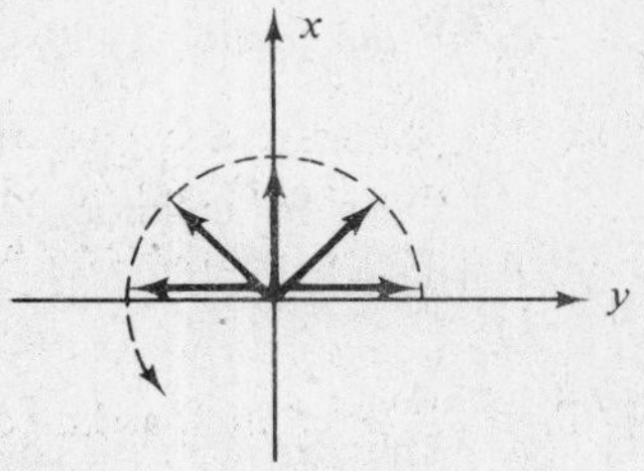

Fig. 14-7

As shown in Fig. 14-7, $\mathbf{E}(z,t)$ is circularly polarized. In addition, the wave travels in the $-\mathbf{a}_z$ direction.

14.6. An **H** field travels in the $-\mathbf{a}_z$ direction in free space with a phase shift constant of 30.0 rad/m and an amplitude of $(1/3\pi)$ A/m. If the field has the direction $-\mathbf{a}_y$ when $t = 0$ and $z = 0$, write suitable expressions for **E** and **H**. Determine the frequency and wavelength.

In a medium of conductivity σ, the intrinsic impedance η, which relates E and H, would be complex, and so the phase of **E** and **H** would have to be written in complex form. In free space this restriction is unnecessary. Using cosines, then,

$$\mathbf{H}(z,t) = -\frac{1}{3\pi}\cos(\omega t + \beta z)\mathbf{a}_y$$

For propagation in $-z$,

$$\frac{E_x}{H_y} = -\eta_0 = -120\pi\ \Omega \qquad \text{or} \qquad E_x = +40\cos(\omega t + \beta z)\quad \text{(V/m)}$$

Thus

$$\mathbf{E}(z,t) = 40\cos(\omega t + \beta z)\mathbf{a}_x \quad \text{(V/m)}$$

Since $\beta = 30$ rad/m,

$$\lambda = \frac{2\pi}{\beta} = \frac{\pi}{15}\ \text{m} \qquad f = \frac{c}{\lambda} = \frac{3\times 10^8}{\pi/15} = \frac{45}{\pi}\times 10^8\ \text{Hz}$$

14.7. Determine the propagation constant γ for a material having $\mu_r = 1$, $\epsilon_r = 8$ and $\sigma = 0.25$ pS/m, if the wave frequency is 1.6 MHz.

In this case,

$$\frac{\sigma}{\omega\epsilon} = \frac{0.25\times 10^{-12}}{2\pi(1.6\times 10^6)(8)(10^{-9}/36\pi)} \approx 10^{-9} \approx 0$$

so that

$$\alpha \approx 0 \qquad \beta \approx \omega\sqrt{\mu\epsilon} = 2\pi f\,\frac{\sqrt{\mu_r\epsilon_r}}{c} = 9.48\times 10^{-2}\ \text{rad/m}$$

and $\gamma = \alpha + j\beta \approx j9.48 \times 10^{-2}\ \text{m}^{-1}$. The material behaves like a perfect dielectric at the given frequency. Conductivity of the order of 1 pS/m indicates that the material is more like an insulator than a conductor.

14.8. Determine the conversion factor between the neper and the decibel.

Consider a plane wave traveling in the $+z$ direction whose amplitude decays according to

$$E = E_0 e^{-\alpha z}$$

From Section 14.10, the power carried by the wave is proportional to E^2, so that

$$P = P_0 e^{-2\alpha z}$$

Then, by definition of the decibel, the power drop over the distance z is $10 \log_{10}(P_0/P)$ dB. But

$$10 \log_{10} \frac{P_0}{P} = \frac{10}{2.3026} \ln \frac{P_0}{P} = \frac{20}{2.3026}(\alpha z) = 8.686(\alpha z)$$

Thus, αz nepers is equivalent to $8.686(\alpha z)$ decibels; i.e.

$$1\ \text{Np} = 8.686\ \text{dB}$$

14.9. At what frequencies may earth be considered a perfect dielectric, if $\sigma = 5 \times 10^{-3}$ S/m, $\mu_r = 1$, and $\epsilon_r = 8$? Can α be assumed zero at these frequencies?

Assume arbitrarily that

$$\frac{\sigma}{\omega\epsilon} \leq \frac{1}{100}$$

marks the cutoff. Then

$$f = \frac{\omega}{2\pi} \geq \frac{100\sigma}{2\pi\epsilon} = 1.13\ \text{GHz}$$

For small $\sigma/\omega\epsilon$,

$$\alpha = \omega \sqrt{\frac{\mu\epsilon}{2}\left(\sqrt{1 + \left(\frac{\sigma}{\omega\epsilon}\right)^2} - 1\right)}$$

$$\approx \omega \sqrt{\frac{\mu\epsilon}{2}\left[\frac{1}{2}\left(\frac{\sigma}{\omega\epsilon}\right)^2\right]} = \frac{\sigma}{2}\sqrt{\frac{\mu}{\epsilon}} = \frac{\sigma}{2}\sqrt{\frac{\mu_r}{\epsilon_r}}(120\pi) = 0.333\ \text{Np/m}$$

Thus, no matter how high the frequency, α will be about 0.333 Np/m, or almost 3 db/m (see Problem 14.8); α cannot be assumed zero.

14.10. Find the skin depth δ at a frequency of 1.6 MHz in aluminum, where $\sigma = 38.2$ MS/m and $\mu_r = 1$. Also find γ and the wave velocity U.

$$\delta = \frac{1}{\sqrt{\pi f \mu \sigma}} = 6.44 \times 10^{-5}\ \text{m} = 64.4\ \mu\text{m}$$

Because $\alpha = \beta = \delta^{-1}$,

$$\gamma = 1.55 \times 10^4 + j1.55 \times 10^4 = 2.20 \times 10^4 \angle 45^\circ\ \text{m}^{-1}$$

and

$$U = \frac{\omega}{\beta} = \omega\delta = 647\ \text{m/s}$$

14.11. Calculate the intrinsic impedance η, the propagation constant γ and the wave velocity U for a conducting medium in which $\sigma = 58$ MS/m, $\mu_r = 1$, at a frequency $f = 100$ MHz.

$$\gamma = \sqrt{\omega\mu\sigma}\,\underline{/45^\circ} = 2.14 \times 10^5\,\underline{/45^\circ} \quad \text{m}^{-1}$$

$$\eta = \sqrt{\frac{\omega\mu}{\sigma}}\,\underline{/45^\circ} = 3.69 \times 10^{-3}\,\underline{/45^\circ} \quad \Omega$$

$$\alpha = \beta = 1.51 \times 10^5 \qquad \delta = \frac{1}{\alpha} = 6.61\ \mu\text{m} \qquad U = \omega\delta = 4.15 \times 10^3 \text{ m/s}$$

14.12. A plane wave traveling in the $+z$ direction in free space $(z < 0)$ is normally incident at $z = 0$ on a conductor $(z > 0)$ for which $\sigma = 61.7$ MS/m, $\mu_r = 1$. The free-space **E** wave has a frequency $f = 1.5$ MHz and an amplitude of 1.0 V/m; at the interface it is given by

$$\mathbf{E}(0, t) = 1.0 \sin 2\pi f t\, \mathbf{a}_y \quad \text{(V/m)}$$

Find $\mathbf{H}(z, t)$ for $z > 0$.

For $z > 0$, and in complex form,

$$\mathbf{E}(z, t) = 1.0e^{-\alpha z} e^{j(2\pi f t - \beta z)} \mathbf{a}_y \quad \text{(V/m)}$$

where the imaginary part will ultimately be taken. In the conductor,

$$\alpha = \beta = \sqrt{\pi f \mu \sigma} = \sqrt{\pi(1.5 \times 10^6)(4\pi \times 10^{-7})(61.7 \times 10^6)} = 1.91 \times 10^4$$

$$\eta = \sqrt{\frac{\omega\mu}{\sigma}}\,\underline{/45^\circ} = 4.38 \times 10^{-4} e^{j\pi/4}$$

Then, since $E_y/(-H_x) = \eta$,

$$\mathbf{H}(z, t) = -2.28 \times 10^3 e^{-\alpha z} e^{j(2\pi f t - \beta z - \pi/4)} \mathbf{a}_x \quad \text{(A/m)}$$

or, taking the imaginary part,

$$\mathbf{H}(z, t) = -2.28 \times 10^3 e^{-\alpha z} \sin(2\pi f t - \beta z - \pi/4) \mathbf{a}_x \quad \text{(A/m)}$$

where f, α, and β are as given above.

14.13. In free space $\mathbf{E}(z, t) = 50 \cos(\omega t - \beta z) \mathbf{a}_x$ (V/m). Find the average power crossing a circular area of radius 2.5 m in the plane $z = \text{const}$.

In complex form,

$$\mathbf{E} = 50e^{j(\omega t - \beta z)} \mathbf{a}_x \quad \text{(V/m)}$$

and since $\eta = 120\pi\ \Omega$ and propagation is in $+z$,

$$\mathbf{H} = \frac{5}{12\pi} e^{j(\omega t - \beta z)} \mathbf{a}_y \quad \text{(A/m)}$$

Then

$$\mathscr{P}_{\text{avg}} = \tfrac{1}{2} \text{Re}(\mathbf{E} \times \mathbf{H}^*) = \tfrac{1}{2}(50)\left(\frac{5}{12\pi}\right) \mathbf{a}_z \text{ W/m}^2$$

The flow is normal to the area, and so

$$P_{\text{avg}} = \tfrac{1}{2}(50)\left(\frac{5}{12\pi}\right) \pi (2.5)^2 = 65.1 \text{ W}$$

14.14. A voltage source, v, is connected to a pure resistor R by a length of coaxial cable, as shown in Fig. 14-8(a). Show that use of the Poynting vector $\mathcal{P}$ in the dielectric leads to the same instantaneous power in the resistor as methods of circuit analysis.

From Problem 7.9 and Ampère's law,

$$\mathbf{E} = \frac{v}{r \ln (b/a)} \mathbf{a}_r \qquad \text{and} \qquad \mathbf{H} = \frac{i}{2\pi r} \mathbf{a}_\phi$$

where a and b are the radii of the inner and outer conductors, as shown in Fig. 14-8(b). Then

$$\mathcal{P} = \mathbf{E} \times \mathbf{H} = \frac{vi}{2\pi r^2 \ln (b/a)} \mathbf{a}_z$$

This is the instantaneous power density. The total instantaneous power over the cross section of the dielectric is

$$P(t) = \int_0^{2\pi} \int_a^b \frac{vi}{2\pi r^2 \ln (b/a)} \mathbf{a}_z \cdot r\, dr\, d\phi\, \mathbf{a}_z = vi$$

which is also the circuit-theory result for the instantaneous power loss in the resistor.

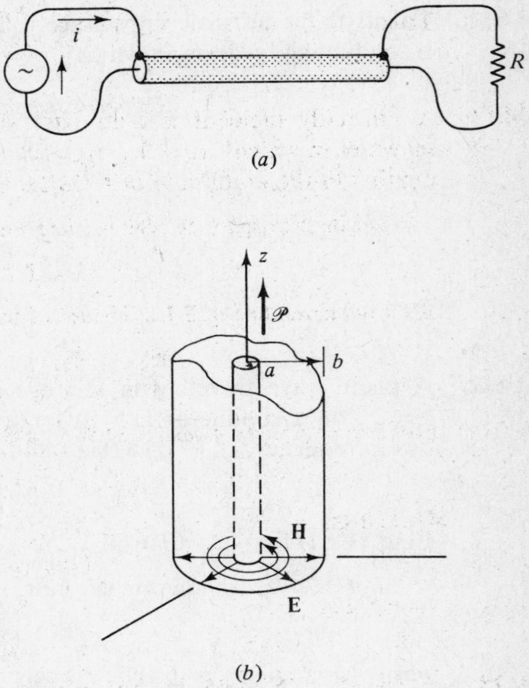

Fig. 14-8

14.15. Determine the amplitudes of the reflected and transmitted **E** and **H** at the interface shown in Fig. 14-9, if $E_0^i = 1.5 \times 10^{-3}$ V/m in region *1*, in which $\epsilon_{r1} = 8.5$, $\mu_{r1} = 1$ and $\sigma_1 = 0$. Region *2* is free space. Assume normal incidence.

$$\eta_1 = \sqrt{\frac{\mu_0 \mu_{r1}}{\epsilon_0 \epsilon_{r1}}} = 129\ \Omega \qquad \eta_2 = 120\pi\ \Omega = 377\ \Omega$$

$$E_0^r = \frac{\eta_2 - \eta_1}{\eta_2 + \eta_1} E_0^i = 7.35 \times 10^{-4}\ \text{V/m}$$

$$E_0^t = \frac{2\eta_2}{\eta_2 + \eta_1} E_0^i = 2.24 \times 10^{-3}\ \text{V/m}$$

$$H_0^i = \frac{E_0^i}{\eta_1} = 1.16 \times 10^{-5}\ \text{A/m}$$

$$H_0^r = \frac{\eta_1 - \eta_2}{\eta_1 + \eta_2} H_0^i = -5.69 \times 10^{-6}\ \text{A/m}$$

$$H_0^t = \frac{2\eta_1}{\eta_1 + \eta_2} H_0^i = 5.91 \times 10^{-6}\ \text{A/m}$$

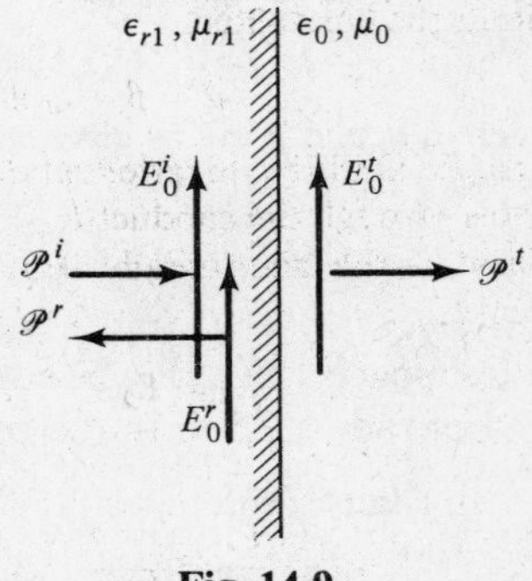

Fig. 14-9

14.16. The amplitude of $\mathbf{E}^i$ in free space (region *1*) at the interface with region *2* is 1.0 V/m. If $H_0^r = -1.41 \times 10^{-3}$ A/m, $\epsilon_{r2} = 18.5$ and $\sigma_2 = 0$, find μ_{r2}.

From

$$\frac{E_0^r}{H_0^r} = -120\pi\ \Omega = -377\ \Omega \qquad \text{and} \qquad \frac{E_0^r}{E_0^i} = \frac{\eta_2 - 377}{377 + \eta_2}$$

$$\frac{E_0^i}{H_0^r} = \frac{1.0}{-1.41 \times 10^{-3}} = \frac{-377(377 + \eta_2)}{\eta_2 - 377} \qquad \text{or} \qquad \eta_2 = 1234\ \Omega$$

Then $$1234 = \sqrt{\frac{\mu_0 \mu_{r2}}{\epsilon_0(18.5)}} \qquad \text{or} \qquad \mu_{r2} = 198.4$$

14.17. A normally incident **E** field has amplitude $E_0^i = 1.0\ \text{V/m}$ in free space just outside of seawater in which $\epsilon_r = 80$, $\mu_r = 1$ and $\sigma = 2.5\ \text{S/m}$. For a frequency of 30 MHz, at what depth will the amplitude of **E** be 1.0 mV/m?

Let the free space be region *1* and the seawater be region *2*.

$$\eta_1 = 377\ \Omega \qquad \eta_2 = 9.73\ \underline{/43.5^\circ}\ \ \Omega$$

Then the amplitude of **E** just inside the seawater is E_0^t.

$$\frac{E_0^t}{E_0^i} = \frac{2\eta_2}{\eta_1 + \eta_2} \qquad \text{or} \qquad E_0^t = 5.07 \times 10^{-2}\ \text{V/m}$$

From $\gamma = \sqrt{j\omega\mu(\sigma + j\omega\epsilon)} = 24.36\ \underline{/46.53^\circ}\ \ \text{m}^{-1}$.

$$\alpha = 24.36 \cos 46.53^\circ = 16.76\ \text{Np/m}$$

Then, from

$$1.0 \times 10^{-3} = (5.07 \times 10^{-2})e^{-16.76z}$$

$z = 0.234$ m.

14.18. A traveling **E** field in free space, of amplitude 100 V/m, strikes a sheet of silver of thickness 5 μm, as shown in Fig. 14-10. Assuming $\sigma = 61.7$ MS/m and a frequency $f = 200$ MHz, find the amplitudes E_2, E_3 and E_4.

For the silver at 200 MHz, $\eta = 5.06 \times 10^{-3}\ \underline{/45^\circ}\ \ \Omega$.

$$\frac{E_2}{E_1} = \frac{2(5.06 \times 10^{-3}\ \underline{/45^\circ})}{377 + 5.06 \times 10^{-3}\ \underline{/45^\circ}} \qquad \text{or} \qquad E_2 = 2.68 \times 10^{-3}\ \text{V/m}$$

Within the conductor,

$$\alpha = \beta = \sqrt{\pi f \mu \sigma} = 2.21 \times 10^5$$

Thus, in addition to attenuation there is phase shift as the wave travels through the conductor. Since E_3 and E_4 represent maximum values of the sinusoidally varying wave, this phase shift is not involved.

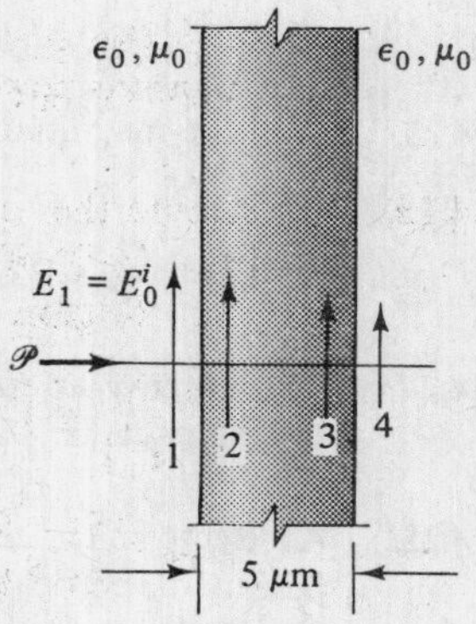

Fig. 14-10

$$E_3 = E_2 e^{-\alpha z} = (2.68 \times 10^{-3})e^{-(2.21 \times 10^5)(5 \times 10^{-6})} = 8.88 \times 10^{-4}\ \text{V/m}$$

and

$$\frac{E_4}{E_3} = \frac{2(377)}{377 + 5.06 \times 10^{-3}\ \underline{/45^\circ}} \qquad \text{or} \qquad E_4 = 1.78 \times 10^{-3}\ \text{V/m}$$

Supplementary Problems

14.19. Given

$$\mathbf{E}(z, t) = 10^3 \sin(6 \times 10^8 t - \beta z)\mathbf{a}_y \quad (\text{V/m})$$

in free space, sketch the wave at $t = 0$ and at time t_1 when it has traveled $\lambda/4$ along the z axis. Find t_1, β and λ. *Ans.* $t_1 = 2.62\ \eta\text{s}$, $\beta = 2\ \text{rad/m}$, $\lambda = \pi\ \text{m}$. See Fig. 14-11.

14.20. In free space,

$$\mathbf{H}(z, t) = 1.0e^{j(1.5\times 10^8 t + \beta z)}\mathbf{a}_x \quad \text{(A/m)}$$

Obtain an expression for $\mathbf{E}(z, t)$ and determine the propagation direction.
Ans. $E_0 = 377$ V/m, $-\mathbf{a}_z$

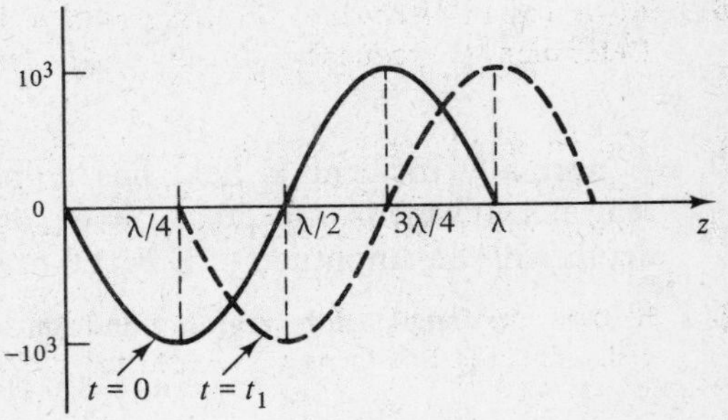

Fig. 14-11

14.21. In free space,

$$\mathbf{H}(z, t) = 1.33 \times 10^{-1} \times \cos(4 \times 10^7 t - \beta z)\mathbf{a}_x \quad \text{(A/m)}$$

Obtain an expression for $\mathbf{E}(z, t)$. Find β and λ.
Ans. $E_0 = 50$ V/m, (4/30) rad/m, 15π m

14.22. A traveling wave has a velocity of 10^6 m/s and is described by

$$y = 10\cos(2.5z + \omega t)$$

Sketch the wave as a function of z at $t = 0$ and $t = t_1 = 0.838$ μs. What fraction of a wavelength is traveled between these two times?
Ans. 1/3. See Fig. 14-12.

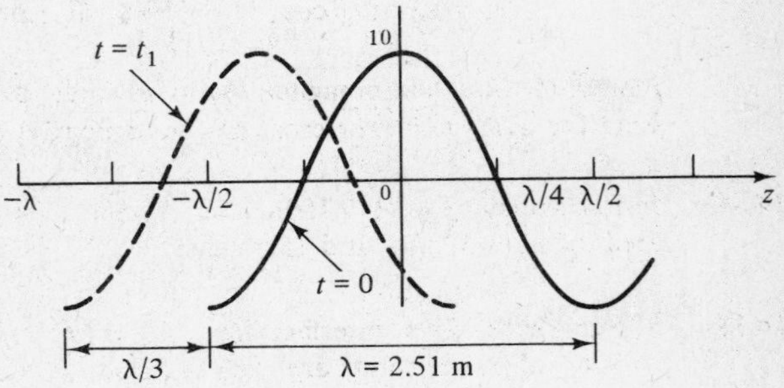

Fig. 14-12

14.23. Find the magnitude and direction of

$$\mathbf{E}(z, t) = 10\sin(\omega t - \beta z)\mathbf{a}_x - 15\sin(\omega t - \beta z)\mathbf{a}_y \quad \text{(V/m)}$$

at $t = 0$, $z = 3\lambda/4$. *Ans.* 18.03 V/m, $0.555\mathbf{a}_x - 0.832\mathbf{a}_y$

14.24. Determine γ at 500 kHz for a medium in which $\mu_r = 1$, $\epsilon_r = 15$, $\sigma = 0$. At what velocity will an electromagnetic wave travel in this medium? *Ans.* $j4.06 \times 10^{-2}\,\text{m}^{-1}$, 7.74×10^7 m/s

14.25. An electromagnetic wave in free space has a wavelength of 0.20 m. When this same wave enters a perfect dielectric, the wavelength changes to 0.09 m. Assuming that $\mu_r = 1$, determine ϵ_r and the wave velocity in the dielectric. *Ans.* 4.94, 1.35×10^8 m/s

14.26. An electromagnetic wave in free space has a phase shift constant of 0.524 rad/m. The same wave has a phase shift constant of 1.81 rad/m upon entering a perfect dielectric. Assuming that $\mu_r = 1$, find ϵ_r and the velocity of propagation. *Ans.* 11.9, 8.69×10^7 m/s

14.27. Find the propagation constant at 400 MHz for a medium in which $\epsilon_r = 16$, $\mu_r = 4.5$ and $\sigma = 0.6$ S/m. Find the ratio of the velocity U to the free-space velocity c. *Ans.* $99.58\,\underline{/60.34^\circ}\ \text{m}^{-1}$, 0.097

14.28. In a partially conducting medium, $\epsilon_r = 18.5$, $\mu_r = 800$ and $\sigma = 1$ S/m. Find α, β, η and the velocity U, for a frequency of 10^9 Hz. Determine $\mathbf{H}(z, t)$, given

$$\mathbf{E}(z, t) = 50.0e^{-\alpha z}\cos(\omega t - \beta z)\mathbf{a}_y \quad \text{(V/m)}$$

Ans. 1130 Np/m, 2790 rad/m, $2100\,\underline{/22.1^\circ}\ \Omega$, 2.25×10^6 m/s,
$2.38 \times 10^{-2}e^{-\alpha z}\cos(\omega t - 0.386 - \beta z)(-\mathbf{a}_x)$ (A/m)

14.29. For silver, $\sigma = 3.0$ MS/m. At what frequency will the depth of penetration δ be 1 mm?
Ans. 84.4 kHz

14.30. At a certain frequency in copper ($\sigma = 58.0$ MS/m) the phase shift constant is 3.71×10^5 rad/m. Determine the frequency. *Ans.* 601 MHz

14.31. The amplitude of **E** just inside a liquid is 10.0 V/m and the constants are: $\mu_r = 1$, $\epsilon_r = 20$ and $\sigma = 0.50$ S/m. Determine the amplitude of **E** at a distance of 10 cm inside the medium for frequencies of (*a*) 5 MHz, (*b*) 50 MHz, and (*c*) 500 MHz. *Ans.* (*a*) 7.32 V/m; (*b*) 3.91 V/m; (*c*) 1.42 V/m

14.32. In free space, $\mathbf{E}(z, t) = 1.0 \sin(\omega t - \beta z)\mathbf{a}_x$ (V/m). Show that the average power crossing a circular disk of radius 15.5 m in a $z = \text{const.}$ plane is 1 W.

14.33. In spherical coordinates, the *spherical wave*

$$\mathbf{E} = \frac{100}{r}\sin\theta\cos(\omega t - \beta r)\mathbf{a}_\theta \quad (\text{V/m}) \qquad \mathbf{H} = \frac{0.265}{r}\sin\theta\cos(\omega t - \beta r)\mathbf{a}_\phi \quad (\text{A/m})$$

represents the electromagnetic field at large distances r from a certain dipole antenna in free space. Find the average power crossing the hemispherical shell $r = 1$ km, $0 \le \theta \le \pi/2$. *Ans.* 55.5 W

14.34. In free space, $\mathbf{E}(z, t) = 150 \sin(\omega t - \beta z)\mathbf{a}_x$ (V/m). Find the total power passing through a rectangular area, of sides 30 mm and 15 mm, in the $z = 0$ plane. *Ans.* 13.4 mW

14.35. A free space–silver interface has $E_0^i = 100$ V/m on the free-space side. The frequency is 15 MHz and the silver constants are $\epsilon_r = \mu_r = 1$, $\sigma = 61.7$ MS/m. Determine E_0^r and E_0^t at the interface. *Ans.* -100 V/m, $7.35 \times 10^{-4} \angle 45°$ V/m

14.36. A free space–conductor interface has $H_0^i = 1.0$ A/m on the free-space side. The frequency is 31.8 MHz and the conductor constants are $\epsilon_r = \mu_r = 1$, $\sigma = 1.26$ MS/m. Determine H_0^r and H_0^t and the depth of penetration of $\mathbf{H}^t$. *Ans.* 1.0 A/m, 2.0 A/m, 80 μm

14.37. A traveling **H** field in free space, of amplitude 1.0 A/m and frequency 200 MHz, strikes a sheet of silver of thickness 5 μm with $\sigma = 61.7$ MS/m, as shown in Fig. 14-13. Find H_0^t just beyond the sheet. *Ans.* 1.78×10^{-5} A/m

14.38. A traveling **E** field in free space, of amplitude 100 V/m, strikes a perfect dielectric, as shown in Fig. 14-14. Determine E_0^t. *Ans.* 59.7 V/m

14.39. A traveling **E** field in free space strikes a partially conducting medium, as shown in Fig. 14-15. Given a frequency of 500 MHz and $E_0^i = 100$ V/m, determine E_0^t and H_0^t. *Ans.* 19.0 V/m, 0.0504 A/m

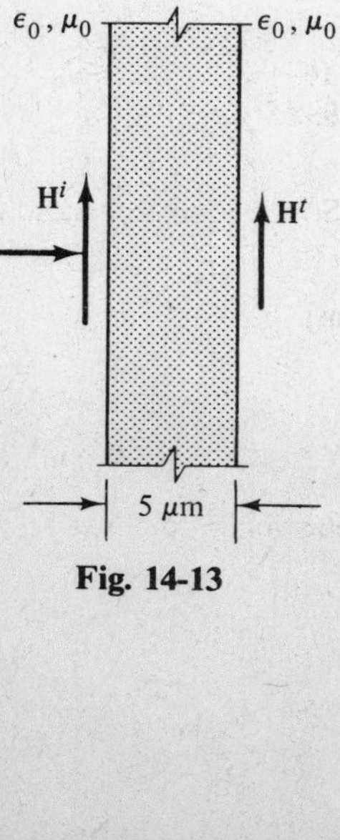

Fig. 14-13

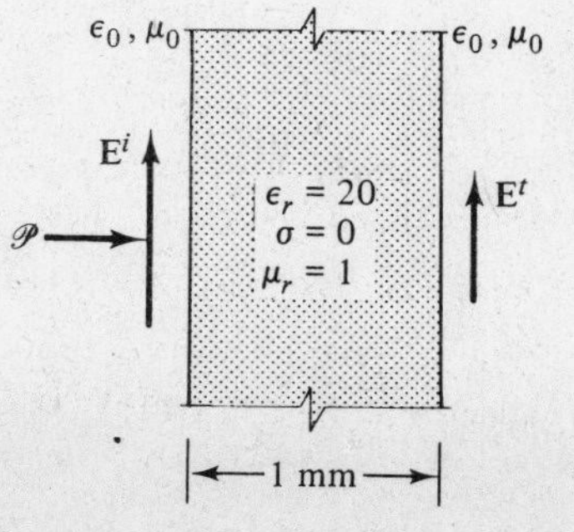

Fig. 14-14

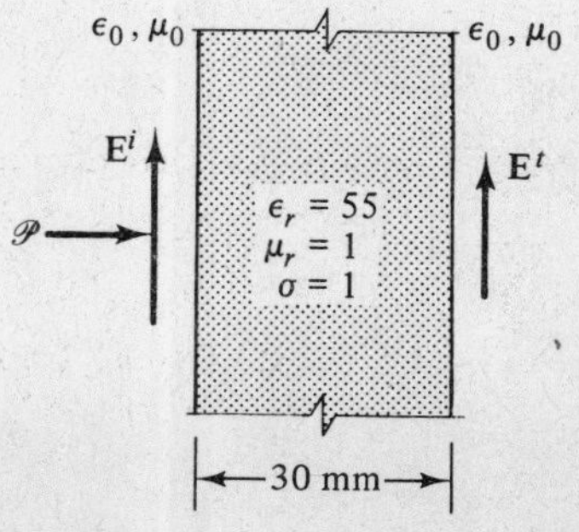

Fig. 14-15

Appendix

SI Unit Prefixes

Factor	Prefix	Symbol	Factor	Prefix	Symbol
10^{18}	exa	E	$(10^{-1}$	deci	d)
10^{15}	peta	P	$(10^{-2}$	centi	c)
10^{12}	tera	T	10^{-3}	milli	m
10^{9}	giga	G	10^{-6}	micro	μ
10^{6}	mega	M	10^{-9}	nano	n
10^{3}	kilo	k	10^{-12}	pico	p
$(10^{2}$	hecto	h)	10^{-15}	femto	f
(10	deka	da)	10^{-18}	atto	a

Divergence, Curl, Gradient, and Laplacian

Cartesian Coordinates

$$\nabla \cdot \mathbf{A} = \frac{\partial A_x}{\partial x} + \frac{\partial A_y}{\partial y} + \frac{\partial A_z}{\partial z}$$

$$\nabla \times \mathbf{A} = \left(\frac{\partial A_z}{\partial y} - \frac{\partial A_y}{\partial z}\right)\mathbf{a}_x + \left(\frac{\partial A_x}{\partial z} - \frac{\partial A_z}{\partial x}\right)\mathbf{a}_y + \left(\frac{\partial A_y}{\partial x} - \frac{\partial A_x}{\partial y}\right)\mathbf{a}_z$$

$$\nabla V = \frac{\partial V}{\partial x}\mathbf{a}_x + \frac{\partial V}{\partial y}\mathbf{a}_y + \frac{\partial V}{\partial z}\mathbf{a}_z$$

$$\nabla^2 V = \frac{\partial^2 V}{\partial x^2} + \frac{\partial^2 V}{\partial y^2} + \frac{\partial^2 V}{\partial z^2}$$

Cylindrical Coordinates

$$\nabla \cdot \mathbf{A} = \frac{1}{r}\frac{\partial}{\partial r}(rA_r) + \frac{1}{r}\frac{\partial A_\phi}{\partial \phi} + \frac{\partial A_z}{\partial z}$$

$$\nabla \times \mathbf{A} = \left(\frac{1}{r}\frac{\partial A_z}{\partial \phi} - \frac{\partial A_\phi}{\partial z}\right)\mathbf{a}_r + \left(\frac{\partial A_r}{\partial z} - \frac{\partial A_z}{\partial r}\right)\mathbf{a}_\phi + \frac{1}{r}\left[\frac{\partial}{\partial r}(rA_\phi) - \frac{\partial A_r}{\partial \phi}\right]\mathbf{a}_z$$

$$\nabla V = \frac{\partial V}{\partial r}\mathbf{a}_r + \frac{1}{r}\frac{\partial V}{\partial \phi}\mathbf{a}_\phi + \frac{\partial V}{\partial z}\mathbf{a}_z$$

$$\nabla^2 V = \frac{1}{r}\frac{\partial}{\partial r}\left(r\frac{\partial V}{\partial r}\right) + \frac{1}{r^2}\frac{\partial^2 V}{\partial \phi^2} + \frac{\partial^2 V}{\partial z^2}$$

Spherical Coordinates

$$\nabla \cdot \mathbf{A} = \frac{1}{r^2}\frac{\partial}{\partial r}(r^2 A_r) + \frac{1}{r \sin\theta}\frac{\partial}{\partial \theta}(A_\theta \sin\theta) + \frac{1}{r\sin\theta}\frac{\partial A_\phi}{\partial \phi}$$

$$\nabla \times \mathbf{A} = \frac{1}{r\sin\theta}\left[\frac{\partial}{\partial \theta}(A_\phi \sin\theta) - \frac{\partial A_\theta}{\partial \phi}\right]\mathbf{a}_r + \frac{1}{r}\left[\frac{1}{\sin\theta}\frac{\partial A_r}{\partial \phi} - \frac{\partial}{\partial r}(rA_\phi)\right]\mathbf{a}_\theta + \frac{1}{r}\left[\frac{\partial}{\partial r}(rA_\theta) - \frac{\partial A_r}{\partial \theta}\right]\mathbf{a}_\phi$$

$$\nabla V = \frac{\partial V}{\partial r}\mathbf{a}_r + \frac{1}{r}\frac{\partial V}{\partial \theta}\mathbf{a}_\theta + \frac{1}{r\sin\theta}\frac{\partial V}{\partial \phi}\mathbf{a}_\phi$$

$$\nabla^2 V = \frac{1}{r^2}\frac{\partial}{\partial r}\left(r^2\frac{\partial V}{\partial r}\right) + \frac{1}{r^2\sin\theta}\left(\sin\theta\frac{\partial V}{\partial \theta}\right) + \frac{1}{r^2\sin^2\theta}\frac{\partial^2 V}{\partial \phi^2}$$

INDEX